Linear and Nonlinear Optics

Materials, Properties, and Applications

Kitsakorn Locharoenrat

JENNY STANFORD
PUBLISHING

Published by

Jenny Stanford Publishing Pte. Ltd.
Level 34, Centennial Tower
3 Temasek Avenue
Singapore 039190

Email: editorial@jennystanford.com
Web: www.jennystanford.com

British Library Cataloguing-in-Publication Data
A catalogue record for this book is available from the British Library.

Linear and Nonlinear Optics: Materials, Properties, and Applications

ISBN 978-981-4877-15-2 (Hardcover)
ISBN 978-1-003-05588-4 (eBook)

Contents

Preface ix

PART I Linear Optics

1. Au-Core Pd-Shell Nanoparticles 3
1.1 Introduction 3
1.2 Principle 4
1.2.1 Surface Plasmon Resonance 4
1.2.2 Computational Method 11
1.3 Methods 15
1.3.1 Absorbance and Fluorescence 15
1.3.2 Fluorescence Quantum Yield 16
1.3.3 Fluorescence Polarization 17
1.4 Applications 19
1.4.1 Fluorescence Enhancer of Mercury Detection 19
1.4.2 Fluorescence Enhancer of Latent Fingerprints Detection 21
1.5 Implementations 23
1.5.1 Absorbance and Fluorescence 23
1.5.2 Fluorescence Quantum Yield 34
1.5.3 Fluorescence Polarization 38
1.5.4 Fluorescence Enhancer of Mercury Detection 40
1.5.5 Fluorescence Enhancer of Image Enhancement of Latent Fingerprints 42

2. ZnO Nanoparticles 47
2.1 Introduction 47
2.2 Fabrications 48
2.2.1 Nanolithography 48
2.2.2 Laser Ablation 49
2.3 Applications 49
2.3.1 Biomedical Uses 49
2.3.2 Industrial Uses 52

2.3.3 Photovoltaic Uses 56
2.4 Method 58
2.5 Implementations 58
2.5.1 ZnO Nanoparticles 58
2.5.2 ZnO/Au Nanocomposites 64
2.5.3 ZnO/Coumarin-153 Nanocomposites 68

3. Fe_3O_4 Magnetic Nanoparticles 73
3.1 Introduction 73
3.2 Principle 73
3.3 Fluorescence Resonance Energy Transfer 75
3.4 Applications 76
3.5 Implementations 78

4. Organic Light-Emitting Diode 87
4.1 Introduction 87
4.2 Principle 88
4.3 Methods 89
4.4 Implementations 90
4.4.1 White OLED 90
4.4.1.1 Single host structures 92
4.4.1.2 White OLED structures 93
4.4.2 Blue OLED 95

5. Pb–Fe-Shielded Material 99
5.1 Introduction 99
5.2 Gamma–Matter Interaction 99
5.2.1 Compton Scattering 101
5.2.2 Photoelectric Absorption 101
5.2.3 Pair Production 101
5.3 Computational Method 102
5.3.1 Interpolation Method 104
5.3.2 Optimization Model 106
5.3.3 Contour Plot Method 109
5.4 Implementation 113

6. Probe Sensors 119
6.1 Introduction 119
6.2 Principle 119
6.2.1 Turbid Tissue 119
6.2.2 Reflection Probe 120

6.2.3 Barcode Scanner 126
6.2.4 Smart Cane 129
6.3 Implementations 130
6.3.1 Reflection Probe 130
6.3.2 Barcode Scanner 141
6.3.3 Smart Cane 144

PART II Nonlinear Optics

7. Nonlinear Medium 151
7.1 Introduction 151
7.2 Component of Electric Fields 151
7.3 Component of Magnetic Fields 153
7.4 Classification of Nonlinear Medium 155

8. Wave Propagation 159
8.1 Introduction 159
8.2 Polarization 159
8.3 Wave Propagation in Nonlinear Medium 160
8.4 Implementations 163
8.4.1 Plane Wave Propagation in Nonlinear Medium at Normal Angle 163
8.4.2 Plane Wave Propagation in Nonlinear Medium at a Tile Angle 164

9. Second Harmonic Generation 169
9.1 Introduction 169
9.2 Second Harmonic Generation 169
9.3 Realistic SHG 185
9.4 Implementations 191
9.4.1 Birefringence Phase Matching 191
9.4.2 Effective Nonlinear Polarization 194
9.4.3 Quasi-Phase Matching 202
9.4.4 Spontaneous Polarization 208

10. Other Frequency Mixing Processes 219
10.1 Introduction 219
10.2 Different Frequency Generation 219
10.3 Sum Frequency Generation 229
10.3.1 Sum Frequency Generation with Phase Matching 229

10.3.2 Sum Frequency Generation without Phase Matching 231
10.4 Optical Parametric Interactions 233
10.4.1 Optical Parametric Amplification 233
10.4.2 Optical Parametric Oscillation 238
10.5 Optical Resonant Oscillator 240
10.5.1 Single Resonant Oscillator 242
10.5.2 Double Resonant Oscillator 246

Suggested Readings **251**

Solutions to Problems **253**

Appendix A **297**

Appendix B **309**

Appendix C **317**

Index 329

Preface

This book emphasizes the underlying principles of the linear and nonlinear optical properties of materials, as well as their implementation and applications to help students and researchers in physics and related fields to develop their physics insight. Several examples presented throughout this book and the problem sets in each chapter will enable readers to develop their skills and to measure their understanding of each topic step-by-step. The book will be a helpful tool to solve problems both in classroom and in real life.

The textbook is divided into two parts. Part I, Linear Optics, is composed of Chapters 1–6. Chapter 1 describes bimetallic nanoparticles in dielectric media that show a noticeable band shift of electromagnetic field around them, owing to the plasmonic effect applicable for enhancing and quenching fluorescence intensities of the dye molecules. Concentration dependence of the fluorescence quantum yield and fluorescence polarization is related to the energy transfer between the metals and dye molecules. The integration of bimetallic nanoparticles to dye molecules shows the selective and sensitive detection of trace amounts of heavy metals such as mercury ion at the nanometer level. Furthermore, the integration of bimetallic nanoparticles to dye molecules during excited state as luminescence enhancer improves image contrasts of artificial latent fingerprints and serves as an information bank of the donor, which has high importance in forensic inspection and biomedical diagnosis. Chapter 2 shows that the dosage of ZnO nanoparticles under UV exposure is related to the levels of the oxidant generated from electron–hole pairs and the human breast cancer cells via phototoxic activity. Integration of bimetallic nanoparticles to the ZnO matrix promotes broadening of the absorption spectrum from the ultraviolet to the visible wavelength. The photocatalytic activity is improved by surface plasmon resonance and Schottky junction effect. The integration of coumarin-153 to the ZnO matrix also shows effective and stable photovoltaic effect. Higher absorbance indicates that coumarin-153 has an important role in light harvesting.

Coumarin-153 absorbs light in higher frequency regions, transferring their electrons to the conductive band of ZnO nanoparticles and enhancing the semiconductive activity of ZnO. Chapter 3 explains that the dye removal kinetics by Fe_3O_4 magnetic nanoparticles under pulsed white light originates from surface adsorption and photocatalytic degradation. A reduction of fluorescence quantum yield of rhodamine 6G conforms to the improvement of dye removal. Chapter 4 describes the uniformity of the electron transport layer stamped on a blue-OLED, serving as an alternative for the deposition of an organic thin film. On the other hand, the white-OLED based on p-i-n structure shows great power performance compared to the traditional structures because the high hole–electron ability maintains an effective electron–hole balance. Chapter 5 introduces the double-layer shielding tank design to safely store radioactive waste. The weight and cost of the tank made from the designed materials are computed via the optimization model and the contour plotting method. Chapter 6 explains photon propagation through the multilayer structures of a human tissue model. A point source and two photo detectors to be set on human skin are dependent on the distribution and penetration depth of the photon. A photon packet transfixed through the skeletal muscle can fairly detect the contraction and relaxation states of the skeletal muscles for clinical tests.

Part II, Nonlinear Optics, is composed of Chapters 7–10. Chapters 7 and 8 cover the basic topics on nonlinear optical effects. They focus on general concepts such as electromagnetic theory, nonlinear medium, and wave propagation. Chapters 9 and 10 focus on more advanced concepts such as second harmonic generation, phase matching, optical parametric interactions, different frequency generation, sum frequency generation, tunable laser, and optical resonant oscillator.

Kitsakorn Locharoenrat

Autumn 2020

Part I
Linear Optics

Chapter 1

Au-Core Pd-Shell Nanoparticles

1.1 Introduction

This chapter deals with the characteristic plasmon bands of noble metal nanostructures (Au-core Pd-shell nanorods) observed in the absorption spectra as they might play a key role in chemical, biological, and optical sensors. Lower plasmon peak wavelength depends on the shell thickness and dielectric indices of dielectric media. It is related to the electron oscillation along the transverse plasmon mode of the rods. A higher plasmon peak wavelength shows band shift by modification of shell thickness of the metal nanostructures and dielectric media. It is related to the electron oscillation along the longitudinal plasmon mode of the rods. Furthermore, this chapter covers the influences of sizes, shapes, and amounts of the noble metal nanostructures on fluorescence signals of the organic dyes such as rhodamine 6G (R6G), owing to a promising energy transfer between them. Fluorescent quantum yield and respective fluorescence polarization are proportional to the concentration ratio of the metal nanoparticles and dye molecules. Enhancement and reduction of fluorescence polarization are originated by shortening of the fluorescence lifetime of dye molecules and temperature effect of metal nanoparticles, respectively. This great photosensitivity is a preliminary step toward the development of fluorescence sensing

Linear and Nonlinear Optics: Materials, Properties, and Applications
Kitsakorn Locharoenrat

ISBN 978-981-4877-15-2 (Hardcover), 978-1-003-05588-4 (eBook)
www.jennystanford.com

in heavy metal pollutants as one of the environmental concerns and latent fingerprint as one of the most valuable information storage platforms for individual for forensic perspective and health assessment. On one hand, a fluorescence sensor is remarkably selective and sensitive to trace amounts of mercury and the limit of detection of mercury ion is low in nanometer scale. The mechanism of fluorescence sensors is described via the fluorescence resonance energy transfer and surface plasmon resonance energy transfer between spirolactam forms of the dye molecules and metal nanoparticles. On the other hand, image enhancement of artificial latent fingerprints is improved by metal-fluorescence molecules because the dye molecules are served as a glue adhered very well to the metal nanoparticles as carrier of any signaling and recognition molecules like α-amylase in the artificial latent fingerprints. The localized surface plasmon resonance of metal nanoparticles coupled with emission of dye molecules results in the electromagnetic development of emission signal and luminescence image.

1.2 Principle

1.2.1 Surface Plasmon Resonance

Nanomaterials such as metal nanoparticles are now important in comparison to bulk materials because of the following reasons:

- Size-, shape-, and concentration-dependence of optical properties is applicable to nanoscale lasers and photonic devices.
- Surface plasmon resonance exists especially for noble metals because a simultaneous vibration of free electrons in the conduction band under the applied electric field introduces a high field near their surfaces.
- Electric field enhancement offers remarkable Raman scattering, absorption, and fluorescence intensities.

Among metal nanoparticles, Au and Ag have been widely reported. However, these nanoparticles have some drawbacks, such as they are sometimes unsuitable to photosensitivity. Herein we focus on bimetallic nanoparticles such as Au-core Pd-shell nanorods because they show unique physical properties in comparison to single metallic nanoparticles. These unique abilities allow their use in two possible applications:

1. Size-, shape-, and concentration-dependence of physical properties of bimetallic nanoparticles is suggested to improve the photocatalytic properties. The total material cost is also reduced when one uses a cheap metal core in place of an expensive metal shell (Table 1.1).
2. The core–shell nanostructure interface keeps much more effective coupling of surface plasmon resonance than alloy. Consequently, it is useful for chemical and biological sensors.

Among the noble metals, Pd is considered a potential shell, whereas Au serves as a potential core for the production of bimetallic nanorods. Not only do they show strong optical absorption in the visible region, they can also show photocatalytic activity under their large surface-to-volume ratio. Surface plasmon resonance of Au-core and Pd-shell nanorods also results in fluorescence intensities of dye molecules.

Table 1.1 Forecast metal market: USD trillion as of January 8, 2020

Metal	Price, USD/ozt
Gold	1587.51
Silver	18.49
Platinum	969.97
Palladium	2075.16

Source: InvestmentMine

Surface plasmon resonance is an optical phenomenon in which the collective oscillations of free electrons of metals couple with the applied polarized light (Fig. 1.1).

First, let us define the following terms as follows:

- m^* is effective mass.
- Γ_0 is damping constant.
- q is electric charge.
- x is displacement.
- N is electron concentration.
- ω is the applied frequency.
- ω_p is the plasma frequency.

The equation of motion of free electron oscillations of metals without the restoring force is written as

$$\vec{x}(t) = \vec{x}_0 e^{-i\omega t} \tag{1.1}$$

$$\vec{E}(t) = \vec{E}_0 e^{-i\omega t} \tag{1.2}$$

$$\sum \vec{F} = \vec{F}_q + \vec{F}_{\text{electron collisions}} \tag{1.3}$$

$$\sum \vec{F} = q\vec{E}(t) - m^* \Gamma_0 \frac{d\vec{x}(t)}{dt} \tag{1.4}$$

$$\sum \vec{F} = q\vec{E}(t) - m^* \Gamma_0 \frac{d\vec{x}(t)}{dt} = m^* \vec{a} = m^* \frac{d^2\vec{x}(t)}{dt^2} \tag{1.5}$$

$$q\vec{E}_0 e^{-i\omega t} = m^* \left[-\omega^2 \vec{x}_0 - i\omega \vec{x}_0 \Gamma_0 \right] e^{-i\omega t} \tag{1.6}$$

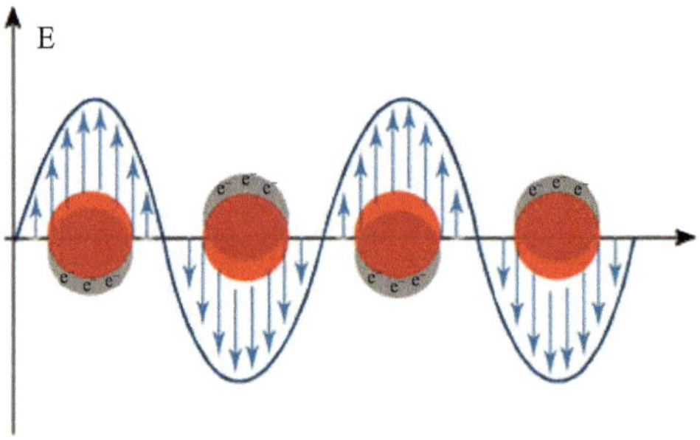

Figure 1.1 Surface plasmon resonance due to coupling of free electrons of metals with an external electromagnetic field.

The resultant amplitude of free electron oscillations becomes

$$\vec{x}_0 = \frac{q\vec{E}_0}{m^* \left[-\omega^2 - i\omega\Gamma_0 \right]} \tag{1.7}$$

The dielectric constant of metals due to a polarization term is written as

$$\vec{P} = Nq\vec{x}_0 e^{-i\omega t} = (\varepsilon_m - \varepsilon_o)\vec{E}_0 e^{-i\omega t} \tag{1.8}$$

$$\varepsilon_m = \varepsilon_o - \frac{Nq^2}{m^* \omega(\omega + i\Gamma_0)} \tag{1.9}$$

The dielectric constant of metals due to $\omega >> \Gamma_0$ is written as

$$\varepsilon_m = \varepsilon_o - \frac{Nq^2}{m^* \omega^2} \tag{1.10}$$

Defining a resonance condition as $\omega = \omega_p$, the dielectric constant of metals becomes

$$\omega_p^2 = \frac{Nq^2}{m^* \varepsilon_o} \tag{1.11}$$

$$\varepsilon_m = \varepsilon_o \left[1 - \frac{\omega_p^2}{\omega^2} \right] \tag{1.12}$$

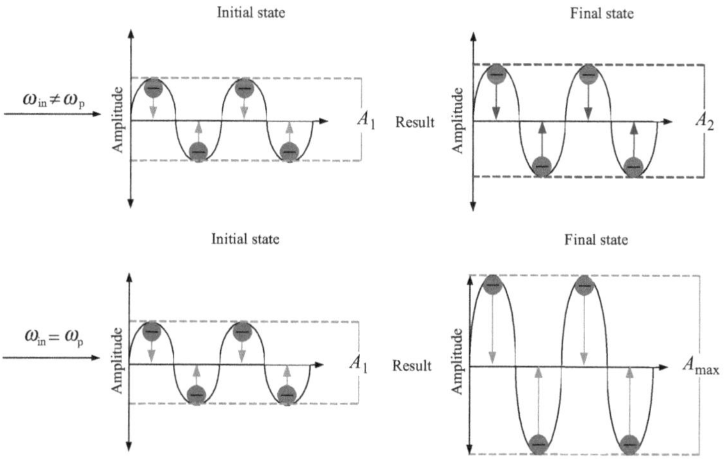

Figure 1.2 Resultant amplitude of free electron oscillations as the applied frequency is not equivalent to the plasma frequency (top) and as the applied frequency is equivalent to the plasma frequency (bottom).

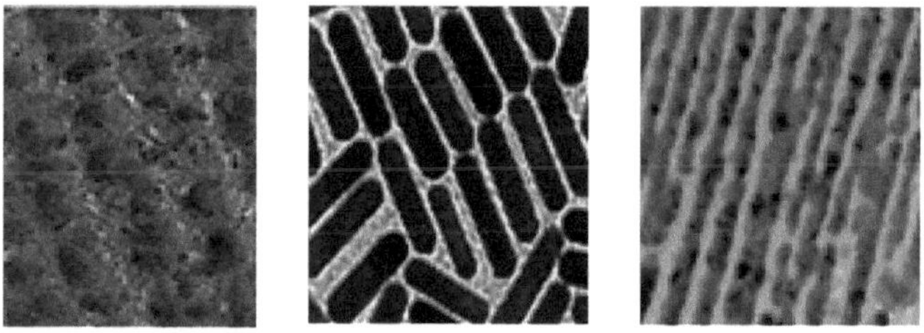

Figure 1.3 Transmission electron microscopic images of nanodots (left), nanorods (middle), and nanowires (right).

The amplitude of free electron oscillations of metals, as shown in Fig. 1.2, is dependent on four factors as follows:

(a) Aspect ratio a/b (Fig. 1.3 and Table 1.2) where a is the long axis and b is the short axis. The higher the aspect ratio, the stronger the plasmonic effect.

Table 1.2 Detailed geometrical specifications of the representative metal nanostructures

Nanostructure	Shape	Long axis (a), nm	Short axis (b), nm	Aspect ratio (a/b)
Nanodots	Sphere	5	5	1.0
Nanorods	Ellipse	96	40	2.4
Nanowires	Wire	—	14	∞

(b) Electromagnetic field (Fig. 1.4)

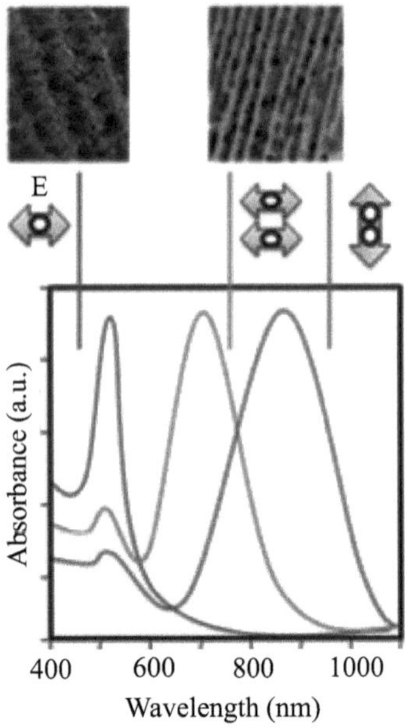

Figure 1.4 Absorbance with respect to the electric field distributions of Cu nanodots (left), Cu nanowires under s-polarized light (middle), and Cu nanowires under p-polarized lights (right).

(b1) One mode for the single nanoparticle offers poor plasmonic effect.

(b2) Double modes for the multiple nanoparticles offer high plasmonic effect.

- Transverse magnetic (TM) mode or coupled mode is preferred because the applied electric field is parallel to the long axis of the metal nanoparticles.
- Transverse electric (TE) mode or uncoupled mode is not recommended because the applied electric field is perpendicular to the long axis of the metal nanoparticles.

(c) Dielectric constant (Table 1.3): As the imaginary part of the dielectric constant of metals is low, the tendency of spectral width of surface plasmon becomes sharp.

(d) Symmetric structures: Electric field enhancement is strongly dependent on the asymmetric nanostructures especially for the circular shapes in comparison to triangle and square ones. The higher the symmetric structures, the larger the plasmonic effect.

Table 1.3 Representative dielectric constants of metal nanoparticles

Metal	Wavelength (nm)	Dielectric constant
Ag	633	−18.22 + 0.48i
Cu	633	−14.67 + 0.72i
Au	633	−10.92 + 1.49i
Pd	620	−14.40 + 14.60i
Pt	640	−11.10 + 15.70i
Fe	633	−1.02 + 17.81i

The surface plasmon resonances of the metal nanoparticles are also extensively used to modify the optical responses of the dye molecules. Integration of metal nanoparticles to organic dyes such as R6G applicable to fluorescent probe is concerned due to the following reasons:

- R6G offers good photostability.
- It offers an excellent optical absorption coefficient.
- It offers a broad fluorescence in the visible region.

- It is inexpensive.
- It is environmentally nontoxic.
- It offers a high fluorescent quantum yield (0.95).
- It offers great fluorescence properties.
- Fluorescence and absorption appear in the visible light.
- Spirolactam ring opening increases in the fluorescence intensity.
- Spirolactam-type rhodamine derivatives are non-fluorescent.
- Further fluorescence intensity of rhodamine 6G possibly originates by plasmon resonance energy transfer.
- Enhancement of fluorescence intensity resolves the limitations of chemical, physical, and biomedical devices.
- R6G is possibly applicable to luminescent solar concentrators (Fig. 1.5).

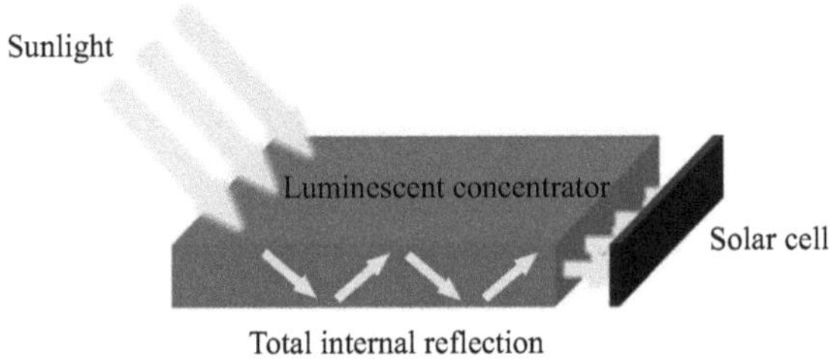

Figure 1.5 Organic dyes often used in luminescent solar concentrators.

Metal nanoparticles can either quench or enhance the intrinsic luminescence of emitters like dye molecules relying on the following reasons:

- Distance between the donor and the emitter.
- Fluorescence enhancement comes from the surface plasmon resonance when the incident light hits on the surface of metal nanoparticles. This point increases the excitation decay rate and radiative quantum efficiency.
- The dipole energy at the metal surface suppresses the ratio of radiative to nonradiative decay rate through the energy transfer mechanism. This point quenches the intrinsic fluorescence from the dye molecules.

1.2.2 Computational Method

Centered finite difference is the first-order differential equation developed from the Taylor series as

$$f(x+\Delta x/2)=f(x)\pm\frac{\partial f(x)}{\partial x}\frac{\Delta x}{2}\frac{1}{1!}+\frac{\partial^2 f(x)}{\partial x^2}\left(\frac{\Delta x}{2}\right)^2\frac{1}{2!} \\ \pm\frac{\partial^3 f(x)}{\partial x^3}\left(\frac{\Delta x}{2}\right)^3\frac{1}{3!}+\ldots \tag{1.13}$$

Assuming that Δx becomes zero, Eq. (1.13) is approximated as

$$\frac{f(x+\Delta x/2)-f(x-\Delta x/2)}{\Delta x}=\frac{\partial f(x)}{\partial x}+(\Delta x)^2\frac{\partial^3 f(x)}{\partial x^3}\frac{1}{24}+\ldots \tag{1.14}$$

$$\frac{\partial f(x)}{\partial x}\approx\frac{f(x+\Delta x/2)-f(x-\Delta x/2)}{\Delta x} \tag{1.15}$$

Defining a relationship of the differential equation upon time and space as

$$\frac{\partial f(x,t)}{\partial x}=\eta\left(\frac{\partial f(x,t)}{\partial t}\right) \tag{1.16}$$

$$\frac{f(x+\Delta x/2,t)-f(x-\Delta x/2,t)}{\Delta x} \\ =\eta\left(\frac{f(x,t+\Delta t/2)-f(x,t-\Delta t/2)}{\Delta t}\right) \tag{1.17}$$

The electromagnetic field with respective plasma frequency of free electron oscillation of metal in a time and space derivative is, therefore, calculated as

$$\nabla\times\vec{H}=\frac{\partial\vec{D}}{\partial t} \tag{1.18}$$

$$\vec{D}=\varepsilon_{\mathrm{m}}\vec{E} \tag{1.19}$$

$$\frac{\partial\vec{H}}{\partial x}=\varepsilon_{\mathrm{o}}\left[1-\frac{\omega_{\mathrm{p}}^2}{\omega^2}\right]\frac{\partial\vec{E}}{\partial t} \tag{1.20}$$

$$\frac{\vec{H}(x+\Delta x/2,t)-\vec{H}(x-\Delta x/2,t)}{\Delta x} = \varepsilon_{\mathrm{o}}\left[1-\frac{\omega_{\mathrm{p}}^{2}}{\omega^{2}}\right]\left(\frac{\vec{E}(x,t-\Delta t/2)-\vec{E}(x,t+\Delta t/2)}{\Delta t}\right) \tag{1.21}$$

$$\frac{\vec{H}^{t_i}_{x+\Delta x/2}-\vec{H}^{t_i}_{x-\Delta x/2}}{\Delta x} = \varepsilon_{\mathrm{o}}\left[1-\frac{\omega_{\mathrm{p}}^{2}}{\omega^{2}}\right]\left(\frac{\vec{E}^{t_f}_{x}-\vec{E}^{t_i}_{x}}{\Delta t}\right) \tag{1.22}$$

In order to converge the first-order approximation of a differential equation to the exact solution, the finite-difference time-domain (FDTD) method was introduced for the first time by Kane S. Yee in 1966. It is a numerical analysis using centered finite difference operators for modeling the electrodynamics via Yee cells served as rectangular unit cells of Cartesian computational grids (Figs. 1.6 and 1.7).

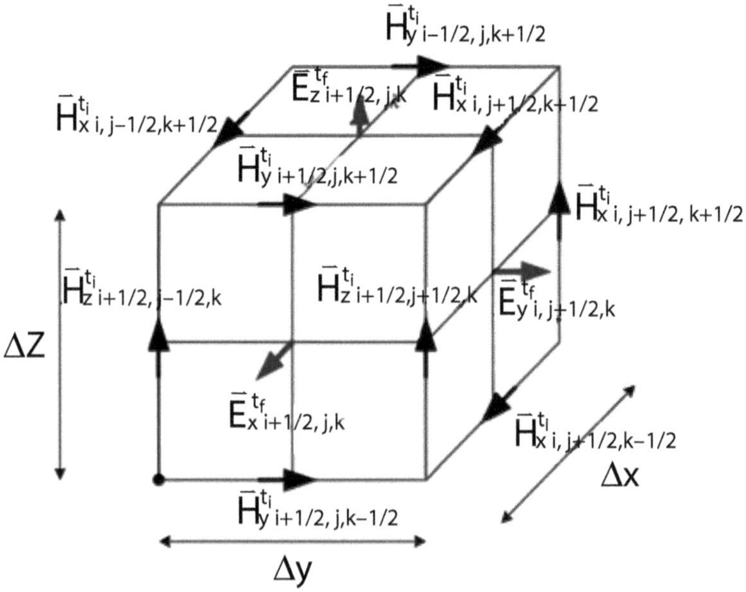

Figure 1.6 Yee cells composed of rectangular unit cells of Cartesian computational grids in time and space of each electromagnetic component in Maxwell's equations.

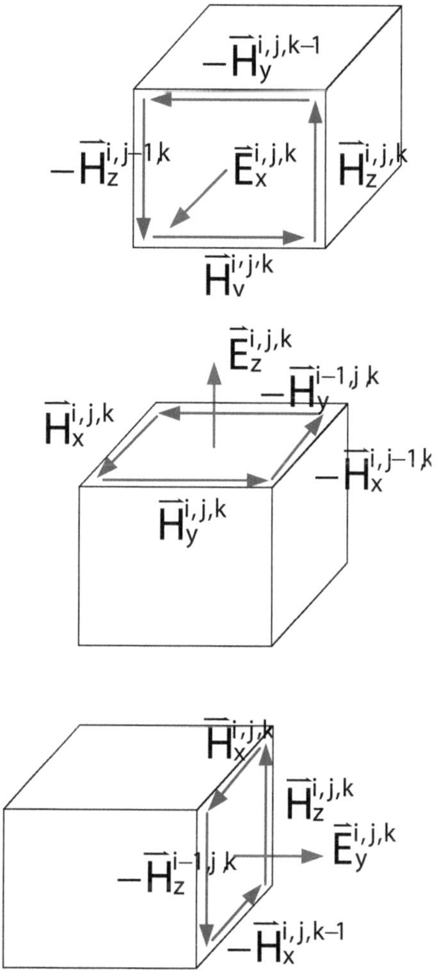

Figure 1.7 Change in each electromagnetic field upon time and space interval in each Yee cell.

Using the FDTD method, the electromagnetic field with respective plasma frequency of free electron oscillation of metal is calculated as follows:

Electromagnetic field in the x-direction is written as

$$\varepsilon_o\left(1-\frac{\omega_p^2}{\omega^2}\right)\frac{\partial \vec{E}_x}{\partial t}=\frac{\partial \vec{H}_y}{\partial z}-\frac{\partial \vec{H}_z}{\partial y} \quad (1.23)$$

$$\varepsilon_o\left(1-\frac{\omega_p^2}{\omega^2}\right)\left(\frac{\vec{E}_x^{t_f}-\vec{E}_x^{t_i}}{\Delta t}\right)=\left(\frac{\vec{H}_y^{i,j,k}-\vec{H}_y^{i,j,k+1}}{\Delta z}\right)-\left(\frac{\vec{H}_z^{i,j,k}-\vec{H}_z^{i,j+1,k}}{\Delta y}\right) \tag{1.24}$$

$$\vec{E}_x^{t_f}=\vec{E}_x^{t_i}+\frac{\Delta t}{\varepsilon_o(1-\omega_p^2/\omega^2)}\left[\left(\frac{\vec{H}_y^{i,j,k}-\vec{H}_y^{i,j,k+1}}{\Delta z}\right)-\left(\frac{\vec{H}_z^{i,j,k}-\vec{H}_z^{i,j+1,k}}{\Delta y}\right)\right] \tag{1.25}$$

Electromagnetic field in the y-direction is written as

$$\varepsilon_o\left(1-\frac{\omega_p^2}{\omega^2}\right)\frac{\partial\vec{E}_y}{\partial t}=\frac{\partial\vec{H}_z}{\partial x}-\frac{\partial\vec{H}_x}{\partial z} \tag{1.26}$$

$$\varepsilon_o\left(1-\frac{\omega_p^2}{\omega^2}\right)\left(\frac{\vec{E}_y^{t_f}-\vec{E}_y^{t_i}}{\Delta t}\right)=\left(\frac{\vec{H}_z^{i,j,k}-\vec{H}_z^{i+1,j,k}}{\Delta x}\right)-\left(\frac{\vec{H}_x^{i,j,k}-\vec{H}_x^{i,j,k+1}}{\Delta z}\right) \tag{1.27}$$

$$\vec{E}_y^{t_f}=\vec{E}_y^{t_i}+\frac{\Delta t}{\varepsilon_o(1-\omega_p^2/\omega^2)}\left[\left(\frac{\vec{H}_z^{i,j,k}-\vec{H}_z^{i+1,j,k}}{\Delta x}\right)-\left(\frac{\vec{H}_x^{i,j,k}-\vec{H}_x^{i,j,k+1}}{\Delta z}\right)\right] \tag{1.28}$$

Electromagnetic field in the z-direction is written as

$$\varepsilon_o\left(1-\frac{\omega_p^2}{\omega^2}\right)\frac{\partial\vec{E}_z}{\partial t}=\frac{\partial\vec{H}_y}{\partial x}-\frac{\partial\vec{H}_x}{\partial y} \tag{1.29}$$

$$\varepsilon_o\left(1-\frac{\omega_p^2}{\omega^2}\right)\left(\frac{\vec{E}_z^{t_f}-\vec{E}_z^{t_i}}{\Delta t}\right)=\left(\frac{\vec{H}_y^{i,j,k}-\vec{H}_y^{i+1,j,k}}{\Delta x}\right)-\left(\frac{\vec{H}_x^{i,j,k}-\vec{H}_x^{i,j+1,k}}{\Delta y}\right) \tag{1.30}$$

$$\vec{E}_z^{t_f}=\vec{E}_z^{t_i}+\frac{\Delta t}{\varepsilon_o(1-\omega_p^2/\omega^2)}\left[\left(\frac{\vec{H}_y^{i,j,k}-\vec{H}_y^{i,j+1,k}}{\Delta x}\right)-\left(\frac{\vec{H}_x^{i,j,k}-\vec{H}_x^{i,j+1,k}}{\Delta y}\right)\right] \tag{1.31}$$

After getting the electromagnetic field intensity upon a solution of Maxwell's equations, the absorbance of metals is calculated in terms of the absorption intensity per unit volume as follows:

The absorption intensity with respective potential energy of electromagnetic field is written as

$$P_{\text{abs}} = \frac{\partial U}{\partial t} = -\frac{1}{2}\nabla \cdot \vec{S} - \frac{1}{2}\vec{E} \cdot \vec{J} \tag{1.32}$$

Substituting the divergence of the Poynting vector in Eq. (1.32), we have

$$\nabla \cdot \vec{S} = \vec{H}(\nabla \times \vec{E}) - \vec{E}(\nabla \times \vec{H}) \tag{1.33}$$

$$P_{\text{abs}} = -\frac{1}{2}\left[\vec{H}(\nabla \times \vec{E}) - \vec{E}(\nabla \times \vec{H})\right] - \frac{1}{2}\vec{E} \cdot \vec{J} \tag{1.34}$$

Substituting the rotation of the electromagnetic field in Eq. (1.34), we get

$$\nabla \times \vec{E} = -i\omega\vec{B} \tag{1.35}$$

$$\nabla \times \vec{H} = i\omega\vec{D} + \vec{J} \tag{1.36}$$

$$P_{\text{abs}} = -\frac{1}{2}\left[\vec{H}(-i\omega\vec{B}) - \vec{E}(i\omega\vec{D} + \vec{J})\right] - \frac{1}{2}\vec{E} \cdot \vec{J} \tag{1.37}$$

$$P_{\text{abs}} = -\frac{i\omega}{2}(\vec{H} \cdot \vec{B}) + \frac{i\omega}{2}(\vec{E} \cdot \vec{D}) \tag{1.38}$$

$$P_{\text{abs}} \simeq \frac{i\omega}{2}(\vec{E} \cdot \vec{D}) \tag{1.39}$$

Substituting the displacement field in Eq. (1.39), the absorption intensity reaches

$$\vec{D} = \varepsilon\vec{E} \tag{1.40}$$

$$P_{\text{abs}} \simeq \frac{i\omega}{2}(\vec{E} \cdot \varepsilon\vec{E}) \tag{1.41}$$

$$P_{\text{abs}} \simeq \frac{i\omega}{2}\left|\vec{E}\right|^2 \varepsilon \tag{1.42}$$

1.3 Methods

1.3.1 Absorbance and Fluorescence

Spectroscopy is an optical method to detect the light–matter interaction via a spectroscope with a prism. This method is used to detect the absorption spectra upon the photon energy stimulated electrons from the ground state and to detect the fluorescence spectra upon the photon energy of electrons released from the excited state as well. Absorbance and fluorescence of the prepared samples are generally measured using a spectrometer (Fig. 1.8). A tungsten lamp is usually used as the excitation source of absorption measurement. A 3 W green LED at the excitation wavelength of 516–525 nm is suggested as the excitation light of R6G, whereas 3 W UV LEDs at the excitation wavelength of 398 nm are suggested as the excitation light of coumarin-153.

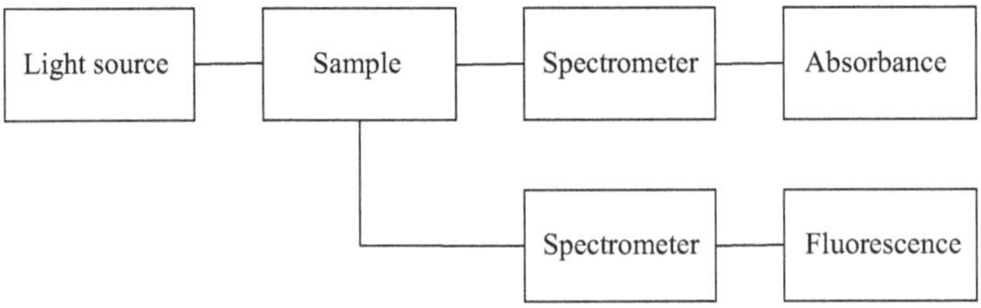

Figure 1.8 Block diagrams of the absorption and fluorescence measurements.

1.3.2 Fluorescence Quantum Yield

Interactions between metal nanoparticles and fluorophore (e.g., Ag and rhodamine B, Ag and OLED, etc.) have been a subject of many studies, owing to their applications in biosensors, cell imaging, and photovoltaics. Enhancement and quenching of the fluorescence depend on two conditions:

1. When the fluorophore is in a proximity greater than 10 nm to the metal nanoparticles, enhancement of fluorescent signal is dominant due to the excitation of surface plasmon on the metal surface known as metal-enhanced fluorescence.
2. When the fluorophore is in a proximity less than 10 nm to the metal nanoparticles, fluorescence resonance energy transfer is dominant, leading to quenching of the fluorescent signal.

The fluorescence effect of metal nanoparticles and fluorophore on respective fluorescent quantum yield under identical conditions is defined as

$$\mathrm{FQY} = \mathrm{FQY_R}\frac{I}{I_\mathrm{R}}\frac{A_\mathrm{R}}{A}\left(\frac{n}{n_\mathrm{R}}\right)^2 \tag{1.43}$$

where I is the integrated fluorescence intensity, A is absorbance, n is the refractive index of solvent, and R is reference fluorophore.

The ratio of emitted to absorbed photons of the fluorescence dyes theoretically determines the efficiency of fluorophores via the response surface method with the following benefits:

- The response surface method is a combination of statistical and mathematical approaches to optimize any properties.
- It is capable of simultaneously determining the individual and interactive effects of any factors of interests.
- It requires a smaller number of experiments in comparison to the conventional method.
- It can create an analytical model for the amount of metal on fluorescent quantum yield of dye molecules of interest.
- It is validated with experimental measurement.
- The total number of test samples for experiment is designed as

  $$2^k + 2k + \mathrm{n}$$

  where k is the number of input factors and n is replicated runs.
- Analysis of variance (ANOVA) can check the reliability of a prediction model of a response surface method as

  $$\mathrm{FQY} = \beta_0 + \sum_{i=1}^{k} \beta_i X_i + \sum_{i=1}^{k} \beta_{ii} X_i^2 + \sum_{i=1}^{k} \beta_{ij} X_i X_j \tag{1.44}$$

  where FQY is the forecast fluorescent quantum yield, β_0 is the constant coefficient of intercept, β_{i} is the constant coefficient of linear variable, β_{ii} is the constant coefficient of quadratic variable, β_{ij} is the constant coefficient of interaction effect, and X_{i} and X_{j} are independent variables.

1.3.3 Fluorescence Polarization

Fluorescence quenching and enhancement of metal nanoparticles and dye molecules are possible to detect by the fluorescence polarization process applicable to molecular sensing and imaging (Fig. 1.9). The concept of this method is as follows:

- The polarization degree of molecule is concerned with its rotation arising from Brownian dynamic.
- When a big molecule is excited by polarized light, it rotates slowly and hence emits more polarized light.
- When a small molecule is excited by polarized light, it rotates quickly and hence emits less polarized light.
- A small polarization degree reveals that the fluorescent molecules move easily in the surroundings.
- A high polarization degree reveals the complex of molecules.

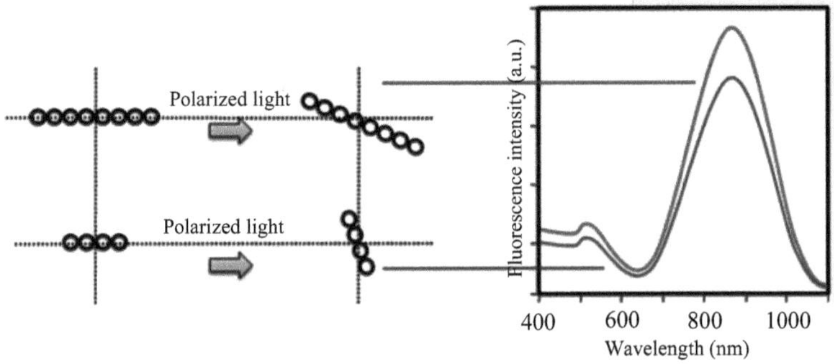

Figure 1.9 High and low fluorescence polarizations from slow and fast rotations of any molecules during the excited state, respectively.

Fluorescence polarization (FP) is experimentally carried out as follows (Fig. 1.10):

- Monochromatic light passes through the polarizer to hit the sample.
- Fluorescence of the sample is detected with horizontal and vertical planes.
- The degree of fluorescence polarization of the sample is defined as follow:

$$\mathrm{FP} = \frac{F_{//} - F_{\perp}}{F_{//} + F_{\perp}} \tag{1.45}$$

where $F_{//}$ is the fluorescence intensity parallel to the excitation plane and $F_{\perp}$ is the fluorescence intensity normal to the excitation plane.

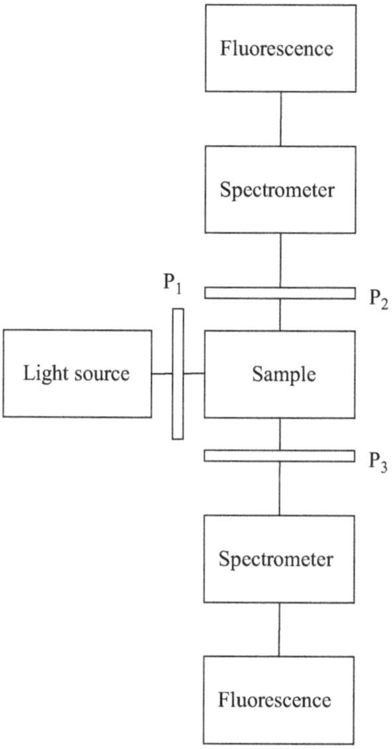

Figure 1.10 Block diagram of fluorescence polarization measurement.

1.4 Applications

1.4.1 Fluorescence Enhancer of Mercury Detection

The electronics industry, as one of the users of heavy metals, is now facing protests for heavy metal disposal causing environmental problems. The spread of heavy metals such as mercury in the human body causes severe problems, for instance nephritic syndrome, Minamata disease, cyanosis, pulmonary edema, and intelligence quotient loss. As a result, many instruments, including inductively coupled plasma-atomic absorption spectrometry, atomic fluorescence spectrometry, mass spectrometry, and gas chromatography, have been introduced so that mercury can be easily detected. However, these tools are expensive, complicated, not easy to operate, and time consuming. Thus, there is a huge demand to have sensitive and selective sensing devices for the detection of mercury and their uses in onsite tests. The fluorescence method is an alternative tool because of the following reasons:

- Tracking a trace amount of heavy metal ion is possible.
- The fluorescence method is low cost.
- It offers great sensitivity.
- It offers a quick response time for real-time observation.
- Fluorophores offer a moderate to high sensitivity to sensors.
- The fluorescence method offers an alternative approach to instant mercury determination with very simple and easy preparation of test solution without complicated instrumentation.
- The fluorescence method of mercury detection can be a designing start for improving a compact and portable instrument for onsite mercury determination.
- A fluorescence enhancer for the detection of mercury is alternatively employed for industrial applications.

Improvement in fluorescence sensing is possible through the metal-enhanced fluorescence mechanism. Although there are many monometallic nanoparticles and organic dyes suitable for sensing mercury ion, no literature has been reported on bimetallic nanoparticles and organic dyes served as the fluorescence enhancer. Herein we represent R6G hydrazine derivative and Au-core Pd-shell nanorods (Au@Pd) served as the fluorescence sensor with the following reasons:

- Fluorescence resonance energy transfer exists.
- Plasmonic phenomena between R6G and Au@Pd exist when the spirolactum-type rhodamine derivative (e.g. R6G-hydrazine) is covered on the surface of Au@Pd.
- Au@Pd and R6G behave like a donor for Hg(II) probe at the surface and hence drastically increase the fluorophoric radiative rate.
- Au@Pd and R6G offer excellent metal sorption sensitivity and selectivity of detection of Hg(II).
- The limit of detection of Hg(II) is developed to the nanometer levels if the concentration ratio of R6G:Au@Pd is well prepared.

- The limit of detection, LOD, is calculated as

$$\text{LOD} = \frac{k}{a}\sqrt{\frac{SD^2}{n} + \frac{R^2}{12}} \tag{1.46}$$

where k is the uncertainty constant, a is the slope of linear approximation, SD is the standard deviation of blank, n is the measurement cycle of fluorescence intensity, and R is the instrument resolution.

1.4.2 Fluorescence Enhancer of Latent Fingerprints Detection

Enhancement methods of latent fingerprints have been the subject of many studies due to potential applications in forensic and biomedical sciences. A latent fingerprint is the impression of the ridge pattern left on the substrate. It contains a mixture of substances from epidermis to dermis with the following details:

- Intrinsic components: metabolites, medication traces, drug usages
- Extrinsic contaminants: blood, dirt, grease, makeup, food contaminants, moisturizers, hair-care products
- Eccrine gland composed of proteins and carbohydrates: albumin, histone, protamine, glucose, lactose, amylose
- Sebaceous gland composed of lipids, sterols, and squalene: fatty acids, fatty acid esters, glycerides, cholesterol, lipid esters

The latent fingerprint constituents of those particular and variable matrices depend on the following conditions:

- Deposition stage disturbed by donor factors (i.e., sex and race), transfer conditions (i.e., contact time, angle, and pressure), substrate nature (i.e., porosity, curvature, and surface texture)
- Aging stage influenced by substrate, environmental variations (i.e., temperature, humidity, and light exposure), enhancement methods (i.e., optical, physical, chemical ways), time elapsed during the deposition stage

Many attempts have focused on physical/chemical changes in fresh and aged latent fingerprints for their image enhancements with different methods as follows:

- **Electromagnetic radiation method:** We can use a chromatic white light to produce the topographic images of fingermarks on a variety of substrates. Also we can use a green laser over blue or yellow one to visualize the untreated fingermarks on steel and plaster walls. We can use UV light to investigate fingermarks on thermal papers.
- **Small particle reagents method:** We can use zinc carbonate and eosin-dye-based small particle reagents to improve fingerprints on nonporous surfaces. Also we can use zinc carbonate and crystal-violet-based small particle reagents to detect fingermarks on nonporous color surfaces.
- **Fluorescent dyes method:** We can use fuchsin-dye-based small particle reagents to develop fingermarks on nonporous surfaces. Also we can use Nile-red to develop fingermarks on wet papers. We can use ninhydrin to increase blood-contaminated fingerprints on color-porous papers. We can use Nile-blue to observe fingermarks on papers. We can use Oil Red O to improve fingermarks on moist porous surfaces. We can use curcumin to clearly see the increase in ridge details of fingermarks on wet papers. We can use Blue B to develop cannabis-contaminated fingerprints on porous and nonporous surfaces.

Among the above-mentioned methods, most novel studies have focused on the fluorescence method as it responses very well even when the treatment time is longer. We can see a degree of red shift from fresh to prolonged latent fingerprints. For instance, fresh latent fingerprints might show green or yellow fluorescence in comparison to the orange one in old latent fingerprints. However, the fluorescence intensity of a single fluorescence dye is unlikely to provide the highest contrast mode of images of the latent fingerprints. To obtain a fairly reliable expression of data via fluorescence method, a combination of single fluorescence dye with metal nanoparticles is therefore introduced. The appropriate band gap structure of metal nanoparticles allows for additional changes in fluorescence intensity of single dye molecule under visible light. Herein we introduce Au-core Pd-shell nanorods and R6G effectively transferred on the target artificial latent fingerprints under green LED in comparison to white LED exposition.

The optical image contrast of the artificial latent fingerprints is further confirmed using the ImageJ program via the following steps:

- Image preprocessing (i.e., noise reduction and background subtraction)
- Image detection (i.e., thresholds and regions of interests)
- Image measurement (i.e., quantify intensity values)
- Gray value is counted with respect to the particular distance of sample selection in which

$$\text{Image contrast} = \text{Maximum gray value} - \text{Minimum gray value} \tag{1.47}$$

1.5 Implementations

1.5.1 Absorbance and Fluorescence

The Au-core in cetyltrimethylammonium bromide is produced via a seed-mediated growth technique. $PdCl_2$ is, hence, added in the prepared Au-core so that Pd-shells at different thicknesses are formed. For instance, the weight ratios of Pd:Au are 0.20:1, 0.10:1, 0.05:1, and 0:1. The actual amounts of Au and Pd are detected by inductively coupled plasma mass spectrometry (ICP-MS). The diameters of Au-core and the thickness of Pd-shell are characterized by transmission electron microscopy (TEM). The absorbance of Au-core Pd-shell nanorods is observed by a UV-Vis spectrometer. Xenon lamp is used as a light source. The absolute magnitudes of the sample absorption intensity are achieved by the ratio of the sample absorbance and quartz reference absorbance response to the plasmon mode. The dependence of the plasmon peak of the Au-core Pd-shell nanorods on the refractive index of the surrounding medium is carried out by dispersing them in, at least, two organic solvents, such as ethanol and toluene. For instance, the refractive indices are 1.33, 1.36, 1.41, 1.43, 1.45, and 1.49.

TEM images (Fig. 1.11) display uniformity of Au-core and Au-core Pd-shell with the following details:

- Long axis a of Au-core is 96±6 nm.
- Short axis b is 40±3 nm.

- Aspect ratio a/b is 2.4±2 nm.
- Thickness of Pd-shell is 3±0.2 nm.

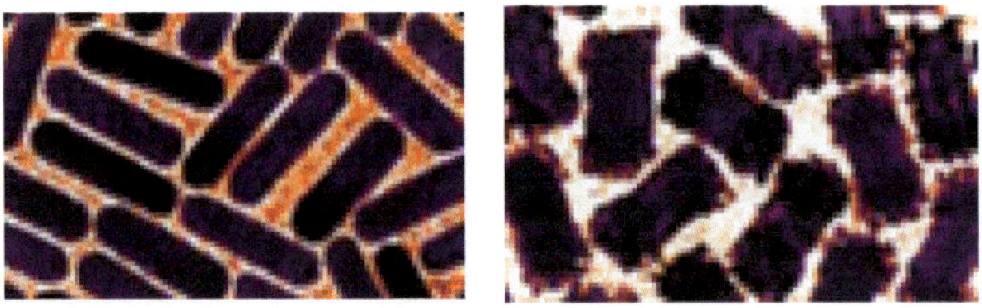

Figure 1.11 TEM images of Au-core nanorods (left) and Au-core Pd-shell nanorods (right).

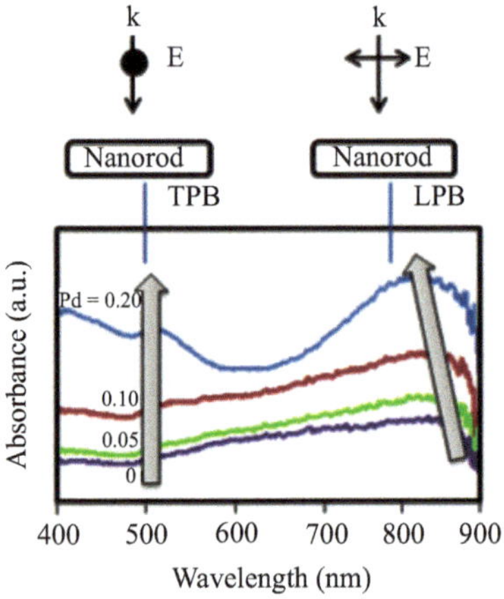

Figure 1.12 Absorbance of Au-core Pd-shell nanorods dispersed in DI water with different weight ratios of Pd:Au. TPB and LPB represent the transverse plasmon band and longitudinal plasmon band, respectively.

Two main absorption bands of plasmon resonances of Au-core Pd-shell nanorods (Fig. 1.12) are observed:

1. The left peak reveals the electron oscillations perpendicular to the long axis of rods, so-called the transverse plasmon band (TPB). Transverse oscillations of electrons provide the plasmon peak at the shorter wavelength (525 nm). Since the short axis a is predominant as sphere-like segments, the aspect ratio a/b is unity. Light polarization is less effective on such centrosymmetric structures. This plasmon mode is,

therefore, independent of the thickness of Pd-shell. Absorption maximum results in the undetectable shift in plasmon band.

2. The right peak reveals the conduction electron oscillations parallel to the long axis of rods, so-called the longitudinal plasmon band (LPB). Longitudinal oscillations of electrons provide the plasmon peak at the longer wavelength (800–900 nm). Since the long axis b is predominant as rod-like segments, the aspect ratio a/b is not equivalent to 1. Light polarization is largely effective on such non-centrosymmetric structures. This plasmon mode is, therefore, dependent on the thickness of Pd-shell. Since the real part of the permittivity of Pd (−14.40+14.60i) is less negative than that of Au (−10.92+1.49i), the optical responses of Pd-shell are dominant, whereas the influence of the Au-core on plasmonic sensitivity is perfectly screened. Consequently, the plasmon peak results in a blue-shifted LPB (820–860 nm) as the amount of Pd is increased.

Table 1.4 LPB sensitivity of Pd-shell immersed in different refractive indices of dielectric media

ethanol and toluene fractions	**water**	**100% ethanol**	**67% ethanol +33% toluene**	**50% ethanol +50% toluene**	**33% ethanol +67% toluene**	**100% toluene**
Pd:Au	1.33	1.36	1.41	1.43	1.45	1.49
0:1	860	864	869	872	874	880
0.05:1	840	843	851	858	863	870
0.10:1	830	835	840	847	855	860
0.20:1	820	826	830	835	842	850

Note: It is noted that the TPB sensitivity of Pd-shell is independent of the dielectric media, resulting in unchanged plasmon peak.

Table 1.4 shows the plasmon peak wavelength as Au-core Pd-shell nanorods are dispersed in dielectric media. Under the same dielectric medium, the LPB sensitivity on Pd-shell shows a blue shift in the plasmon peak position. Namely, plasmon peak wavelengths of Pd:Au = 0.20:1, 0.10:1, 0.05:1, and 0:1 under n = 1.33 (DI water) are 820, 830, 840, and 860 nm, respectively. Under different dielectric media, the LPB sensitivity on constant Pd-shell shows a red shift. Namely, plasmon peak wavelengths of Pd:Au = 0.20:1, 0.10:1, 0.05:1, and 0:1 under n = 1.33–1.49 are 820–850, 830–860, 840–870, and 860–880 nm, respectively. The high refractive index of the organic solvents is definitely more polarizable of the local electromagnetic field and hence couples more easily with the conduction electron oscillations on the spectral changes. The electron excitation energy is, therefore, decreased.

Comparative Study 1.1

We show a few sets of the noble metal nanoparticles as follows (Table 1.5):

- Au-core Pd-shell nanorods (Au@Pd NR)
- Au nanobipyramids (Au NBP)
- Single-crystalline porous Pd nanocrystals (Pd PNC)

Table 1.5 (a) Detailed preparation of test solution set A (R6G + Au@Pd NR)

Sample No.	Reference solution (R6G), μM	Test solution (R6G + Au@Pd NR)	
		R6G, μM	Au@Pd, mM
A1	1.5000	1.5000	3.0500
A2	0.7500	0.7500	1.5250
A3	0.3750	0.3750	0.7625
A4	0.1880	0.1880	0.3812
A5	0.0940	0.0940	0.1906

Note: R6G and C153 are dissolved in DI water and ethanol, respectively. M is mole/L.

Table 1.5 (b) Detailed preparation of test solution set B (R6G + Au NBP)

Sample No.	Reference solution (R6G), μM	Test solution (R6G + Au NBP)	
		R6G, μM	Au, mM
B1	1.0000	1.0000	83.8000
B2	0.5000	0.5000	41.9000
B3	0.2500	0.2500	20.9000
B4	0.1250	0.1250	10.5000
B5	0.0630	0.0630	5.24000

Table 1.5 (c) Detailed preparation of test solution set C (C153 + Pd PNC)

Sample No.	Reference solution (C153), μM	Test solution (C153 + Pd PNC) C153, μM	Pd, mM
C1	2.7083	2.7083	14.6800
C2	2.6875	2.6875	7.3400
C3	2.6770	2.6770	3.6700
C4	2.6719	2.6719	1.8300
C5	2.6692	2.6692	0.9100

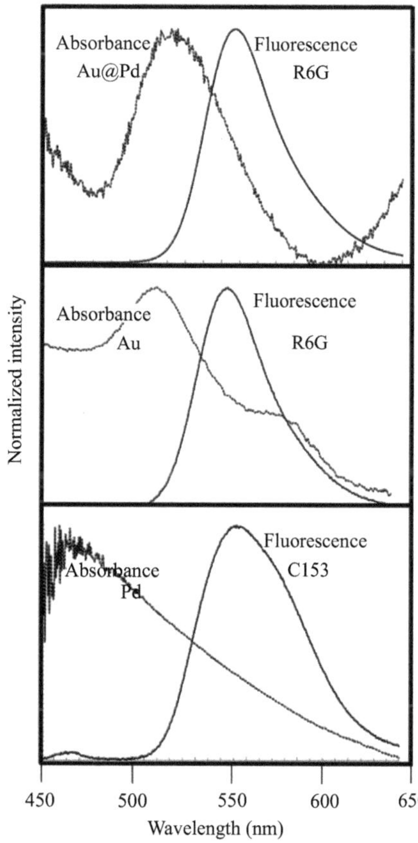

Figure 1.13 Absorbance and fluorescence of Au@Pd NR (top), Au NBP (middle), and Pd PNC (bottom) in comparison to the fluorescence of corresponding dye molecules (R6G and C153).

Their absorbance and fluorescence show some interesting points as follows:

- Au@Pd NR, Au NBP, and Pd PNC appropriately show their absorption spectra overlapping with the corresponding fluorescence spectra of R6G and C153 (Fig. 1.13).
- Fluorescence signals of such dye molecules can enhance/ quench upon the addition of metal nanoparticles without any changes in spectral shapes (Fig. 1.14), except for Au@Pd NR and R6G. The fluorescence peak of Au@Pd NR and R6G is blue-shifted because the collision of dye molecules on Au@Pd NR possibly changes the local dielectric environment of Au@ Pd NR.

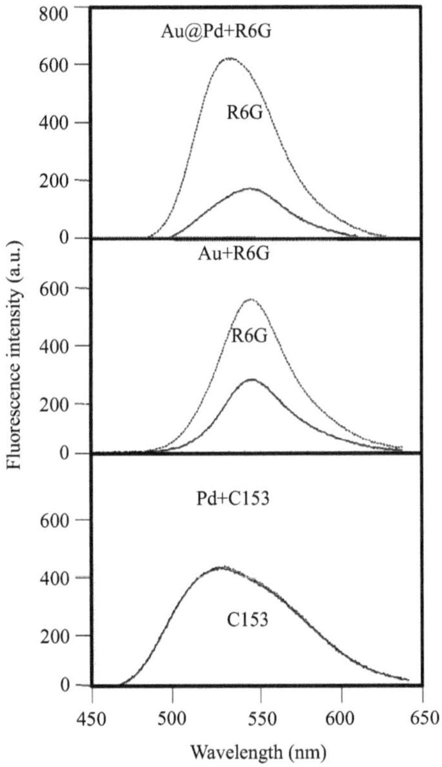

Figure 1.14 Representative fluorescence of Au@Pd NR (sample A3, top), Au NBP (sample B2, middle), and Pd PNC (sample C2, bottom) in comparison to the fluorescence of corresponding dye molecules (R6G and C153).

- The fluorescence enhancement mechanism is explained by the following reasons:
  - Increased adsorption rate of dye molecules results in the plasmonic effect easily coupling with metal nanostructures.
  - The orientation of metal nanostructures well controls the plasmonic interactions between the metal nanostructures and dye molecules.

The concentration dependence of the fluorescence intensity of dye-metal molecules is confirmed by the ratio of fluorescence intensity according to Eq. (1.48), and they reveal four aspects (Table 1.6) as follows:

$$\text{Fluorescence intensity ratio, R} = \frac{\text{Peak intensity of dye+metal}}{\text{Peak intensity of dye}} \tag{1.48}$$

Table 1.6 Fluorescence intensity ratios of metal–dye molecules for sample sets A, B, and C

	Fluorescence intensity ratio of sample number				
Sample set	**1**	**2**	**3**	**4**	**5**
A	1.95	2.75	3.50	3.20	2.70
B	2.05	2.10	2.20	2.15	1.60
C	0.98	1.05	1.10	1.05	0.80

Note: At the optimal condition, the fluorescence intensity is enhanced by 3.5, 2.2, and 1.1 folds for R6G+Au@Pd NR, R6G+Au NBP, and C153+Pd PNC, respectively.

1. As the concentrations of the test solution are high (samples A1, B1, C1), dye molecules are very close to metal nanoparticles (so-called the fluorescence quenching process), due to the high nonradiative transfer of the carriers to the metal nanostructures.
2. As the concentrations of test solution are low (samples A2, B2, C2), a gap between metal nanoparticles and dye molecules tends to increase (so-called fluorescence enhancement process). A small number of dye molecules are not easy to

couple well with metal nanoparticles due to less nonradiative transfer mechanism.

3. There is optimal concentration for plasmonic effect in each set of sample (samples A3, B3, C3), and the fluorescence intensity ratio of R6G+Au@Pd NR is the highest among them.
4. Beyond the optimal conditions (samples A5, B5, C5), the fluorescence intensity ratios decrease once again with a reduction in the concentration of metal–dye molecules, due to the re-quenching effect of the intrinsic fluorescence from the mismatching concentration of the solution.

This study concluded that Au-core Pd-shell nanorods and R6G are the best for potential applications.

Comparative Study 1.2

We show the two sets of Au-core Pd-shell nanorods and R6G in two forms: One set is prepared in solution, and another set is prepared in film (Table 1.7). The solution is mixed with 0.4 g polymethyl methacrylate (PMMA) and thus spin-coated on the glass substrate at 3500 rpm for 40 s to obtain the film thickness of 3.58 μm. The absorption band of Au-core Pd-shell nanorods in the solution form at the wavelength range of 490–550 nm reveals a good overlapping with the fluorescence band of R6G in the solution form at the wavelength range of 523–645 nm (Fig. 1.15). This point confirms a possible promotion of metal-enhanced fluorescence signal.

The fluorescence peak intensity of each sample in Figs. 1.16 and 1.17 is converted to a ratio of fluorescence intensity in Table 1.8. The intensity ratios of the solution show similar tendency to the film as follows:

- The high concentration of sample A gives the lowest intensity ratio because of high interactions of nearby molecules.
- The optimal concentration of sample B gives the highest intensity ratio because of plasmon resonance energy transfer of free electrons from Au-core Pd-shell nanorods to dye molecules.
- The low concentrations of samples C and D give a reduction in intensity ratio because the coupling between Au-core Pd-shell nanorods and dye molecules disappears.

- The maximum fluorescence intensity ratios are 1.6 and 1.4 for samples in solution and film, respectively.
- The free movement of molecules is more possible in solution than in a film embedded by a polymer matrix.

This comparative study concluded that Au-core Pd-shell nanorods and R6G in the solution form are the best for potential applications.

Table 1.7 Detailed preparation of test solution of Au-core Pd-shell nanorods and rhodamine 6G

Sample No.	Reference solution (R6G in DCM), μM	Test solution (R6G+Au@Pd) R6G, μM	Au@Pd, mM	Molar ratio of Au@Pd:R6G
A	2.0000	2.0000	0.1206	60.3
B	3.6000	3.6000	0.0516	14.3
C	4.6000	4.6000	0.0241	5.24
D	5.3000	5.3000	0.0117	2.21

Note: DCM is dichloromethane. R6G is rhodamine 6G. Au@Pd is Au-core Pd-shell nanorods. M is mole/L.

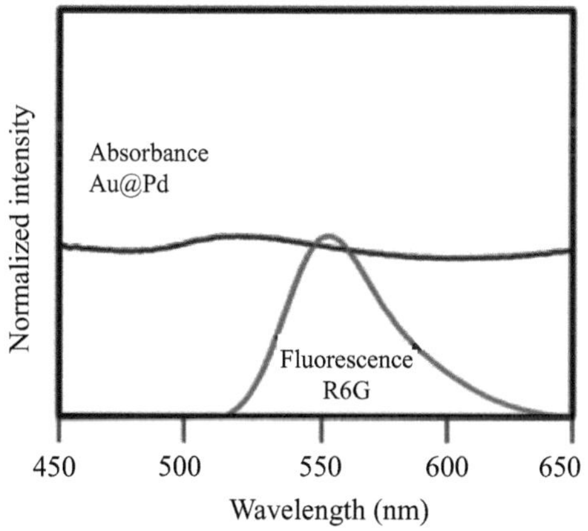

Figure 1.15 Absorbance of Au-core Pd-shell nanorods and fluorescence of rhodamine 6G in solution form. The excitation wavelength is 524 nm.

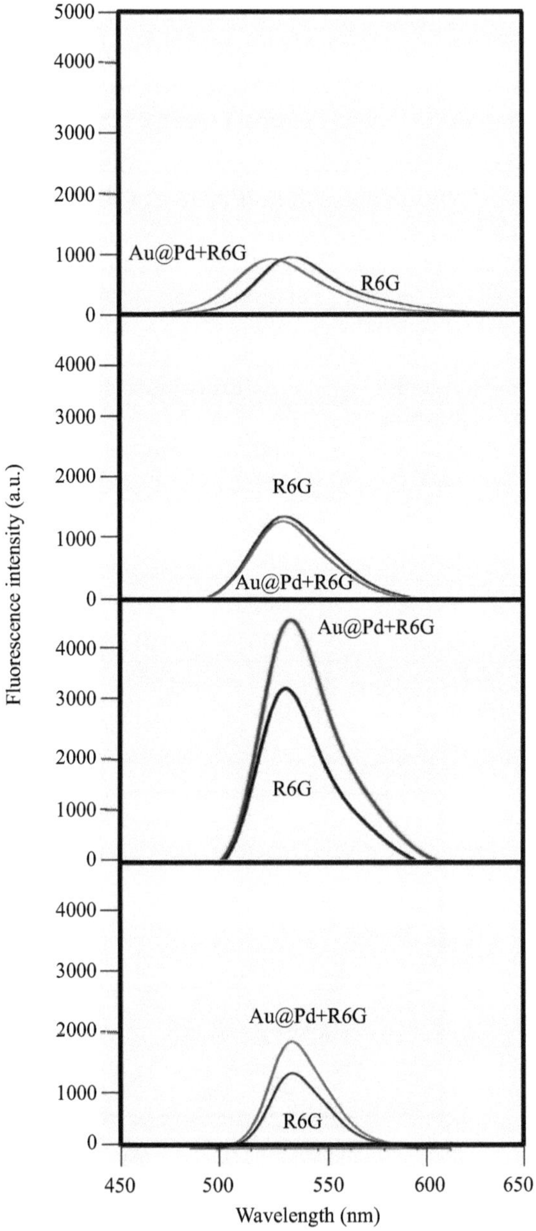

Figure 1.16 Fluorescence of Au-core Pd-shell nanorods and rhodamine 6G in the solution form from sample A–D.

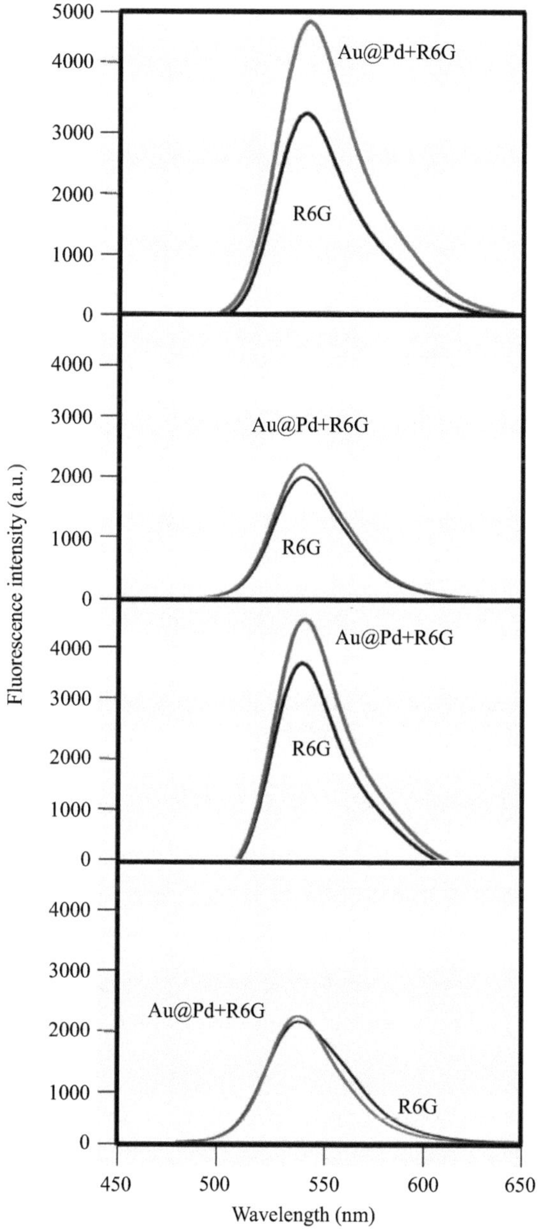

Figure 1.17 Fluorescence of Au-core Pd-shell nanorods and rhodamine 6G in the film form from sample A–D.

Table 1.8 Fluorescence intensity ratios of Au-core Pd-shell nanorods and rhodamine 6G in solution and film

	Fluorescence intensity ratio of sample number			
Sample	**A**	**B**	**C**	**D**
Solution	1.00	1.60	1.50	1.25
Film	0.95	1.40	1.10	1.00

1.5.2 Fluorescence Quantum Yield

This section covers the fluorescence spectra with respective fluorescence quantum yield of Au-core Pd-shell nanorods and R6G. The concentration of R6G stock solution is 6μM, which is then diluted to 0.35–1.70 μM, whereas the concentration of Au-core Pd-shell nanorods (ϕ = 40±3 nm, length = 96±6 nm) is 7.06 μM–0.14 mM (Table 1.9).

The R6G concentration of 1 μM is used as reference fluorophore with the following spectroscopic information:

- Fluorescent quantum yield is 0.95.
- Absorbance is 0.055.
- Integrated fluorescence intensity is 149435.761.
- Refractive index is 1.3320.

Using the response surface method, the input factors are composed of

- Concentration of Au-core Pd-shell nanorods
- Concentration of R6G

The total number of test samples for the experiment is

$$2^k + 2k + n = 13$$

where k is the number of input factors (k = 2) and n is replicated runs (n = 5).

It has been found that both absorbance and fluorescence are increased upon the addition of Au-core Pd-shell nanorods. The absorbance peak is centered at the wavelength of 526 nm, whereas the fluorescence peak is centered at the wavelength of 556 nm (Fig. 1.18). The fluorescence intensities are, hence, used to calculate the fluorescent quantum yield and showed in Table 1.10. The scattering effect by metal particles reduces the light path through a cuvette and then results in the quenching of the fluorescent quantum yield.

Table 1.9 Detailed design parameters using Design Expert Software 9.0.3 and their experimental results

	Concentration of		Absorbance, *A*		Integrated fluorescence, *I*		Refractive index, *n*		Measured FQY	
Sample No.	R6G, μM	Au@Pd, μM	R6G	Test solution	R6G	Test solution	R6G	Test solution	R6G	Test solution
1	0.35	71.70	0.022	0.119	50302.005	228739.129	1.3312	1.3319	0.78	0.67
2	0.55	26.00	0.032	0.071	82603.744	224184.172	1.3322	1.3316	0.90	1.00
3	0.55	117.00	0.032	0.166	83819.022	223775.028	1.3319	1.3321	0.90	0.47
4	1.03	7.060	0.058	0.075	159959.841	241897.851	1.3324	1.3322	0.97	1.00
5	1.03	71.70	0.058	0.155	158155.123	275404.869	1.3319	1.3318	0.95	0.62
6	1.03	71.70	0.059	0.158	161241.487	278442.149	1.3316	1.3319	0.95	0.61
7	1.03	71.70	0.057	0.157	158536.007	281573.074	1.3319	1.3317	0.97	0.63
8	1.03	71.70	0.058	0.157	161413.131	279859.644	1.3322	1.3324	0.97	0.63
9	1.03	71.70	0.067	0.159	180939.336	279066.648	1.3326	1.3316	0.94	0.61
10	1.03	136.00	0.058	0.216	156218.158	215292.498	1.3315	1.3318	0.94	0.35
11	1.50	26.00	0.082	0.116	224084.840	313036.022	1.3335	1.3326	0.96	0.95
12	1.50	117.00	0.081	0.225	221412.541	255871.570	1.3323	1.3339	0.96	0.40
13	1.70	71.70	0.096	0.194	260328.150	320301.143	1.3317	1.3321	0.95	0.58

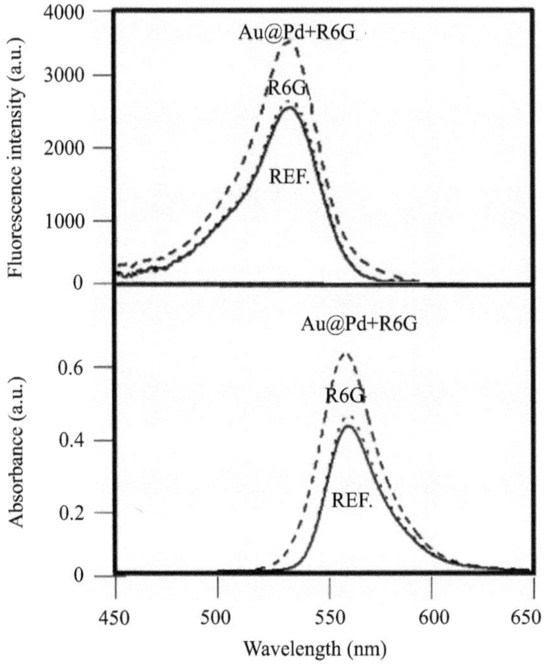

Figure 1.18 Absorbance (bottom) and fluorescence (top) of reference fluorophore, rhodamine 6G, Au-core Pd-shell nanorods, and rhodamine 6G from sample 4 as a representative example.

The response surface method offers a mathematical model, showing a relationship of the concentration dependence of the fluorescent quantum yield, FQY, as

$$\begin{aligned} \text{FQY} = {} & 1.48345 - (2.93524 \times 10^5 \times A) - (11949.31203 \times B) \\ & + (9.20837 \times 10^8 \times A \times B) + (6.64820 \times 10^{10} \times A^2) \\ & + (3.34804 \times 10^7 \times B^2) \end{aligned} \tag{1.49}$$

where A is the concentration of R6G and B is the concentration of Au-core Pd-shell nanorods.

ANOVA, as shown in Table 1.10, confirms that a mathematical model via the response surface method is reliable because the p-value is less than 0.0001 together with the F-value of 129.45. It is also found that the concentration of Au-core Pd-shell nanorods is inversely proportional to the fluorescent quantum yield

(Fig. 1.19). The integration of Au-core Pd-shell nanorods in the dye molecules decreases the fluorescent quantum yield of dye molecules in comparison to a typical value of a reference fluorophore (0.95), due to the nonradiative energy transfer of the metal–dye molecules. Therefore, this case study concluded that Au-core Pd-shell nanorods are served as the fluorescence quencher.

Table 1.10 Analysis of variance (ANOVA) of the response surface method

Source	Sum of squares	Degree of freedom	Mean square	F-value	p-value
Model	0.6900	5	0.1400	129.4500	<0.0001
A = R6G	0.0150	1	0.0150	14.0500	0.0072
B = Au@Pd	0.6400	1	0.6400	599.7300	<0.0001
AB	0.0016	1	0.0016	1.4900	0.2616
A^2	0.0015	1	0.0015	1.4600	0.2664
B^2	0.0340	1	0.0340	31.7700	0.0008
Residual	0.0075	7	0.0012	—	—
Lack of fit	0.0071	3	0.0024	23.7000	0.0052
Pure error	0.0004	4	0.0001	—	—
Core total	0.7000	12	—	—	—

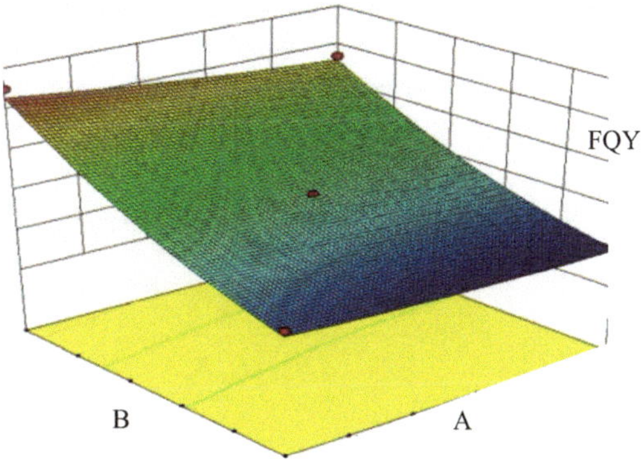

Figure 1.19 Response surface plot of concentration dependence of the fluorescent quantum yield, FQY, of Au-core Pd-shell nanorods and rhodamine 6G.

1.5.3 Fluorescence Polarization

This section covers the fluorescence spectra with respective fluorescence polarization of Au-core Pd-shell nanorods and R6G (Table 1.11).

Table 1.11 Detailed preparation of Au-core Pd-shell nanorods (ϕ = 90±10 nm) and rhodamine 6G

Sample No.	R6G, μM	Au@Pd, mM	Concentration ratio of Au@Pd:R6G
1	5	0	—
2	5	0.62	124
3	5	1.86	372
4	5	3.10	619
5	5	4.34	867
6	5	5.57	1115

The fluorescence polarization setup is modified a little bit as follows:

- The sample is excited with the linear polarized light via the 3 mW green LED at the wavelength of 532 nm and dichroic polarizer.
- Fluorescence is detected at a normal angle to the incident light.
- The fluorescence intensities parallel, $F_{//}$, and perpendicular, $F_{\perp}$, to the excitation plane are simultaneously detected using two UV/Vis spectrometers.
- The fluorescence polarization is, therefore, modified as

$$\mathrm{FP} = \frac{\int F_{\lambda,//} d\lambda - \int C_{\lambda} F_{\lambda,\perp}}{\int F_{\lambda,//} d\lambda + \int C_{\lambda} F_{\lambda,\perp}} \tag{1.50}$$

where

$$\text{Correction factor, } C = \frac{\text{Fluorescence intensity from spectrometer 1}}{\text{Fluorescence intensity from spectrometer 2}} \tag{1.51}$$

The addition of Au-core Pd-shell nanorods to R6G results in fluorescence quenching (Fig. 1.20). This is originated by the nonradiative energy transfer process from the excited fluorophores to the metal nanoparticles when the fluorescence of dye molecules as donor and the absorption spectra of metal as acceptor overlap. The fluorescence polarization as a function of the test solution is summarized in Table 1.12 with the following details:

- Fluorescence polarization shows a small change when a small concentration ratio of Au@Pd:R6G is applied (sample 2).
- Fluorescence polarization increases when a high concentration ratio of Au@Pd:R6G is applied (samples 3–5) because of the shortening of the fluorescence lifetime of dye molecules.
- Fluorescence polarization decreases when the concentration ratio of Au@Pd:R6G goes over 867 (sample 6) because metal nanoparticles behave like a heat source. Additional molecules of metal nanoparticles increase the temperature of the solution system and hence result in faster rotation of fluorescence molecules.

This study concluded that the fluorescence intensity ratio is inversely proportional to fluorescence polarization, which can predict the degree of fluorescence quenching.

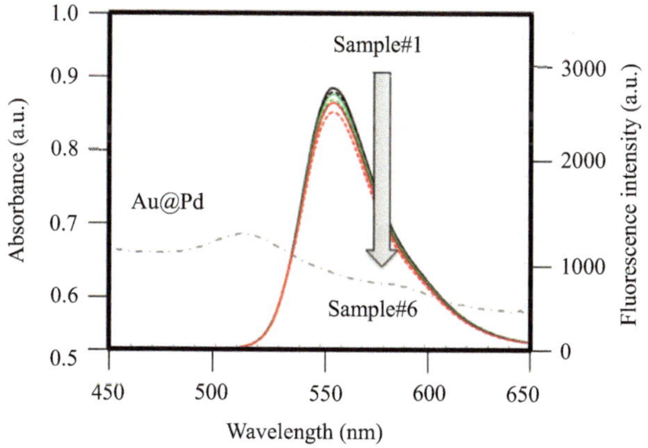

Figure 1.20 Absorbance and fluorescence of Au-core Pd-shell nanorods and rhodamine 6G.

Table 1.12 Fluorescence polarization and fluorescence intensity ratio of Au-core Pd-shell nanorods and rhodamine 6G

Sample No.	Fluorescence polarization	Fluorescence intensity ratio	Concentration ratio of Au@Pd:R6G
1	0.048	0.986	—
2	0.051	0.985	124
3	0.067	0.978	372
4	0.073	0.960	619
5	0.078	0.948	867
6	0.075	0.945	1115

1.5.4 Fluorescence Enhancer of Mercury Detection

One hundred milligram of R6G and 0.1 mL of triethylamine are mixed in 5 mL of absolute ethanol under Ar atmosphere for 30 min. Then 0.1 mL of hydrazine monohydrate is mixed with the solution. This mixture is heated under Ar gas to reflux for 24 h. After cooling down, the residue is filtered and dried under room temperature. Finally, 67 mg R6G–hydrazine derivative is achieved. The developed R6G–hydrazine is mixed with Au–Pd core–shell nanorods (Au@Pd, 40±3 nm diameter, 96±6 nm length) according to Table 1.13. Mix metal ions such as Na(I), K(I), Ag(I), Cu(II), Pb(II), Ca(II), Hg(II), and Fe(III) at concentrations of 0–15.5 μM are studied as representative metal ions in tap water. The fluorescence of each sensor on metal ion is detected by a luminescence spectrometer. Excitation and emission wavelengths are 500 and 550 nm, respectively.

Table 1.13 Detailed preparation of different sensors for metal detection (M is mole/L)

Sensor No.	R6G, μM	Au@Pd, M
A	4.40	—
B	4.40	10.00
C	4.40	1.00
D	4.40	0.01

Sensor B reveals the highest fluorescence intensity centered at the wavelength of 550 nm due to the following reasons (Fig. 1.21):

- The transfer of fluorescence energy of dye molecules to Au–Pd core–shell nanorods (Au@Pd) dampens the dipole oscillations by a nearby metal.
- The conducting Au@Pd increases the local incident field on dye molecules via plasmonic effect and hence increases the excitation rate.
- The nearby Au@Pd enhances the intrinsic radiative decay rate of dye molecules and thus raises the emission rate.

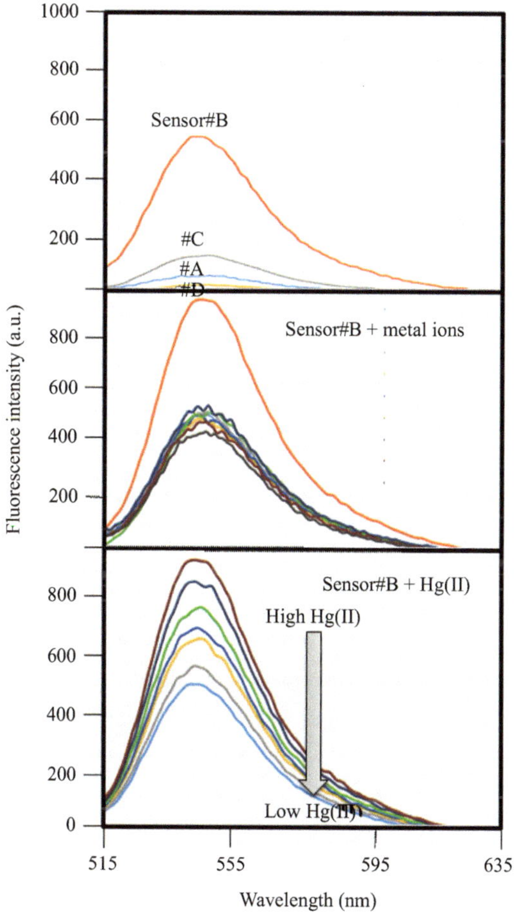

Figure 1.21 Fluorescence spectra of sensors A–D (top), selective sensor B under various metal ions with the identical concentration of 15.5 μM (middle), and sensitive sensor B with different concentrations of Hg(II) (bottom).

Not only does sensor B offer the highest selectivity to analyte Hg(II), but also it is sensitive to Hg(II) dosage.

The response of fluorescence of sensor B on Hg(II) dosages is fitted as

$$y = 2.4372x^2 - 9.2612x - 550.72$$

Since the second term is neglected as sensor B functions close to zero metal concentration, the approximation is modified as

$$Y = 9.2612x - 550.72$$

The limit of detection (LOD) is calculated together with the following numbers:

- k is the uncertainty constant ($k = 3$).
- a is the slope of linear approximation ($a = 9.2612$ 1/μM).
- SD is the standard deviation of blank (SD = 0.11).
- n is the measurement cycle of fluorescence intensity ($n = 3$).
- R is the instrument resolution ($R = 0.01$).

The limit of detection of sensor B is 20.6 nM. This loading amount is significantly reduced in comparison to the harmless level for tap water set by the Environmental Protection Agency (Fact Sheet EPA-823-F-01-011), USA.

1.5.5 Fluorescence Enhancer of Image Enhancement of Latent Fingerprints

Artificial latent fingerprints are produced by the following steps:

- The cartoon fingerprint model available for free download is transferred onto the transparent card.
- The gelatin solution serving as artificial fingerprints is deposited on the substrate and then left under room temperature for 24 h.
- The α-amylase serving as latent print residue is dropped on the artificial fingerprints and then put in the fridge to dry.
- Artificial latent fingerprints are ready for test.

Rhodamine 6G of concentration 1.5 μM is prepared from 6.0 μM stock dye solution and then mixed with Au@Pd ($\phi = 40 \pm 3$ nm, length

= 96±6 nm) (Table 1.14). Test solutions are divided into two sets. The first set is added to a cuvette to detect fluorescence spectra via a UV-Vis spectrometer. The second set is applied on the well-prepared artificial latent fingerprints after transferring them on optical plastic type under 10 N force for 20 s. The luminescence image quality of the treated artificial latent fingerprints is exposed by a 3 W green LED attached on the optical microscope at a large oblique angle of 60° to highlight a scattering light of fingermarks and is, hence, recorded by a CCD camera. Images of artificial latent fingerprints covered with metal-fluorescence dye molecules are simultaneously detected under white light and LED so that all images have the same camera gain at the start of light irradiations. Note that the plain fingerprint marks of untreated metal-fluorescence dyes are less visible to the naked eye.

Table 1.14 Detailed preparation of test solution prior to application on artificial latent fingerprints

Sample No.	Au-core Pd-shell nanorods, mM	Molar ratio of Au@ Pd:R6G
A	0.004	3: 1
B	0.009	6: 1
C	0.014	9: 1
D	0.018	13: 1
E	0.023	16: 1
F	0.070	47: 1
G	0.117	78: 1
H	0.164	110: 1

Note: Concentration of rhodamine 6G is 1.50 μM. M is mol/L.

The recorded fluorescence of the test solution before and after application on artificial latent fingerprints is shown in Fig. 1.22. Fluorescence enhancement existed for samples A–D, while fluorescence quenching appeared for samples D–H. The fluorescence quencher and enhancer of the test solution reveal the dual-mode imaging for concurrent visualization of the artificial latent fingerprints.

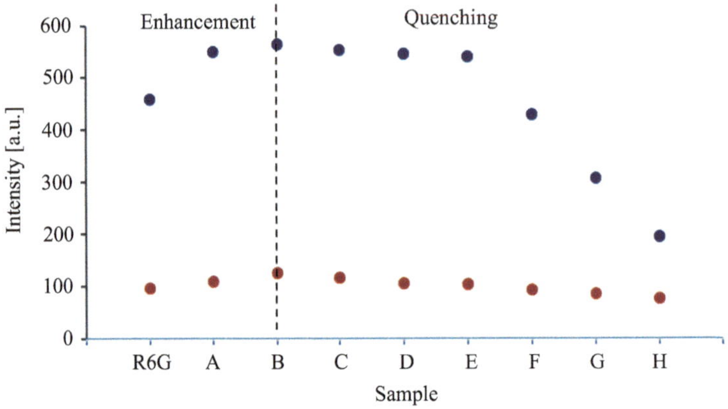

Figure 1.22 Fluorescence intensity ratio of test solution before (upper plots) and after (lower plots) application on the artificial latent fingerprints.

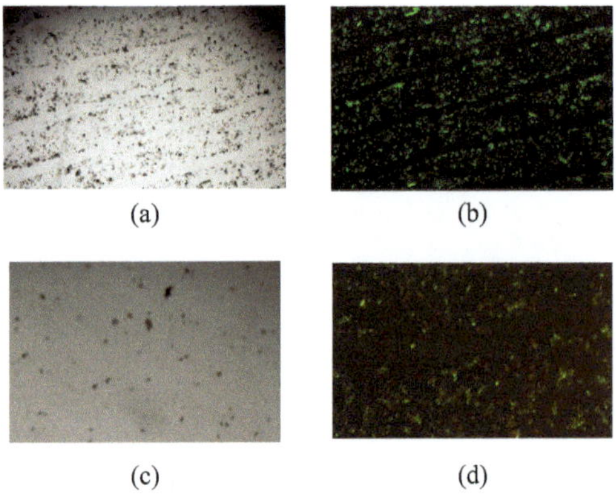

Figure 1.23 Luminescence images of the treated artificial latent fingerprints by representative samples C (a, b) and H (c, d). Excitation is white light (a, c) and LED (b, d). The excitation wavelength of the green LED is 500 nm. Optical magnification is 4×. Black dots are observed under a bright field, while green dots are observed under a dark field.

Metal-fluorescence dye embedded fingerprint marks are highly visible to the unaided eye. Figure 1.23 reveals the fluorescence enhancement and quenching on artificial latent fingerprints as they are both seen under incident white light and LED expositions. Latent fingerprints marked by fluorescence enhancer under LED exposition are most visible than others. The deposited dye is noticeably visible, whereas a contrast reagent such as α-amylase shows luminescence that differs from those of the background compounds.

The optical image contrast of the treated artificial latent fingerprints is further confirmed using the ImageJ program (Fig. 1.24). Image contrasts of the treated artificial latent fingerprints are divided into two sets: one is for fluorescence enhancement and another one is for fluorescence quenching (Table 1.15).

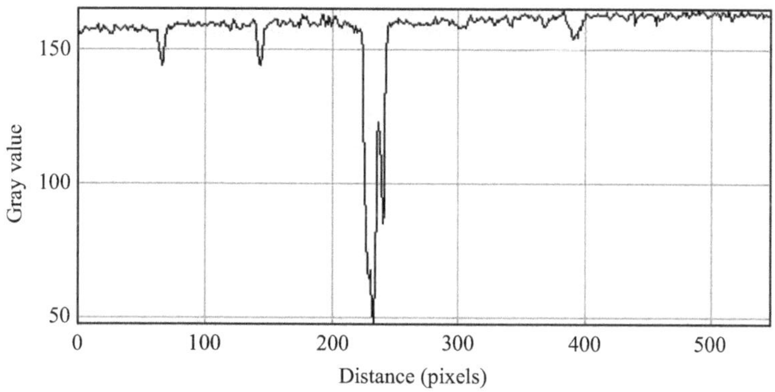

Figure 1.24 Gray value with a specific distance of artificial latent fingerprint image covered with metal-fluorescence dye molecules when a representative sample C is excited by LED. Image contrasts are easily compared in terms of difference of gray value from the valley and ridge patterns of the fingerprints.

Table 1.15 Concentration dependence of the luminescence image contrast excited by white light and LED of the treated artificial latent fingerprints by the representative samples A–C and F–H

Sample No.	Green LED	White LED	Molar ratio of Au@Pd:R6G
A	60	50	3: 1
B	80	70	6: 1
C	100	90	9: 1
F	30	25	47: 1
G	20	15	78: 1
H	10	5	110: 1

Enhancement of image contrasts of the treated artificial latent fingerprints by metal-fluorescence enhancer under LED light is still better than others. The difference of image contrasts of the treated artificial latent fingerprints by metal-fluorescence quencher under different expositions is further improved. Interestingly, the contrasts of luminescence images of the treated latent fingerprints from the fluorescence enhancement are fourfold higher than those from the fluorescence quenching. The degrees of image contrasts are dependent on the following factors:

- They depend on how successfully the amount of metal-dye molecules map on the artificial latent fingerprints.
- The larger the surface area of the metal, the bigger the binding sites for the functional molecules such as α-amylase, similar to human saliva, to notice the targets in the latent fingerprints.
- As the localized surface plasmon resonance of the metal is concurrent coupled with the emission of the dye molecules by a single-excitation mode, the electromagnetic enhancement facilitates the improvement in the emission mode in the latent fingerprints.

Problems

1.1 Write Maxwell's equations in the form of integral and differential equations.

1.2 Show a solution to the current density with respect to the electric field based on the Drude model.

1.3 Draw the surface plasmon resonance of metal nanowires due to the difference in applied electric field.

1.4 Draw a relationship between electric field distribution and different configurations of metal nanostructures due to the plasmonic effects.

1.5 Explain the optical line spectra of hydrogen and sodium atoms.

1.6 Show the database to obtain the dielectric constants of some materials.

Chapter 2

ZnO Nanoparticles

2.1 Introduction

This chapter deals with ZnO nanoparticles with narrow size distributions produced by the pulsed laser ablation method. The laser-fluence dependence of the absorption intensity of ZnO nanoparticles displays a single and sharp absorbance peak. ZnO nanoparticles with different dosages under UV exposition, therefore, possibly hold the promising phototoxic effect on human breast cancer cells. The intimate contact between the cancer cell wall and ZnO nanoparticles upon UV exposition causes changes in the environment of the cell–nanoparticles contact area. The electron–hole pair generated by ZnO nanoparticles produces intracellular reactive oxidation species to destroy the nucleus of the cancer cells. For other purposes, ZnO/Au nanocomposites under UV exposition improve the plasmon-enhanced photocatalytic efficiency of the film via plasmonic and Schottky barrier effects, leading to a reduced amount of heavy metal in wastewater. Moreover, the photovoltaic efficiency of ZnO/coumarin-153 nanocomposites under the UV-Visible regime is applicable to the low-cost per unit of power generation of the dye-sensitized solar cells.

Linear and Nonlinear Optics: Materials, Properties, and Applications
Kitsakorn Locharoenrat

ISBN 978-981-4877-15-2 (Hardcover), 978-1-003-05588-4 (eBook)
www.jennystanford.com

2.2 Fabrications

Because of the size- and shape-dependence of the physical properties of nanomaterials in comparison to the bulk, many studies have sought to develop the nanomaterial production processes. Nanomaterials such as ZnO nanoparticles can be produced by the following methods:

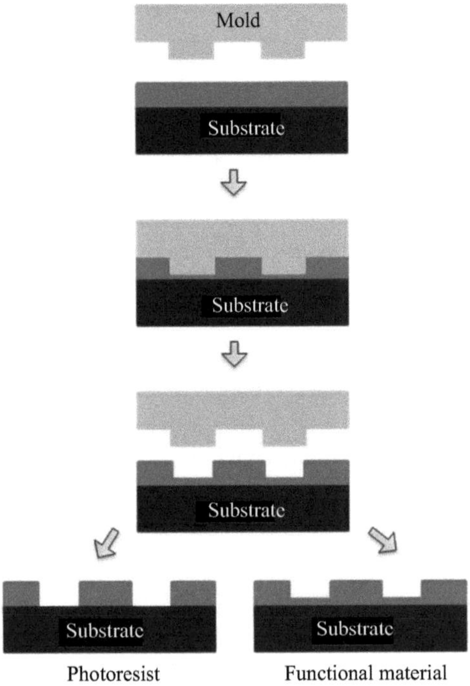

Figure 2.1 Illustration of the nanolithography techniques of the quantum wires.

2.2.1 Nanolithography

The nanolithography method (Fig. 2.1) offers self-assembled nanostructures with a high throughput and low production cost. Not only do they show intrinsic bistability in transport characteristics exploited to realize the nonvolatile quantum dot memory, they can also show signatures of Coulomb blockade and staircase under ambient temperature in quantum wires. This kind of structure is proposed to seek utilization in single-electron transistors. However, there are a few weak points such as technical complexity, low production yields, and contamination caused by organic/inorganic solvents.

2.2.2 Laser Ablation

Pulsed laser ablation (Fig. 2.2) in the solution medium produces size-selected nanoparticles. When the high energy of the pulsed laser, such as Nd:YAG, focuses on the surface of the metal plate, it vaporizes the metal atoms, reacting with the dissolved O_2 in the liquid medium to generate metal oxide nanoparticles. A massive number of metal oxide nanoparticles arrange themselves in precise geometry, whereas they are coated by a layer of the liquid medium to avoid further coalescence and agglomeration and to reduce particle growth. To avoid the dissolved O_2 in the liquid medium, an inert gas such as argon is suggested to be exposed into the liquid medium to form metallic nanoparticles without oxygen component.

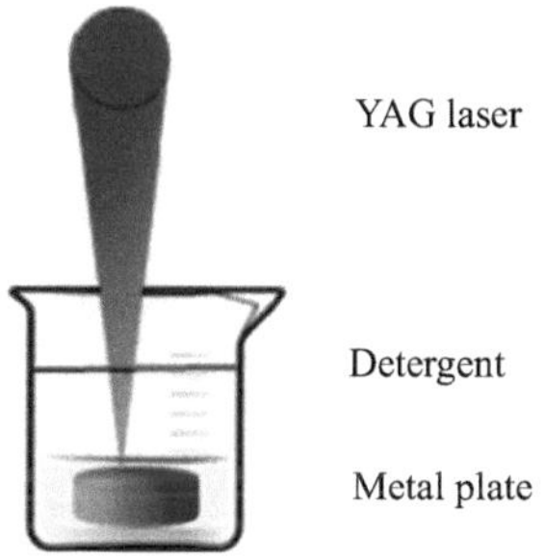

Figure 2.2 Illustration of pulsed laser ablation of the metal oxide nanoparticles.

2.3 Applications

2.3.1 Biomedical Uses

Uncommon DNA results in the speedy increment of cell growth and proliferation, and hence normal human cells become cancer cells. As reported by the National Cancer Institute, breast cancer is the most common cause of human death and mostly found in women worldwide (www.cancerresearchuk.org/health-professional/

cancer-statistics). The following representative nanoparticles bounded with chemo drug with a small dosage are suggested to get better results for killing human breast cancer cells.

- Nanodiamonds offer the following benefits and drawbacks:
  - They have low impurity.
  - They have low toxicity.
  - They are expensive.
- Metal oxide nanoparticles have the following benefits and drawbacks:
  - They kill bacteria.
  - They are used in healthcare products (i.e., sunscreens).
  - They have good biocompatibility.
  - They have poor toxicity.
  - They have enough energy comparable to UV-Vis light (Fig. 2.3).
  - They are cheap.

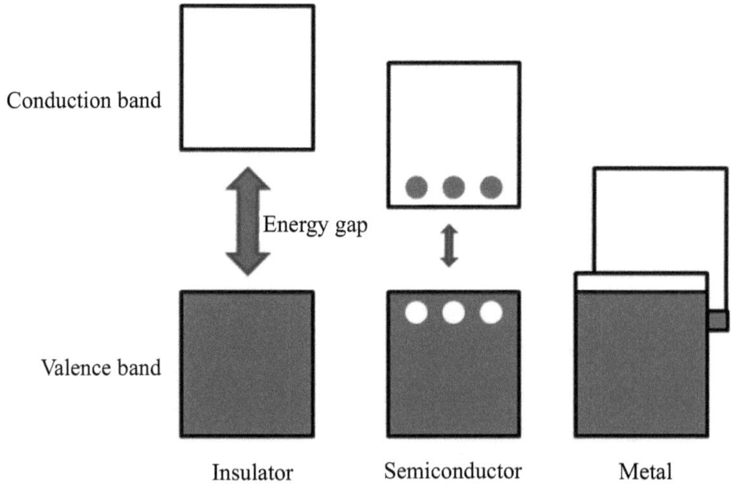

Figure 2.3 Energy gap of semiconductor in comparison to insulator and metal.

Herein we introduce the potential application of ZnO nanoparticles in human breast cancer therapy because of the following reasons:

- Size- and shape-dependence of the optical properties of ZnO nanoparticles in comparison to the biological molecules are suggested to improve the photocatalytic properties, making them suited for the biological applications.
- ZnO nanoparticles under the diameter of 10–150 nm efficiently move through the cells and thus selectively interact with the biomolecules inside the cells.
- The development of the cell-killing approach gives us the scientific understanding of interactions between nanomaterials with biological systems.
- The phototoxic effect of ZnO nanoparticles upon appropriate light exposition increases the cytotoxic reactive oxygen species in the aqueous media, such as superoxide and hydrogen peroxide, so that the cell morphologies, for instance chromatin condensation and fragmentation, are deviated.
- The dosage of ZnO nanoparticles of less than 2.56 mg/mL does not lead to destruction of the red blood cells served as normal cells.

The anticancer activity of ZnO nanoparticles under UV light is improved through the following mechanisms (Table 2.1 and Fig. 2.4):

Table 2.1 (a) Representative work functions of metals

Metal	Work function (eV)
Pt	5.65
Ni	5.15
Pd	5.12
Au	5.10
Mo	4.60
W	4.55
Cr	4.50
Ti	4.33
Al	4.28
Ag	4.26

Table 2.1 (b) Representative work functions of semiconductors

Semiconductor	Work function (eV)
GaAs	4.07
AlAs	3.50
TiO_2	3.20
CuO	4.10
ZnO	3.37

- UV light has enough energy comparable to ZnO nanoparticles. It then excites electrons to the conduction band and leaves holes in the valence band.
- A series of phototoxic effects exist. The electrons interact with oxygen to produce the superoxide radial anion, whereas the holes interact with hydroxyl ions or water to generate the hydroxyl radical.
- Not only does the generated reactive oxygen species lead to DNA damage, they also damage the extensive cytotoxic membrane via lipid peroxidation or protein denaturation.

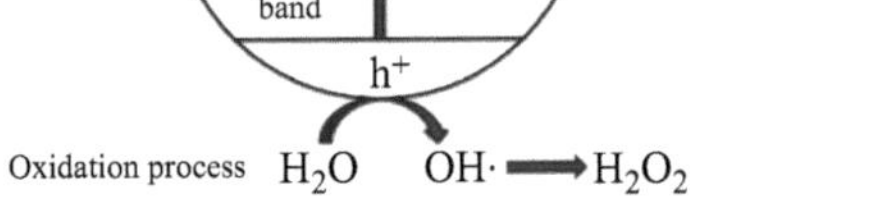

Figure 2.4 Formation of the reactive oxygen species (ROS) via electron/hole pairs of ZnO nanoparticles under UV light.

2.3.2 Industrial Uses

Heavy industries, such as automobile, are now encountering the problem of heavy metal disposal (Table 2.2 and Fig. 2.5). The remediation of heavy-metal-contaminated wastewater using photocatalytic materials under sunlight is, therefore, a greener solution. Nano-sized semiconductor photocatalysts such as ZnO have been introduced in place of the existing TiO_2 and WO_3 due to the following reasons:

Table 2.2 Global automotive market trend

Year	Manufacturing industry value (billion dollars)
2013	1700
2014	1850
2015	1900
2016	2050
2017	2250
2018	2400

Source: ReportLinker

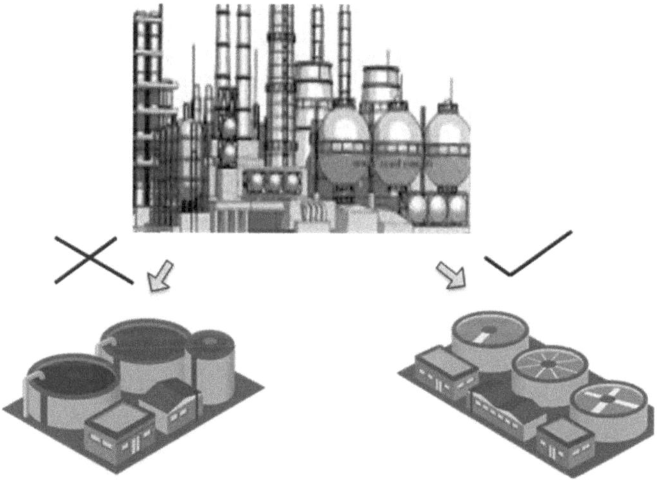

Figure 2.5 Heavy industries with treated and untreated wastewater.

- Large surface-to-volume ratio allows higher adsorption in the target molecules.
- The nanoparticles–pollutant interaction is sensitive and selective to their fluorescence spectra.
- ZnO nanoparticles can remove both chemical and biological pollutants.

- ZnO nanoparticles absorb light efficiently in visible and near UV spectra.

As the noble metal, the so-called plasmonic photocatalyst, is dispersed in the semiconductor, the high amount of electrons in vacant states further inhibit recombination of electron–hole pairs under specific light exposition. Although many plasmonic photocatalysts such as ZnO/Au have been introduced to eliminate most organic dyes (e.g., rhodamine B, rhodamine 6G, thionine), no study has been reported on ZnO/Au applicable in eliminating inorganic pollutants such as heavy metals. The mechanisms of plasmon-enhanced photocatalyst efficiency of ZnO/Au on heavy metal removal are as follows:

- Coupling of electric field and free electrons of Au nanoparticles causes the surface plasmon resonance (SPR).

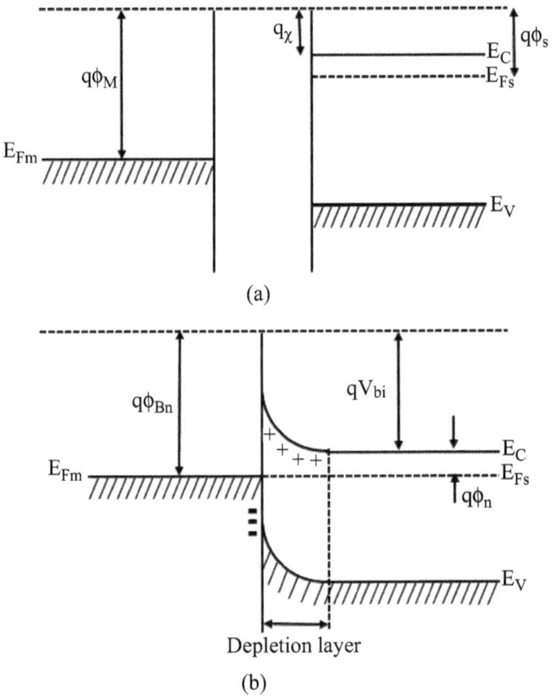

Figure 2.6(a) Energy band diagram at a junction of metal and n-type semiconductor before contact with each other (a) and after contact with each other (b) ($\Phi_S < \Phi_M$). Φ is the Schottky barrier height, V_{bi} is the build-in potential, E_C is the conduction band energy, E_F is the Fermi energy, E_V is the valence band energy, and q is electric charge.

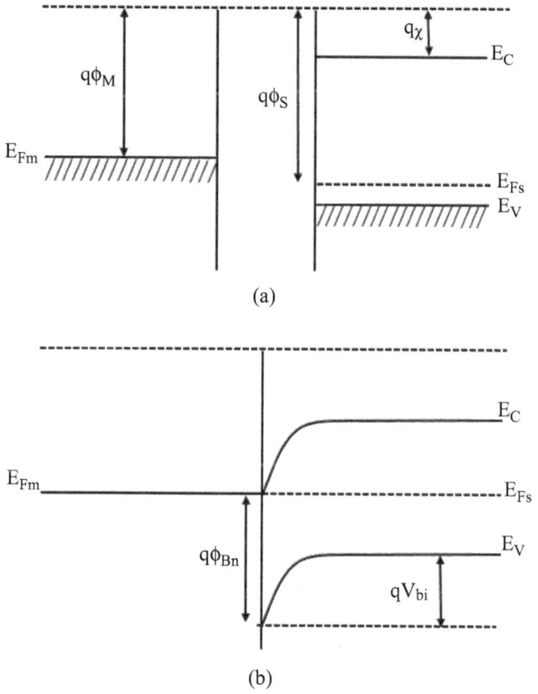

Figure 2.6(b) Energy band diagram of metal–p-type semiconductor junction (top) before contact and (bottom) after contact of the metal-semiconductor ($\Phi_M < \Phi_S$).

- Electrons from ZnO nanoparticles are excited from the valence to the conduction bands to produce electron–hole pair as

$$Au + h\nu \rightarrow SPR \quad (2.1)$$

$$ZnO + h\nu \rightarrow e^- + h^+ \quad (2.2)$$

- Both Au and ZnO strongly induce separation of the electron–hole pair of ZnO as

$$SPR + h^+ + 2H_2O \rightarrow H_2O_2 + 2H^+ \quad (2.3)$$

$$SPR + e^- + O_2 \rightarrow \bullet O^-_2 \quad (2.4)$$

- As the Fermi energy level of ZnO is higher than that of Au, electrons at the conduction band of ZnO are transferred to the hot electrons of the in-contact Au (Fig. 2.6).

- The localized electrons from Au are unable to cross over the Schottky barrier height of ZnO, and the confined electrons of Au are, hence, trapped at the ZnO–Au interface, leading to the rapid separation rate of the electron–hole pair.
- A high amount of electrons existing at the ZnO–Au interface continuously react with Fe(II) served as pollutant as

$$2H^+ + \bullet O^-_2 + e^- \rightarrow H_2O_2 \quad (2.5)$$

$$Fe^{2+} + H_2O_2 \rightarrow Fe^{3+} + \bullet OH + OH^- \quad (2.6)$$

- The photocatalytic performance of ZnO/Au film is finally improved as a result of a reduced amount of ferrous ions.

2.3.3 Photovoltaic Uses

Dye-sensitized solar cells (DSSCs) have been the subject of numerous studies as an example of photovoltaic technology and an alternative to silicon photovoltaic cells since their discovery in 1991. Their markets are now steadily growing due to high efficiency, low cost, and ease of manufacturing under ambient conditions (Table 2.3).

Table 2.3 Global solar photovoltaic market trend

Year	Manufacturing industry value (billion dollars)
2014	108.3
2015	113.1
2016	118.5
2017	123.2
2018	128.7
2019	133.9
2020	148.5

Source: Market research store

DSSCs are constructed with the following main materials:

- Metal oxide film
- Photosensitive dye
- Electrolyte-hole transporter
- Counter electrode

Among them, dyes serve as a sensitizer and play a vital role in the DSSC system. The main function of dyes is to absorb light in the visible and near-infrared parts of the spectra, considering that metal oxide only absorbs in the UV part of the spectrum. In the past, the necessity of an alternative to ruthenium-based dyes appeared because of the high cost of rare metals and the difficult process of synthesizing ruthenium-based dyes, as well as complicated purification processes. Nowadays, organic dyes are an alternative due to the following reasons:

- Organic dyes are easily synthesized and sometimes extracted from the plants.
- Structural modifications are easily obtained by a simple synthesis process in order to tune position and intensity of the charge transfer transition.
- Organic dyes have a higher molar extinction coefficient in comparison to ruthenium-based dyes.
- Among the metal-free organic dyes, the most common system is donor–π-bridge–acceptor (D-π-A), which ensures effective charge transport and separation.
- Since the electron donor affects the molecular energy level, it affects the absorption spectra of the dye molecule.
- The electron-donor component is in the geometric structure of the molecule, providing efficient charge transfer.
- Red-shifting absorption is achieved with the π-conjugated component of the dye molecule, which affects the HOMO/LUMO energy level.

Many metal-free organic dyes show promising results in photovoltaic applications. These are derivatives of coumarin, triphenylamine, carbazole, and indoline. Coumarin-153 is a derivative of coumarin, with additional electron-donor and electron-acceptor groups. The electron-donor group greatly lowers the energy of $S_{\pi\pi^*}$ and $T_{\pi\pi^*}$. In addition, the electron-acceptor group shifts the fluorescence and the absorption maximum peak to a longer wavelength, resulting in the absorbance peak of coumarin-153 at the wavelength of 430 nm in comparison to the absorbance peak of coumarin at the wavelength of 365 nm. Herein we introduce a combination of ZnO and coumarin-153 in order to enhance the efficiency of sunlight absorption.

2.4 Method

We introduce a laser ablation method to produce ZnO nanoparticles with the following procedures:

- The zinc plate in a quartz cell is filled with 5 mL of 1 mM sodium dodecyl sulfate under ambient temperature, resulting in a liquid thickness of 1 mm apart from Zn surface.
- Pulse and photon energies of the Nd:YAG laser at a wavelength of 1064 nm are operated at 1.17 eV and 100 mJ/pulse, respectively.
- Laser light (ϕ = 2–3 mm) irradiates on the surface of Zn plate.
- The solution is stirred slowly during the ablation.
- The solution is then centrifuged and redispersed in DI water.
- The TEM image of ZnO nanoparticles shows mostly spherical shape and good uniformity (ϕ = 45–60 nm).

2.5 Implementations

2.5.1 ZnO Nanoparticles

One milliliter of ZnO nanoparticles with different concentrations (0–2560 μg/mL) is added to 1 mL of human breast cancer cell lines (mid-log phase, MCF-7 = 1.0×10^5 cells/mL) in a 24-well plate under a UV box for 24 h after the cells are cultured in cell culture media (DMEM supplied with 10% fetal bovine serum, 1% penicillin-streptomycin) at a standard culture condition (37°C, 5% CO_2, 24 h). We introduced ferrous ion served as a visible absorption probe to check the potential phototoxic activity of ZnO nanoparticles on the ferrous ion chelating to the cells (ferrous ion is prepared by iron (II) chloride tetrahydrate mixed with PDT disulfonate. One set of the treated solution is then detected by a UV-Vis spectrometer. Another set of the treated solution remaining in the 24-well plate is washed by phosphate buffer saline (0.1 M PBS, pH = 7.4) and hence isolated by the trypsinization method (0.25% trypsin/EDTA) for 5–10 min, followed by redispersion with the cell culture media. The 400 μL cell suspension is transferred to the microcentrifuge tube and hence added with 100 μL of 0.4% trypan blue. The stained solution is loaded on the hemacytometer to count the cells under the optical microscope. Ferrous ion chelating and cell viability are calculated as follows:

$$\text{Ferrous ion chelating, \%} = \left[1 - \frac{\text{Absorbance of tested cells}}{\text{Absorbance of controlled cells}}\right] \times 100 \quad (2.7)$$

$$\text{Cell viability, \%} = \frac{\text{Amount of tested cells}}{\text{Amount of controlled cells}} \times 100 \quad (2.8)$$

The absorbance of ZnO nanoparticles is centered at the wavelength of 375 nm, and it is equivalent to the energy gap of 3.37eV, which is the energy interval between the valence and conduction bands (Fig. 2.7). The polarization dependence of resonance at 3.37 eV is undetectable due to the spherical shapes of ZnO nanoparticles. The absorbance peak positions are independent of the ablation times, showing that the size of nanoparticles is insensitive to the ablation time. However, the magnitudes of the absorption intensities of ZnO nanoparticles are dependent on the ablation times due to the effect of the density of ZnO nanoparticles. For instance, the absorbance values of 0.8, 1.5, and 2.8 relate to the ablation times of 10, 15, and 20 min, respectively.

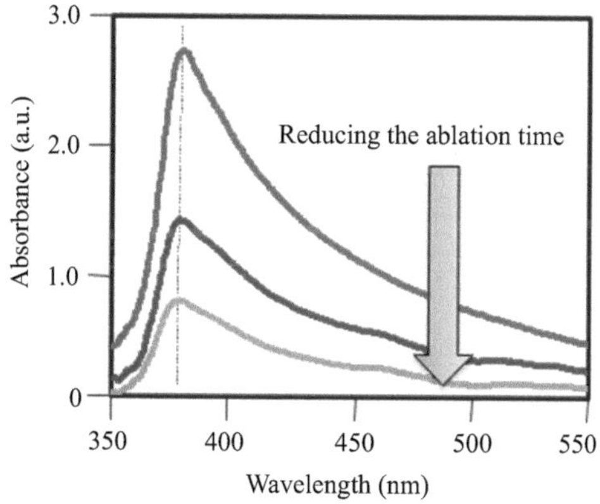

Figure 2.7 Absorbance of ZnO nanoparticles with respect to the ablation times of 10–20 min. The absorbance peak is detected at the wavelength of 375 nm.

The absorbance of a visible absorption probed by ferrous ion shows a reduction in the absorption intensity upon the integration of ZnO nanoparticles into the human breast cell lines under UV light

(Fig. 2.8). This point shows that a small amount of ZnO nanoparticles is left outside the cells, conforming to the phototoxic activity of ZnO nanoparticles on ferrous ion chelating to the cells (Table 2.4). Thus, ZnO nanoparticles possibly pass through the cell membrane and attack the cells.

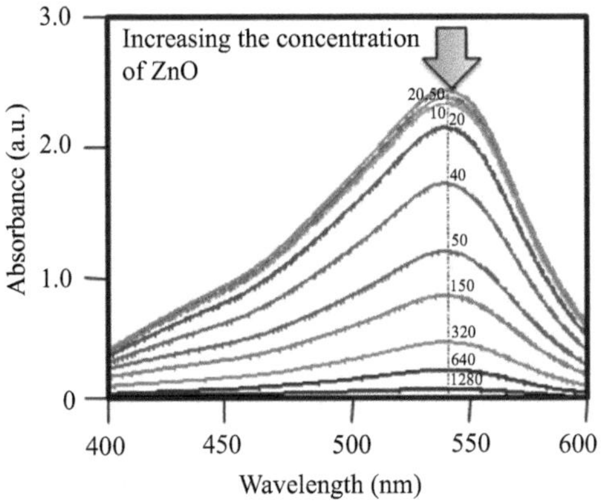

Figure 2.8 Absorbance of a visible absorption probed by ferrous ion upon the addition of ZnO nanoparticles to human breast cancer cell lines under UV light for 24 h. The number presented on each curve shows the concentration of ZnO nanoparticles in µg/mL.

Table 2.4 Ferrous ion chelating and cell viability of the human breast cancer cell lines treated by ZnO nanoparticles under UV light for 24 h

ZnO, µg/mL	Ferrous ion chelating, %	Cell viability, %
0	0	100.00
2.5	5.01	81.25
5.0	11.17	62.19
10	15.05	55.46
20	38.59	50.34
40	72.10	40.49
80	88.05	36.68
160	93.28	30.16
320	95.13	25.11
640	97.95	20.05
1280	98.50	10.06
2560	100.00	0

Since the cell-killing approach is ineffective in controlled cells, such living cells show smooth surface (Fig. 2.9). However, dead cells show condensed or fragmented chromatin. Cell death is proportional to a variety of concentrations of ZnO nanoparticles. For instance, the survival rate is likely to reduce with higher dosages of ZnO nanoparticles.

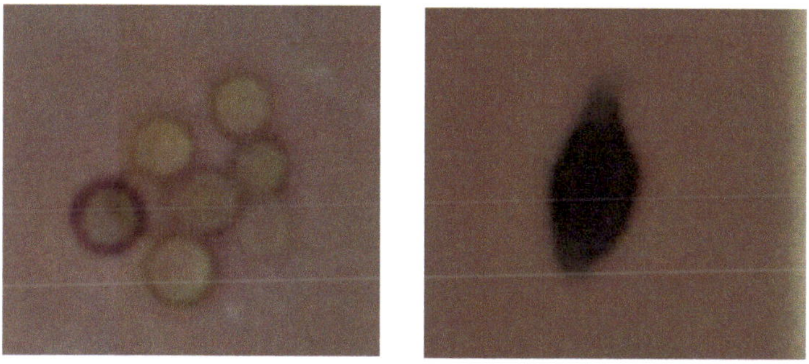

Figure 2.9 Morphological changes in living cells (left) and dead cell (right) of the human breast cancer cell lines.

Case Study 2.1

We show the contribution of phototoxic effect on human breast cancer cells by other functional metal oxide nanoparticles under UV light. Human breast cancer cell lines, MCF-7, served as tested cells and are added in well plates 1–12, while cell culture medium served as controlled cells and are added in well plates 13–24 (Fig. 2.10). Each well plate contains 100,000 cells/1 mL. They are then treated with a set of metal oxide nanoparticles, ZnO, TiO_2, and CuO, under UV box for 24 h. The 0.9 mL DI water, 0.033 mL of 2 mM ferric chloride, and 0.067 mL of 5 mM ferrozine are filled in a 24-well plate. After 40 min, the first set of treated solution is used for absorbance check. The second set of the treated solution is disposed and hence added with 0.5 mL PBS buffer to hold the reaction between cells and metal oxide nanoparticles. The PBS buffer is removed after 1 min, and 0.2 mL trypsin is then added to acquire the non-adherent cells. Trypsin is removed after 10 min, and the 0.4 mL cell culture medium is hence filled, followed by 0.1 mL Trypan blue. The final strained solution is loaded on the hemacytometer for cell count.

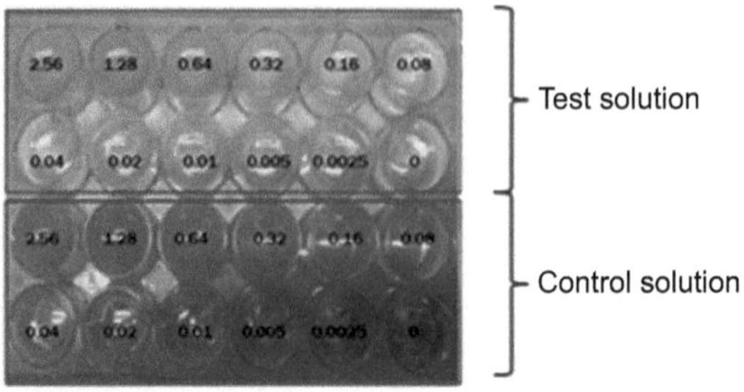

Figure 2.10 Detailed preparation of human breast cancer cell lines treated with a set of metal oxide nanoparticles: ZnO (ϕ = 100±10 nm), TiO_2 (ϕ = 100±10 nm), and CuO (ϕ = 50±5 nm), under UV box for 24 h. The number written on each well plate indicates the series of concentrations of metal oxide nanoparticles in mg/mL.

Owing to the size-dependence of the optical properties of metal oxide nanoparticles comparable to the cell membranes, metal oxide nanoparticles under this study are transferred to the cells via the pinocytosis process. During the cell treatment under UV light, metal oxide nanoparticles simultaneously generate the electron/hole pairs and the oxidative stress is thus induced by the generation of reactive oxygen species between cells and metal oxide nanoparticles (Fig. 2.11). This point is confirmed by ferrous ion chelating and cell viability. That is, each metal oxide nanoparticle under UV light reveals the unique phototoxic activities of metal oxide nanoparticles on the ferrous ion chelating to the cells (Table 2.5). The "initial" reactive oxygen species generated by metal oxide nanoparticles follow the order: TiO_2 > CuO > ZnO. The generation of reactive oxygen species is related to the electronic structures characterized by the energy gap. The excited electrons in the conduction band donate electrons to oxygen for the production of superoxide radial anion, whereas the holes in the valence band accept electrons from water for the production of hydroxyl radicals. The "final" reactive oxygen species generated by metal oxide nanoparticles transported into the cells thus follow the order ZnO > CuO > TiO_2 if the metal oxide nanoparticles have the same diameter (100±10 nm). Since the cell pore of nucleus (39 nm) is smaller than the diameter of metal oxide nanoparticles (50–100 nm), such free radical compounds are slowly agglomerated around the nucleus of cells. The intracellular reactive oxidation species damages the nucleus of cells. However, the

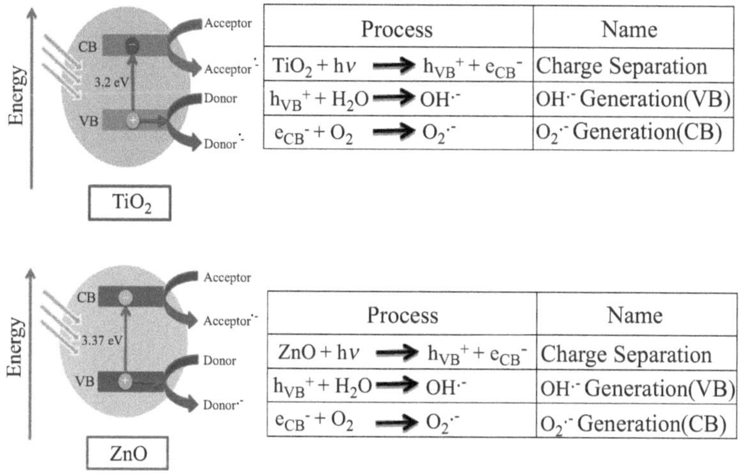

Figure 2.11 Representative formation of reactive oxygen species (ROS) via electron/hole pairs of metal oxide nanoparticles under UV light.

Table 2.5 Ferrous ion chelating and cell viability with respect to metal oxide nanoparticles

Concentration of metal oxide, μg/mL	ZnO		CuO		TiO_2	
	Ferrous ion chelating	Cell viability	Ferrous ion chelating	Cell viability	Ferrous ion chelating	Cell viability
0	0	100.00	0	100.00	0	100.00
2.5	5.01	81.25	10.10	64.22	15.18	85.45
5.0	11.17	62.19	17.50	55.45	27.15	71.15
10	15.05	55.46	22.30	52.78	32.22	60.67
20	38.59	50.34	42.40	42.10	45.96	52.81
40	72.10	40.49	75.65	38.65	77.45	45.44
80	88.05	36.68	89.45	35.44	91.47	38.69
160	93.28	30.16	94.58	27.16	96.11	34.47
320	95.13	25.11	96.64	22.05	96.02	27.91
640	97.95	20.05	98.89	15.00	98.05	22.06
1280	98.50	10.06	99.45	5.05	99.01	10.50
2560	100.00	0	100.00	0	100.00	0

size-dependence of metal oxide nanoparticles affects the enhanced reactive oxidation species generation power. Since the diameter of CuO (50±5 nm) in this study is smaller than the diameters of TiO_2 (100±10 nm) and ZnO (100±10 nm), CuO nanoparticles possibly give more available active sites or large surface areas and thus diffuse into the nucleus of cells so that the structures of DNA of the cells are destroyed. This point firmly causes cell death. Hence, the cell-killing effects via the phototoxicity process of metal oxide nanoparticles on human breast cancer cells are modified in the order: CuO > ZnO > TiO_2.

2.5.2 ZnO/Au Nanocomposites

Gold (ϕ = 100±10 nm) and ZnO (ϕ = 50±5 nm) are mixed under sonication for 2 h and spin coated on a glass substrate at 3500 rpm for 5 s (Table 2.6). ZnO/Au film is immersed in 16 mL of 0.2 mM Fe(II) as a representative sample of heavy metal in wastewater (Fig. 2.12). This amount is well below the standard from wastewater treatment. For economic reasons, it is not necessary to add the excess amount of Au in the waste water when the detected Fe(II) is lower than standard value of 0.2 mM. The height of test solution is 1 mm beyond the film's surface. A fluorescent lamp, as a representative of sunlight, exposits at 80 cm above the solution surface. After 90 min, the treated Fe(II) solution and the dipped film are separately transferred to the sample holder to detect their absorbance using a UV-Vis spectrometer. A tungsten lamp is used as the light source for the absorption measurement. Fe(II) stock solution is used as the calibration curve for Fe(II) concentration determination according to the following equation:

$$y = -0.396x^2 + 1.490x + 2.497 \tag{2.9}$$

where x is Fe(II) concentration, and y is the absorbance peak centered at the wavelength of 560 nm.

Reduction in the concentration of Fe(II) is defined as

$$\text{Reduction of concentration of Fe(II)} = \frac{C}{C_0} \tag{2.10}$$

where C_0 is the Fe(II) initial concentration before dipping the film in the test solution (C_0 = 0.2 mM) and C is the Fe(II) final concentration after dipping the film in the test solution.

Table 2.6 Detailed preparation of ZnO/Au film

PMMA, mg	ZnO, mg	Au, mg	DCM, mL	Molar ratio of ZnO:Au
200	100	0	1	1:0
200	100	100	1	1:0.5
200	100	200	1	1:1
200	100	400	1	1:2

Note: DCM is dichloromethane and PMMA is polymethyl methacrylate.

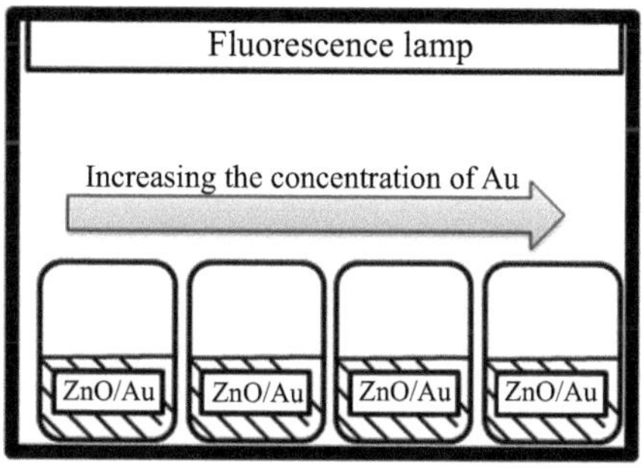

Figure 2.12 Test of photocatalytic performance of ZnO/Au film under the illumination atmosphere for 90 min using the fluorescent lamp with the light intensity of 529.67 Lux as a representative of sunlight.

The absorbance peak of ZnO nanoparticles is centered at the UV wavelength of 380 nm, whereas the broadening peak of Au nanoparticles is seen at the visible wavelength of 550 nm (Fig. 2.13). Fe(II) is further formed on the ZnO/Au film upon the integration of Au nanoparticles in the ZnO matrix (Table 2.7). The molar ratio of ZnO:Au is 1:2, which reveals the highest metal absorber, corresponding to the morphological changes seen under the optical microscope (Fig. 2.14). Au clusters are noticeably seen due to their concentration dependence of the size distributions. For instance, an additional amount of Au at the molar ratio of 1:2 increases the Au cluster aggregations from small to big size. The mechanism of the photocatalytic efficiency of ZnO/Au film is explained as follows (Fig. 2.15):

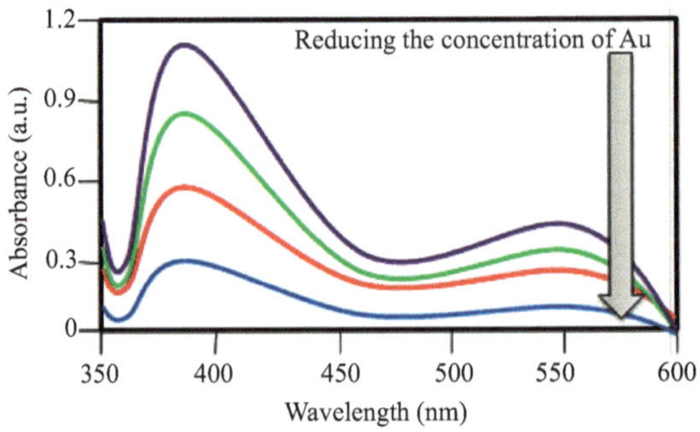

Figure 2.13 Absorbance ZnO/Au films after treatment in the test solution.

Table 2.7 A reduction in Fe(II) concentration, C/C_0

Molar ratio of ZnO:Au	C/C_0
1:0	0.80
1:0.5	0.75
1:1	0.62
1:2	0.55

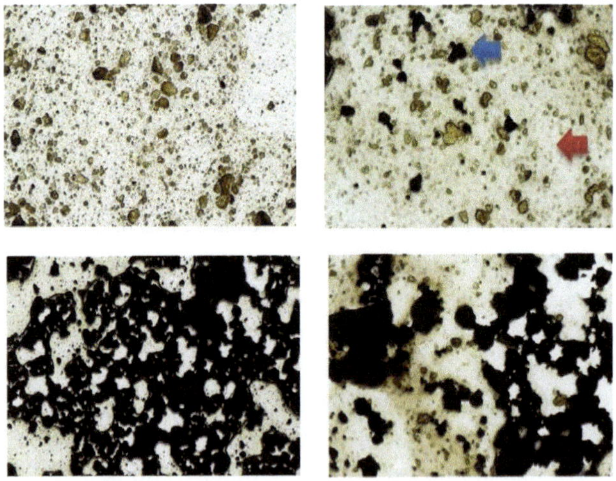

Figure 2.14 Morphological changes in ZnO:Au films at the molar ratio of 1:0 (top left), 1:0.5 (top right), 1:1 (bottom left), and 1:2 (bottom right) after treatment in the test solution. Optical magnification is 10×. The smaller size of ZnO nanoparticles shows uniformity (red arrow), whereas additional amounts of higher size of Au nanoparticles show uniformity (blue arrow). Clay-like color scattering in the films is caused by Fe(II) formation.

- ZnO has an optical band gap in the UV-Visible regime.
- The localized surface plasmonic resonance of Au is contributed to the strong absorption of visible light and excitation of active charge carriers.
- The plasmonic effect of Au enhances the excitation energy of the electron–hole pair of ZnO.
- The Schottky junction is of benefit to the separation of charges.
- The scattering of resonant photon increases the length of the average photon path.

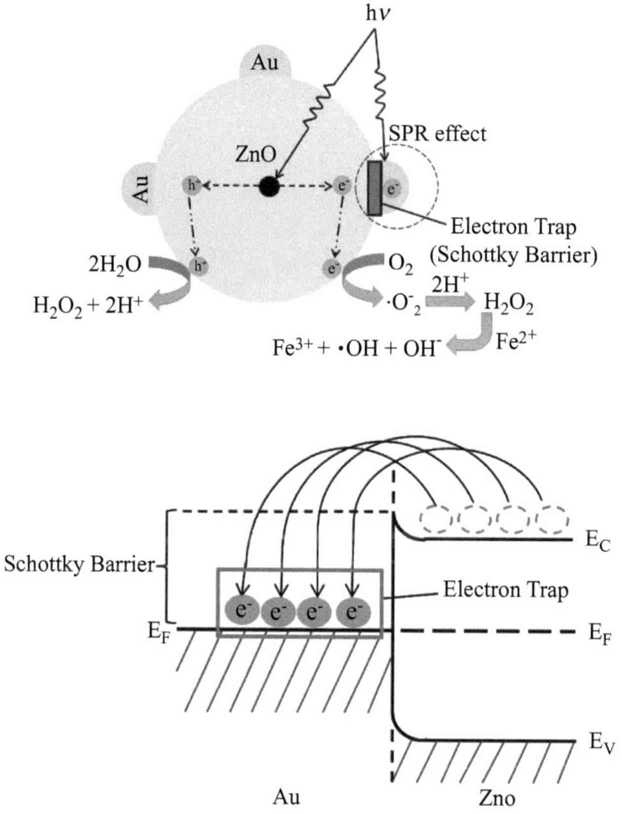

Figure 2.15 (Top) Possible mechanism of photocatalytic efficiency of ZnO/Au film. Note that $\bullet O^-_2$ is the superoxide radical and •OH is the hydroxyl radical. (Bottom) Schottky effect of ZnO/Au film. E_C is the conduction band energy, E_F is the Fermi energy, E_V is the valence band energy, and *e* is the electric charge.

2.5.3 ZnO/Coumarin-153 Nanocomposites

Ethanol solution of 10 mM coumarin-153 is added after polymethyl methacrylate (PMMA) is dissolved in dichloromethane (DCM), followed by 50 wt.% ZnO solution (Table 2.8). ZnO film is prepared by ethanol instead of coumarin-153 ethanol solution. Finally, coumarin-153 film is prepared by mixing PMMA with coumarin-153 ethanol solution, pure ethanol, and DCM. All cocktails are sonicated under a sonicator for 30 min in order to make a homogeneous solution. After drying the samples (Fig. 2.16), they are exposed to UV light for 180 min. Photostability is detected by measuring the fluorescence at intervals of 20–180 min. As a reference for absorbance, the film is produced by dissolving 0.2 g PMMA in 500 μL DCM and 500 μL ethanol.

Table 2.8 (a) Detailed preparation of sol–gel ZnO/coumarin-153 film

PMMA, g	Coumarin-153, μL	ZnO solution, μL	DCM, μL
0.2	400	117.5	482.5
0.2	400	129.5	470.5
0.2	400	141.0	459.0

Table 2.8 (b) Detailed preparation of ZnO film

PMMA, g	Ethanol, μL	ZnO solution, μL	DCM, μL
0.2	400	117.5	482.5
0.2	400	129.5	470.5
0.2	400	141.0	459.0

Table 2.8 (c) Detailed preparation of coumarin-153 film

PMMA, gm	Coumarin-153, μL	Ethanol, μL	DCM, μL
0.2	400	100.0	500.0

Figure 2.16 Coumarin-153 (left), ZnO (middle), ZnO/coumarin-153 films (right).

Although ZnO is commonly used in photovoltaic, the need for more economical ways of acquiring energy leads to the requirement of enhancing the efficiency of these materials. One way is dye sensitization. ZnO has a bang gap of 3.37 eV, meaning that ZnO absorbs light in the UV part of the spectrum. Dyes with lower band gap energy absorb light in lower frequencies, transferring the electrons to the conductive band of ZnO. Therefore, higher light utilization is reached.

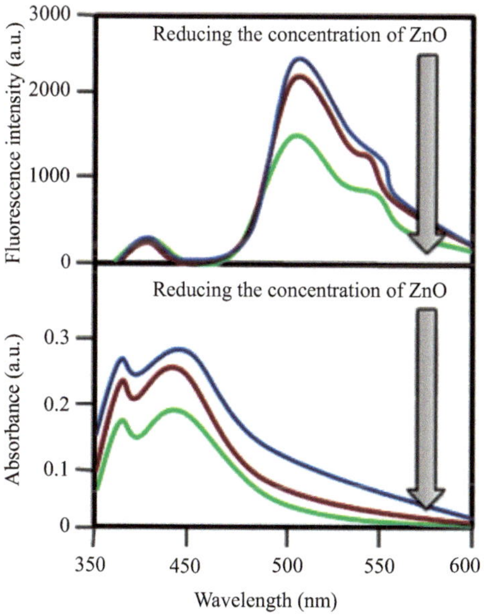

Figure 2.17 Absorbance (bottom) and fluorescence (top) of coumarin-153 mixed with 100–120 mg/mL ZnO nanoparticles. Tungsten lamp is used as a light source for absorption measurement, whereas a 3 W UV-LED at the wavelength of 398 nm is used as the light source for fluorescence measurement.

Figure 2.17 shows the absorbance of ZnO nanoparticles mixed with 4 mM coumarin-153. The first narrow peak at the wavelength of 370 nm belongs to ZnO nanoparticles, whereas the second broadening peak at the wavelength of 420 nm belongs to coumarin-153. It is clearly found that both absorption and fluorescence intensities are proportional to the amount of ZnO.

The integration of ZnO nanoparticles into the coumarin-153 matrix effectively enhances the photostability applicable to photovoltaic (Table 2.9). The photostability of bare coumarin-153 tends to reduce when the UV exposition time is longer. The photostability of ZnO/coumarin-153 tends to be stable when the UV exposition time is longer. Furthermore, ZnO/coumarin-153 shows higher photostability than bare coumarin-153. The improved fluorescence intensity of ZnO/coumarin-153 is, therefore, suitable for making a long-lasting photovoltaic.

Table 2.9 Fluorescence of ZnO/coumarin-153 film and bare coumarin-153 under UV exposition times of 20–180 min

UV exposition time, min	Fluorescence intensity (a.u.)	
	Coumarin-153	Coumarin-153 + 100 mg/mL ZnO
0	1800	2510
20	1754	2502
50	1622	2505
75	1617	2501
110	1605	2507
140	1552	2510
180	1500	2509

Problems

2.1 Describe the band transition of ZnO in comparison to Si and Ge.

2.2 Show the band diagrams of metals and semiconductors.

2.3 Clarify the effect of Schottky barrier at the interface between metals and semiconductors.
2.4 Explain absorbance by the Beer–Lambert law.
2.5 Show the formation of reactive oxygen species via electron/hole pairs of ZnO nanoparticles.

Chapter 3

Fe_3O_4 Magnetic Nanoparticles

3.1 Introduction

This chapter deals with the fluorescence mechanism of Fe_3O_4 magnetic nanoparticles and rhodamine 6G as a dye model under light exposition in a batch mode because there is no systematic report on the relationship between rhodamine 6G adsorption and fluorescence quantum yield. Fe_3O_4 magnetic nanoparticles show active adsorption to quench rhodamine 6G in removing risky water pollutants from dye wastewater in the textile industry, which are otherwise harmful to the surroundings and even human health. Not only does the adsorption characteristic of rhodamine 6G on the Fe_3O_4 surface depend on the adsorbent dosage, but also it relies on the exposition time of pulsed white LED at a variety of frequencies and duty cycles. The fluorescence quantum yield of rhodamine 6G is inversely proportional to dye removal. Upon the addition of pulsed white LED to Fe_3O_4 nanoparticles, dye removal is noticeably improved under a shorten equilibrium state.

3.2 Principle

The textile industry, as the largest user of organic dyes for fiber coloration, is the second biggest polluter in the world. It is now

Linear and Nonlinear Optics: Materials, Properties, and Applications
Kitsakorn Locharoenrat

ISBN 978-981-4877-15-2 (Hardcover), 978-1-003-05588-4 (eBook)
www.jennystanford.com

growing (Table 3.1) and simultaneously encountering protests for the disposal of dyes, one of the sources of environmental pollution worldwide because of the following reasons:

- Some textile factories have low budget and poor technology.
- Effluents originated by some production units are sometimes discharged into water resources.
- Most organic dyes are stable to oxidizers and sunlight.
- Most organic dyes affect the aesthetic merit.
- Most organic dyes reduce light penetration.
- Some dyes are carcinogenic to humans.

Table 3.1 Global textile industrial trend

Year	Manufacturing industry value (billion USD)
2016	748
2017	768
2018	804
2019	847
2020	899
2021	961

Source: Marketline

As a result, there are so many methods available to eliminate dye effluents. Ecofriendly physical methods such as adsorption technique have been introduced so that the dissolved organic and inorganic compounds are turned into nontoxic end products. Although the target pollutants are easily removed by this method, they are still converted from one to another phase. The advanced oxidation process is regarded as a new one-stop system because of the following reasons:

- Advanced oxidation process is needed when both by-product and other contaminant compounds are generated.
- It completes the oxidation of organic compounds to nontoxic products, such as CO_2 and H_2O as donor molecules.
- It is more economically active than traditional primary systems.

The mechanism of advanced oxidation process is explained as follows:

- When incident light hits the adsorbent surface, the electrons in the ground state are excited, which steps them up at the conduction band, whereas holes exist at the valence band.
- The holes will interact with water to produce hydroxyl free radicals as a strong oxidizer responsible for elimination of pollutants.
- The influence of free radicals on existing organic dyes generates biodegradable intermediates and mineralization.

3.3 Fluorescence Resonance Energy Transfer

When dye molecules and solid matters couple, the enhancement or quenching of fluorescence of the single dye molecule originates from the fluorescence energy resonance transfer, depending on d^3, where d is the distance between dye molecules and solid matters (Fig. 3.1). It is considered as a nonradiative energy transfer from the excited donor fluorophore/quencher to the acceptor. This nonradiative dipole–dipole coupling is related to the fluorescence quantum yield by

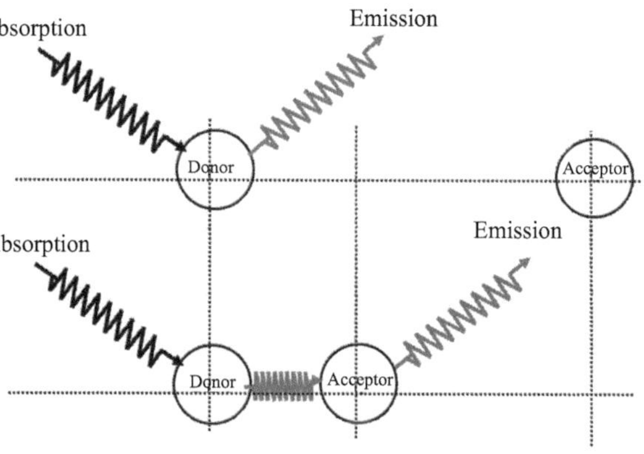

Figure 3.1 Overlapping between donor emission and acceptor absorption results in the active fluorescence resonance energy transfer interaction.

$$\text{Fluorescence quantum yield, } Q_F = Q_R \frac{I_F}{A_F} \frac{A_R}{I_R} \frac{n_F^2}{n_R^2} \tag{3.1}$$

where Q_F is the fluorescence quantum yield of the unknown sample, Q_R is the quantum yield of reference standard (Table 3.2), I_F is the integrated fluorescence intensity of the unknown sample, I_R is the integrated fluorescence intensity of the reference sample, A_F is the respective absorbance value of the unknown sample, A_R is the respective absorbance value of the reference sample, n_F is the solvent refractive index of the unknown sample, and n_R is the solvent refractive index of the reference sample.

Table 3.2 Quantum yield and respective excitation wavelength of the reference standard

Fluorophore	Excitation wavelength, nm	Quantum yield
2-aminopurine	280	0.60
Cresyl violet	550	0.54
Fluorescein	490	0.98
POPOP	365	0.97
Quinine sulfate	350	0.58
Rhodamine 6G	488	0.95
Rhodamine 101	470	1.00

Source: Horiba Scientific

3.4 Applications

Sulfide-based nanocomposites are usually chosen as the adsorbent in the advanced oxidation process. For instance, ZnS/C and ZnS/polyaniline are used under UV light for the removal of rhodamine 6G. The dye removal kinetics is well explained by the Langmuir–Hinshelwood kinetics model. Ag/ZnS/C, under visible light, is used to eliminate rhodamine 6G. The dye removal kinetics is attributed to the photosensitizer and surface plasmonic phenomena of carbon nanotubes and Ag nanoparticles, respectively.

Oxygen-based nanocomposites are now an interesting alternative to substitute the sulfide-based nanocomposites in the advanced oxidation process in order to eliminate effluents with low level of dissolved organic and inorganic compounds. For instance,

TiO_2 nanoparticles under continuous light expositions (UV, visible, and sunlight) are used for dye removal. Although continuous light exposition is sufficiently good for the advanced oxidation process, oxygen-based nanoparticles, on pulsed light exposition of dye pollutants, have not yet been reported elsewhere.

Thus, we introduced Fe_3O_4 magnetic nanoparticles as a magnetic adsorbent in place of TiO_2 nanoparticles because of the following reasons (Fig. 3.2):

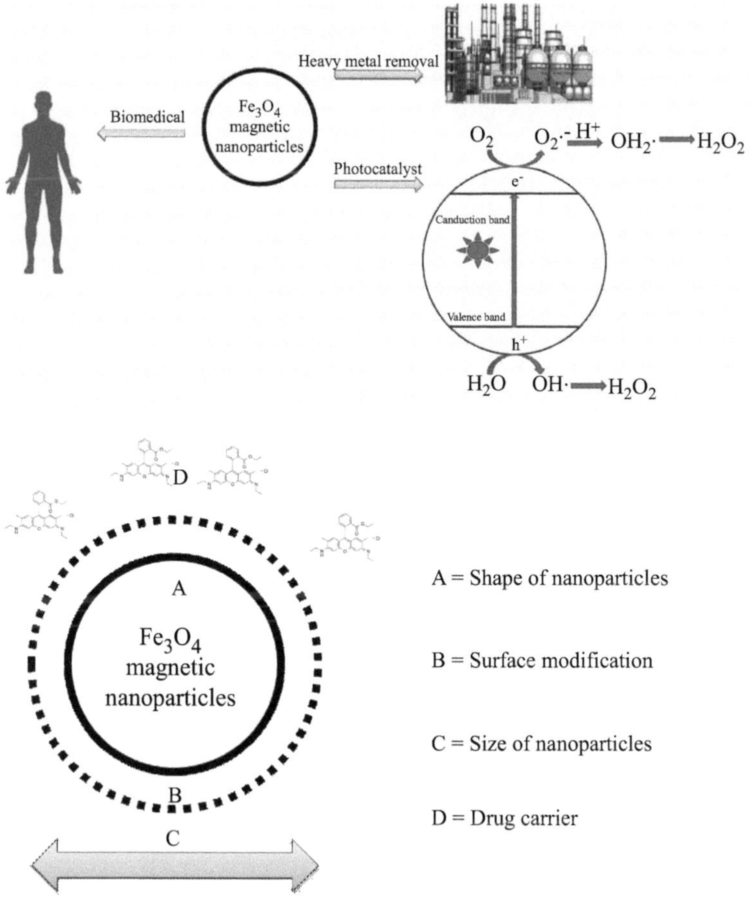

Figure 3.2 Representative applications of Fe_3O_4 magnetic nanoparticles (top) and influence of Fe_3O_4 magnetic nanoparticles in their potential applications (bottom).

- Fe_3O_4 nanoparticles have a high surface-to-volume ratio and suitable pore size for the purpose of adsorption.
- Fe_3O_4 nanoparticles can remove a variety of industrial effluents, such as dyes and heavy metal.
- Fe_3O_4 magnetic nanoparticles effectively eliminate toxic/hazardous adsorbates.
- Fe_3O_4 nanoparticles can treat a high amount of dye wastewater in less time.
- Fe_3O_4 nanoparticles can generate less contaminants.
- Fe_3O_4 nanoparticles are biocompatible.
- Fe_3O_4 nanoparticles offer less toxicity and, hence, are commonly suitable for various biomedical treatments. For example, they are applicable to MRI, hyperthermia, and drug-delivery systems.
- Fe_3O_4 nanoparticles can be easily eliminated from the treated wastewater via the external magnetic separation method.
- Recycling Fe_3O_4 nanoparticles is possible.

In addition to methylene blue and rhodamine widely used in the textile industry, herein we focus on rhodamine in cationic form. This is because not only does it reveal the active adsorption on solid wastes, it is also soluble in water. The influence of Fe_3O_4 magnetic nanoparticles under the pulsed white LED in comparison to the continuous white LED on rhodamine 6G dye removal will be presented to deeply understand the fluorescence phenomena between Fe_3O_4 magnetic nanoparticles and rhodamine 6G. Furthermore, the adsorbent dosage of Fe_3O_4 magnetic nanoparticles plays a key role in their adsorption capacity on the adsorbate rhodamine 6G. Based on the adsorbent Fe_3O_4 dosages, we can find enhancement and quenching of fluorescence of the dye molecules originated from the fluorescence energy resonance transfer.

3.5 Implementations

Rhodamine 6G dye is dissolved in DI water as a stock solution and hence diluted into two sets. The first set is used as a calibration curve to determine the concentration of rhodamine 6G according to the following equation:

$$y = 79289\,x - 0.0032 \tag{3.2}$$

where x is the dye concentration and y is the absorption intensity.

The second set is 1 mL of 3 μM adsorbate rhodamine 6G mixed with 1 mL of 2.5–10 mg/L adsorbent Fe_3O_4 magnetic nanoparticles (ϕ = 50–100 nm) under a sonicator at the ambient temperature for treatment times of 0–150 min. The mixture solution is detected by a UV-Vis spectrometer in every 30 min. Removal of dye and adsorption capacity at the equilibrium state are calculated as follows:

$$\%\,\text{Dye removal} = \frac{\begin{matrix}\text{Initial} \\ \text{concentration of dye}\end{matrix} - \begin{matrix}\text{Equilibrium} \\ \text{concentration of dye}\end{matrix}}{\text{Initial concentration of dye}} \times 100 \tag{3.3}$$

$$\text{Adsorption capacity} = \frac{\begin{matrix}\text{Initial} \\ \text{concentration of dye}\end{matrix} - \begin{matrix}\text{Equilibrium} \\ \text{concentration of dye}\end{matrix}}{\text{Mass of adsorbent}} \times \text{Solution volume} \tag{3.4}$$

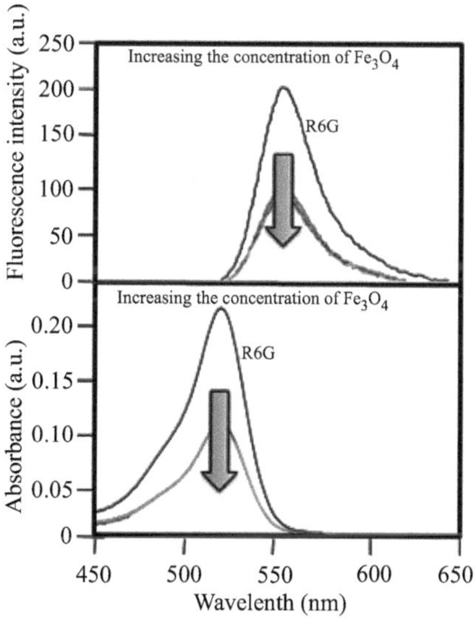

Figure 3.3 Absorbance (bottom) and fluorescence (top) of Fe_3O_4 magnetic nanoparticles (2.5, 5.0, 7.5, and 10.0 mg/L) in rhodamine 6G under treatment time of 0 min as a representative sample. Absorbance and fluorescence peaks are centered at the wavelengths of 525 and 555 nm, respectively.

The absorbance and fluorescence of Fe_3O_4 and rhodamine 6G are noticeably lower than those of bare rhodamine 6G (Fig. 3.3). They are inversely proportional to Fe_3O_4 dosages due to the following reasons:

- Strong adsorption capacity of Fe_3O_4 for rhodamine 6G is caused by high surface-to-volume ratio of Fe_3O_4 magnetic nanoparticles.
- The inner filter effect reduces the fluorescence intensity.
- The quenching of fluorescence of rhodamine 6G by Fe_3O_4 originates from the fluorescence energy resonance transfer.

The decrease in absorbance and fluorescence leads to a reduction in the fluorescence quantum yield (Table 3.2). The fluorescence quantum yield is equivalent to 0.81–0.88. Furthermore, it is inversely proportional to dye removal (Table 3.3). These results concluded that Fe_3O_4 magnetic nanoparticles function as the active fluorescence quencher for rhodamine 6G.

Table 3.3 (a) Absorbance peak of Fe_3O_4 magnetic nanoparticles in rhodamine 6G under treatment times of 0–150 min

Fe_3O_4, mg/L	Absorbance peak under treatment time of (min)					
	0	30	60	90	120	150
0	0.23	0.23	0.22	0.23	0.22	0.22
2.5	0.121	0.124	0.120	0.115	0.101	0.100
5.0	0.124	0.121	0.119	0.109	0.099	0.091
7.5	0.121	0.122	0.112	0.107	0.090	0.090
10.0	0.117	0.115	0.109	0.105	0.089	0.089

Table 3.3 (b) Integrated fluorescence intensity of Fe_3O_4 magnetic nanoparticles in rhodamine 6G under treatment times of 0–150 min

Fe_3O_4, mg/L	Integrated fluorescence intensity under treatment time of (min)					
	0	30	60	90	120	150
0	7500	7450	7325	7202	7114	7000
2.5	4400	4315	4218	4113	4050	4000
5.0	4100	4009	4007	4001	3958	3945
7.5	3900	3894	3877	3829	3814	3750
10.0	3700	3601	3583	3564	3555	3500

Although the influence of Fe_3O_4 magnetic nanoparticles on dye removal reveals a less revolution (Table 3.4), the influence of Fe_3O_4 magnetic nanoparticles on rhodamine 6G uptake reveals a noticeable revolution (Table 3.5). The adsorption capacity is equivalent to 140–580 mg/g. The effect of the treatment time on the adsorption process of Fe_3O_4 magnetic nanoparticles and rhodamine 6G shows interesting tendency:

- The adsorption capacity shows a stable value until the saturation point, due to the available monolayer coverage of dye on the Fe_3O_4 surface.
- It is fast at the beginning (exposition time of 0–30 min) due to large mass-transfer driving force.
- It remains constant after approaching through equilibrium point (treatment time of 60–150 min). This is because most dyes reach a boundary layer and, hence, are hard to diffuse into the tiny pores of Fe_3O_4.
- High dosage means high surface area, resulting in highly active adsorbent sites available to the adsorbate dyes.
- Dye removal is proportional to the adsorbent dosage at the same equilibrium time (i.e., 30 min). The best performance of dye removal of 57% is applicable to 10 mg/L Fe_3O_4 dosage with the adsorption capacity of 580 mg/g.
- The best treatment time of 30 min reveals the lowest operation cost applicable to industry-scale application of adsorbents.

Dye removal does not go over 60% even when greater than 10 mg/L Fe_3O_4 is introduced, partly because of the aggregation of Fe_3O_4 surface site available to dye molecules. To increase the diffusion path length, white LED is additionally recommended to further develop the adsorption capacity of Fe_3O_4 magnetic nanoparticles.

Now, rhodamine 6G dye is dissolved in DI water as a stock solution of 0.1 mM. The stock solution is, hence, diluted into two sets. The first set is used as a calibration curve. The second set is 1 mL of 3 μM adsorbate rhodamine 6G mixed and stirred with 99 mL of 100 mg/L adsorbent Fe_3O_4 (ϕ = 50–100 nm) at 4000 rpm under the pulsed and continuous expositions of 3 W white LED for exposition times of 0–150 min (Fig. 3.4). For the pulsed mode, frequencies and duty cycles are controlled at 0.5–30 kHz and 0–100%, respectively. Prior to every light exposure on rhodamine 6G and Fe_3O_4 for 30 min, this mix solution is centrifuged at 4000 rpm for 15 min.

Table 3.4 Fluorescence quantum yield and dye removal by Fe_3O_4 magnetic nanoparticles

Fe_3O_4, mg/L	Treatment time of (min)											
	0		30		60		90		120		150	
	FQY	Removal	FQY	Removal	FQY	Removal	FQY	Removal	FQY	Removal	FQY	Removal
2.5	0.89	52.00	0.89	52.25	0.88	52.78	0.87	52.99	0.85	53.05	0.82	53.10
5.0	0.88	53.51	0.87	53.59	0.86	53.84	0.85	54.02	0.84	54.56	0.82	54.72
7.5	0.86	54.55	0.86	54.86	0.85	54.99	0.84	55.10	0.83	56.05	0.81	56.54
10.0	0.85	55.83	0.84	55.99	0.84	56.03	0.83	56.15	0.82	57.56	0.81	57.81

Note: Concentration of rhodamine 6G is 3.0 μM.

Table 3.5 Adsorption capacity and dye removal by Fe_3O_4 magnetic nanoparticles

Fe_3O_4, mg/L	Treatment time of (min)											
	0		30		60		90		120		150	
	AC	DR	AC	DR	AC	DR	AC	DR	AC	DR	AC	DR
2.5	0	52.00	575	52.25	580	52.78	580	52.99	580	53.05	580	53.10
5.0	0	53.51	299	53.59	301	53.84	301	54.02	301	54.56	301	54.72
7.5	0	54.55	207	54.86	208	54.99	208	55.10	208	56.05	208	56.54
10.0	0	55.83	140	55.99	145	56.03	145	56.15	145	57.56	145	57.81

Note: Concentration of rhodamine 6G is 3.0 μM. AC is adsorption capacity in mg/g. DR is dye removal in %.

Supernatant is, hence, detected by a UV-Vis spectrometer. The absorbance peak at the wavelength of 525 nm of each spectrum is then used to calculate the initial and final dye concentration via a calibration curve.

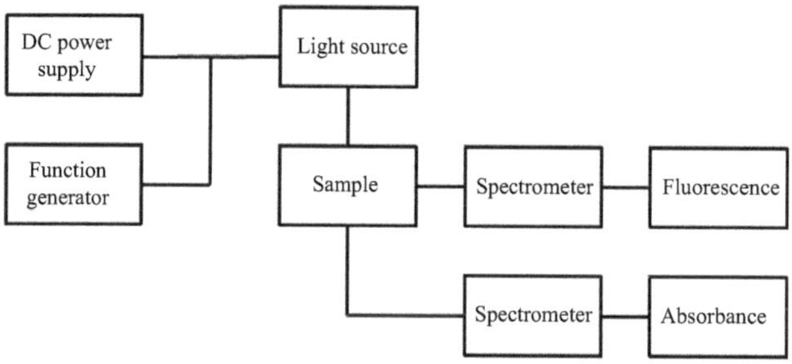

Figure 3.4 Block diagram of Fe_3O_4 magnetic nanoparticles and rhodamine 6G dye under white LED exposition. The absorbance of the supernatant of the mixture sample is detected at the exposition time interval of 30 min.

The duty cycle of the optic pulse is defined as follows:

$$\text{Duty cycle} = \frac{\tau}{T} \times 100\% \tag{3.5}$$

where τ is the optic pulse width during on-LED and T is the optic period during on-LED.

The adsorption characteristics of Fe_3O_4 magnetic nanoparticles and rhodamine 6G under white LED exposition play a key role in further figuring out the dye removal kinetics as follows:

- The absorbance peak of Fe_3O_4 magnetic nanoparticles mixed with rhodamine 6G is centered at the wavelength of 525 nm, whereas the fluorescence peak is centered at the wavelength of 555 nm. This number is comparable to their bandgap of 2.34 eV. Fe_3O_4 magnetic nanoparticles can produce highly active oxidizing and decreasing sites/power so that they oxidize the high amount of dye molecules.
- The transformation of a magnetic dye-adsorbent catalyst to magnetic photocatalyst under white LED originates from a change in the mechanism of dye removal from the surface adsorption to photocatalytic degradation.
- With regard to photocatalyst behavior, Fe_3O_4 magnetic nanoparticles under white LED have a supra-bandgap energy,

leading to the strong association of photo-generated electron–hole pairs. These pairs will spread to the surface, where the electrons adsorb hydroxide, whereas the holes adsorb oxygen molecules to form the free radicals prior to the electron–hole pairs tending to a significant change in the recombination process.

Table 3.6(a) shows that the dye removal kinetics depends on the operated duty cycle of the pulsed white LED and exposition time. It remains a constant value when the exposition time reaches the equilibrium state of 120 min under the pulsed white LED (25%, 50%, and 75% duty cycle), whereas the exposition time reaches the equilibrium state of 60 min under the continuous white LED (100% duty cycle). At the same exposition time, dye removal through the continuous white LED (i.e., 95.99% dye removal at the exposition time of 60 min) is always greater than that from the pulsed white LED (i.e., 85.64% dye removal at the exposition time of 60 min). A reduction in dye removal with a decrease in duty cycle is then confirmed. The reason is that a reduction in the number of photons constrains the formation of electron–hole pairs and hence decreases the generation of hydroxyl and superoxide.

Table 3.6(b) reveals that dye removal kinetics is proportional to the operated frequency of the pulsed white LED and exposition time. Dye removal increases with an increase in both frequency and exposition time. Interestingly, it is found that the dye removal kinetics under the pulsed white LED at the frequency of 30 kHz shows quite similar tendency to that from the continuous white LED. That is, the highest dye removal value reaches at the same equilibrium state of 60 min. Dye removal under this pulsed white LED is 92.25%, whereas dye removal under the continuous white LED is a bit bigger percentage, 95.99%. Therefore, these studies concluded that the pulsed white LED at the frequency of 30 kHz for a constant duty cycle of 25% at the shorten equilibrium state of 60 min shows the optimal condition for the best percentage of dye removal. This is because under this condition, there are still high enough adsorption sites of the unsaturated Fe_3O_4 magnetic nanoparticles during the adsorption process.

Table 3.6 (a) Percentage of dye removal of Fe_3O_4 magnetic nanoparticles mixed with rhodamine 6G under white LED exposition at the frequency of 1 kHz with different duty cycles: 25%, 50%, 75%, and 100%

Duty cycle, %	Exposition time, min				
	30	60	90	120	150
25	60.61	75.29	83.58	89.28	90.36
50	65.33	77.06	86.58	90.86	93.75
75	74.01	85.64	90.79	94.64	95.57
100	91.01	95.99	96.15	96.81	97.25

Table 3.6 (b) Percentage of dye removal of Fe_3O_4 magnetic nanoparticles mixed with rhodamine 6G under white LED exposition at the duty cycle of 25% with different frequencies: 0.5, 1, 10, and 30 kHz

Operated frequency, f (kHz)	Exposition time, min				
	30	60	90	120	150
0.5	48.78	54.58	61.18	63.72	70.97
1.0	60.61	75.29	83.58	89.28	90.36
10.0	81.34	87.11	90.72	90.83	90.84
30.0	88.56	92.25	93.46	93.36	93.44
Continuous	91.01	95.99	96.15	96.81	97.24

Problems

3.1 Show that how the energy (E) consumed by a light source is determined.

3.2 Show the material lists to manipulate 3 W white LED in the pulsed mode.

3.3 Explain the fluorescence quantum yield in nontechnical terms.

3.4 Name other terms describing the fluorescence resonance energy transfer existed via nonradiative dipole–dipole coupling. The energy transfer efficiency is inversely proportional to the third power of a gap distance between donor and acceptor.

Chapter 4

Organic Light-Emitting Diode

4.1 Introduction

This chapter introduces the white organic light-emitting diode (OLED) based on p-i-n structures produced via the vapor deposition method. Tris(8-hydroxyquinolinato) aluminum (Alq_3) doped with 8-hydroxyquinolinato lithium (Liq) is used as n-type, whereas WHI112 doped with molybdenum trioxide (MoO_3) serves as p-type. The great hole-injection capability of WHI112:MoO_3 and the high electron mobility of Alq_3:Liq offer a carrier balance at the electrode–organic layer interface. The power efficiencies of the p-i-n structure is better than those of undoped and n-type structures under small current density, whereas the roll-off of the performance of the p-i-n structure is greater than those of undoped and n-type structures under high current density. Moreover, this chapter introduces a high uniformity of 2,2′,2″-(1,3,5-benzinetriyl)-tris(1-phenyl-1-H-benzimidazole) (TPBi) film as the electron transport layer of the blue OLED. It is transferred on the blue OLED with the help of a polyethylene terephthalate release film via the stamp printing method. The surface morphology of the TPBi film is developed through the manipulation of stamp temperature and stamp time.

Linear and Nonlinear Optics: Materials, Properties, and Applications
Kitsakorn Locharoenrat

ISBN 978-981-4877-15-2 (Hardcover), 978-1-003-05588-4 (eBook)
www.jennystanford.com

4.2 Principle

OLEDs have been the subject of many studies, partly due to the flexibility, high throughput, and low-driving voltage for the display markets, especially for watches and phones (Table 4.1).

Table 4.1 Forecast of OLED market

Year	Manufacturing industry value (billion USD)
2015	3
2016	4
2017	5
2018	9
2019	11
2020	12
2021	14
2022	15

Source: IHS

Herein we introduce the blue OLED and white OLED to understand the influence on carrier transportation, carrier injection, and carrier balance at the electrode–organic layer interface, on the power and current activities of OLED. Structures and materials of OLED are proposed as follows:

- Inserting a thin anode buffer between indium tin oxide (ITO) and hole transport layer (HTL) can decrease the energy barrier and then increase the charge injection at the interface.
- Apply active electron acceptor as dopants in the HTL.
- Apply active electron donor as dopants in the electron transport layer (ETL).
- Most common n-doped ETLs are
  - Li-doped tris(8-hydroxyquinoline) aluminum
  - 4,7-diphyenyl-1,10-phenanthroline-doped tris(8-hydroxyquinoline) aluminum
  - 8-hydroxy-quinolinato lithium

- Most common p-doped HTLs are
  - Molybdenum trioxide
  - Tetrafluoro-tetracyano-quinodimethane
- All organic layers (Table 4.2) are stacked without the intermixing of the ETL.

Table 4.2 Detailed properties of OLED materials

Materials	Properties
Indium tin oxide (ITO)	High conductive transparent Easy to clean Availability on large scale Easy to structure as substrate
Poly(3,4-ethylenedioxythiophene) Poly(styrenesulfonate) (PEDOT:PSS)	Great hole transport layer (HTL) Great work function
2,6-Bis(3-(9H-carbazol-9-yl) phenyl) pyridine (26DCzPPy)	Bipolar host as an emitting layer
Bis(3,5-difluoro-2-(2-pyridyl)phenyl-(2-carboxypyridyl) iridium III (FIrpic)	Blue dopant as emitting layer
Polyethylene terephthalate (PET), 1.5×1.5 cm^2	Resistance up to 210°C Great flexibility
Polydimethylsiloxane (PDMS), 1.5×1.5 cm^2	Glue
2,2′,2″-(1,3,5-Benzinetriyl)-tris(1-phenyl-1-H-benzimidazole) (TPBi)	Good electron transport High mobility
Lithium fluoride (LiF)	Cathode
Aluminum (Al)	Cathode

4.3 Methods

A solution-based OLED structure is produced by some well-known methods as follows:

- Vapor deposition method
- Spin-coating method
- Stamp-printing method

Herein we focus on the stamping method carried out via the release film because of the following reasons:

- Surface polarity and stiffness of stamp are well manipulated.
- Thin film is replicated with good resolution and surface uniformity.
- The release film stamp process is simple in comparison to the current stamping method via PDMS-based polymer, for instance.
- The release film stamp process offers uniform thickness of organic materials.
- The release film stamp is applicable to multilayers devices, partly because of the simple process and economic aspect.
- The release film stamp offers a high-yield process.

Furthermore, we introduce PET release film as the elastomeric material because of the following reasons:

- Layer uniformity is resolved when the organic ink is hardened.
- All organic layers are stacked without the intermixing of electron transport layer.

4.4 Implementations

4.4.1 White OLED

White OLED with an active area of 5×5 mm^2 is produced by the vapor deposition method in an ultrahigh vacuum chamber under the pressure of 4×10^{-6} Torr. White OLED based on the p-i-n structure, using Liq-doped Alq_3 as the n-doped ETL and MoO_3-doped WHI112 as the p-doped HTL based on the energy band diagram (Fig. 4.1), is produced by the following steps:

- Glass slide coated ITO as a substrate is cleaned with acetone, followed by isopropyl alcohol under a sonicator for 15 min, dried with N_2, and exposed with O_2 for 1 min.
- Structures of n-doped ETLs are composed of
  - Al as cathode
  - Tris(2,4,6-trimethyl-3-(pyridin-3-yl)phenyl) borane (3TPYMB) as ETL/hole-blocking layer

  - UBH15 doped with blue fluorescent dopant EB502 as blue EML
  - EPH31 doped with yellow phosphorescent dopant EPY01 as yellow EML
  - LiF as electron injection layer
  - ITO/Alq_3: X wt.% Liq (30 nm)/Al (130 nm)

  where X is 0 wt.% for structure E1, X is 10 wt.% for structure E2, X is 33 wt.% for structure E3, and X is 50 wt.% for structure E4.
- Structures of p-doped HTLs are composed of
  - Al as cathode
  - HTG-1 as HTL
  - UBH15 doped with blue fluorescent dopant EB502 as blue EML
  - EPH31 doped with yellow phosphorescent dopant EPY01 as yellow EML
  - LiF as electron injection layer
  - ITO/WHI112: Y wt.% MoO_3 (50 nm)/HTG-1 (15 nm)/Al (130 nm)

  where Y is 0 wt.% for structure H1, Y is 10 wt.% for structure H2, Y is 20 wt.% for structure H3, and Y is 30 wt.% for structure H4.
- Undoped structure is composed of
  - ITO/WHI112 (55 nm)/HTG-1 (10 nm)/UBH15: 3 wt.% EB502 (10 nm)/EPH31: 3 wt.% EPY01 (25nm)/3TPYMB (10 nm)/Alq_3 (25 nm)/LiF (0.8 nm)/Al (150 nm)
- The n-doped structure is composed of
  - ITO/WHI112 (55 nm)/HTG-1 (10 nm)/UBH15: 3 wt.% EB502 (10 nm)/EPH31: 3 wt.% EPY01 (25nm)/3TPYMB (10 nm)/Alq_3 (25 nm)/LiF (0.8 nm)/Al (150 nm)
  - Alq_3: 33 wt.% Liq (25 nm)
- The p-doped structure is composed of
  - ITO/WHI112 (55 nm)/HTG-1 (10 nm)/UBH15: 3 wt.% EB502 (10 nm)/EPH31: 3 wt.% EPY01 (25nm)/3TPYMB (10 nm)/Alq_3 (25 nm)/LiF (0.8 nm)/Al (150 nm)
  - Alq_3: WHI112: 20 wt.% MoO_3 (55 nm)

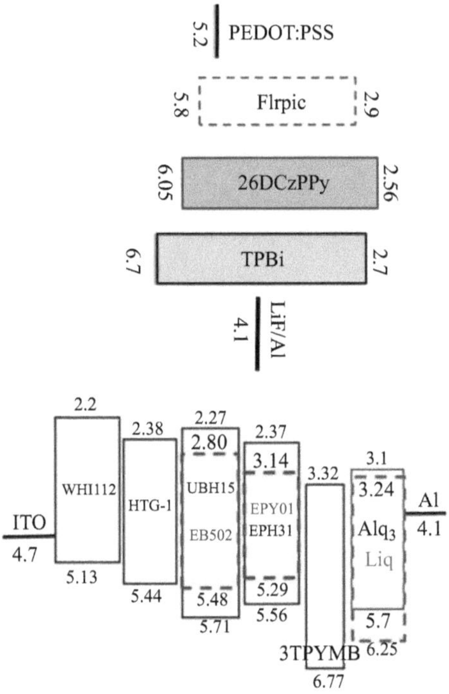

Figure 4.1 Energy band diagrams of blue OLED (top) and white OLED (bottom).

4.4.1.1 Single host structures

For n-doped ETLs, Liq-doped Alq_3 (Table 4.3), current density is proportional to the Liq amount. The highest current density of 33 wt.% Liq clearly improves the electron carrier. The excellent electron conductivity of the Alq_3:Liq matrix originates from the short hopping length in comparison to Alq_3. However, since the LUMO–LUMO difference between Alq_3 (3.10 eV) and Liq (3.24 eV) is very small, electron transport along LUMO shows carrier quenching once Liq goes over 33 wt.% (structure E4).

For the p-doped HTLs, MoO_3-doped WHI112 (Table 4.3), the current density is proportional to the MoO_3 amount. The greatest current density of 20 wt.% MoO_3 leads to the best reduction in potential barrier for hole injection near ITO interface and hence increases the film conductivity. However, since the HOMO–HOMO difference between MoO_3 (5.3 eV) and HTG-1 (5.4 eV) is negligible, subject to their similar HOMO, the holes are easily transferred from the HOMO of WHI112:MoO_3 to the HOMO of HTG-1. High amounts of MoO_3 molecules saturate and, hence, escape to WHI112:MoO_3/HTG-1 interface once MoO_3 goes over 20% (structure H4). Hence, the dipole barrier at the interface results in an increase in the driving voltage similar to structure E4.

Table 4.3 Current density and applied voltage of the n-doped ETL, structures E1–E4. Current density and applied voltage of p-doped HTL, structures H1–H4

Current density, mA/cm^2	Applied voltage, V			
	1	**2**	**3**	**4**
E1	0	2	5	10
E2	5	20	40	60
E3	30	80	120	180
E4	0	2	5	10
H1	0	0	10	30
H2	0	5	20	80
H3	0	7	38	110
H4	0	8	29	92

Note: Driving voltages of E1, E2, E3, and E4 at 100 mA/cm^2 are 10.0, 5.3, 2.6, and 6.7 V, respectively. Driving voltages of H1, H2, H3, and H4 at 100 mA/cm^2 are 5.4, 4.3, 3.8, and 4.2 V, respectively.

4.4.1.2 White OLED structures

In the undoped structures as conventional devices, the number of holes is greater than that of the electrons. Extra holes at the HTL/EML interface increase the EML's cations, leading to device degradation. To produce the white OLED, the carrier balance nearby to recombination and/or emission zone is improved using 3TPYMB as the electron-blocking layer and HTG-1 as the hole-blocking layer. These layers build up further barriers when high carrier concentration is injected. Integration of yellow phosphorescent into blue fluorescent emitters is also suggested.

Herein the undoped and n-type structures are composed of

ITO/WHI112 (55 nm)/HTG-1 (10 nm)/UBH15: 3 wt.% EB502 (10 nm)/EPH31: 3 wt.% EPY01 (25 nm)/3TPYMB (10 nm)/Alq_3: 0–33 wt.% Liq (25 nm)/with or without LiF (0.8 nm)/Al (150 nm)

where Liq is 0 wt.% with LiF for undoped structure and Liq is 33 wt.% without LiF for n-type structure.

High electron injections of n-type structures via Liq help them to move the Fermi level close to the LUMO of Alq (Table 4.4). Thus, current density and luminance are higher than the undoped one. Power efficiencies of undoped and n-type structures are 6.40 and 7.81 Lm/W at 5 mA/cm^2, respectively, with corresponding driving voltages of 11.0 and 7.2 V, respectively.

Table 4.4 Current density, luminance, and power efficiency with respect to the applied voltage of undoped and n-type, and p-i-n structures

Current density, mA/cm^2	Applied voltage, V										
	1	2	3	4	5	6	7	8	9	10	11
Undoped	0	0	1.5	2.5	3.1	3.2	3.5	3.8	4.0	4.5	5.0
n-type	0	0	2	3	4	4	5	10	20	30	50
p-i-n	0	0	2	3	4	4	5	15	25	50	100
Luminance, Cd/m^2											
Undoped	0	0	5	10	20	50	100	190	450	800	1500
n-type	0	0	50	100	300	750	1500	1800	3000	6000	8000
p-i-n	0	0	50	100	300	750	1500	1800	4500	9000	16500
Power efficiency, Lm/W											
Undoped	0	1.5	4.5	6.2	7.5	6.0	5.5	5.1	4.5	4.1	2.2
n-type	0	0	1.0	2.0	9.5	9.0	8.0	7.5	6.2	6.0	5.2
p-i-n	0	0	1.0	5.1	13.5	12.0	8.5	8.5	7.1	6.5	6.1

The p-i-n structure is produced by additional 20 wt.% MoO_3 doped WH112 as follows:

ITO/WHI112: 20 wt.% MoO_3 (55 nm)/HTG-1 (10 nm)/UBH15: 3 wt.% EB502 (10 nm)/EPH31: 3 wt.% EPY01 (25 nm) Alq_3: 33 wt.% Liq (25 nm)/Al (150 nm).

High luminescence and current density of the p-i-n structure give higher power efficiency than the n-type one. Efficiencies of the p-i-n structure, 17.2 cd/A and 5.1 Lm/W, are reduced by about 7% and 21%, respectively, when it is operated at 100 mA/cm^2. Efficiencies of the n-type structure, 14.4 cd/A and 3.8 Lm/W, are reduced by about 20% and 40%, respectively, when it is operated at 100 mA/cm^2. The p-i-n device, integration of the hole transporter of 20 wt.% MoO_3-doped WHI112 into the electron transporter of 33 wt.% Liq-doped Alq_3, is attributed to the hole–electron balance in EML, and thus results in the stable roll-off of current and power efficiencies. The electroluminescence spectra of p-i-n structures are confirmed to be higher than n-type because high electron injection easily shifts the recombination process in the emission zone.

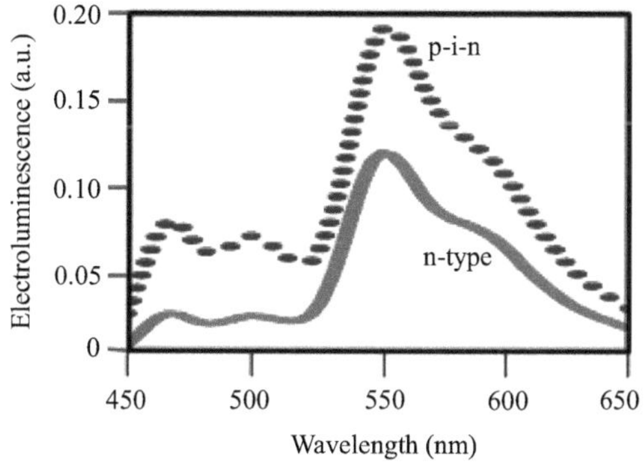

Figure 4.2 Electroluminescence of the white OLED based on n-type and p-i-n devices.

4.4.2 Blue OLED

There is one problem with regard to organic solvent dissolution between the preceding layer and the present layer. A printing method like the stamp transfer technique has been proved to overcome this problem so that each layer is free when stacking multilayer structures all together. ITO/spin-PEDOT:PSS (100 nm)/spin-TCTA:26DCzPPy:FIrpic (45 nm)/RF stamping-TPBi (35 nm)/LiF/Al (0.8/150 nm) based on the energy band diagram (Fig. 4.1) is produced by the following steps (Fig 4.3):

- The ITO substrate is washed under a sonicator for 15 min using acetone, followed by ethanol and dried with N_2 gas.
- PEDOT:PSS is spin coated on the ITO substrate (3500 rpm, 15 s).
- The coated substrate is annealed at 120°C for 20 min and then cooled down to room temperature.
- 26DCzPPy together with FIrpic 15wt.% in chlorobenzene and 0.15 wt.% TPBi in chloroform are stirred for 30 min to obtain a white and clear solution of EML.
- EML is spin coated on the PEDOT:PSS layer to obtain a film thickness of 45 nm.
- PEDOT:PSS/EML is put on a hot plate at 100°C for 15 min.
- 0.15 wt.% TPBi in chloroform is spin coated (3500 rpm, 5 s) on PDMS/PET.
- The PDMS/PET/TPBi film is transferred to EML under certain temperature, physical driving force, and time.

- The film is cooled down to room temperature.
- The PDMS/PET stamping is peeled away, whereas the uniform TPBi is transferred to EML.
- LiF and Al are evaporated and serve as cathode components.

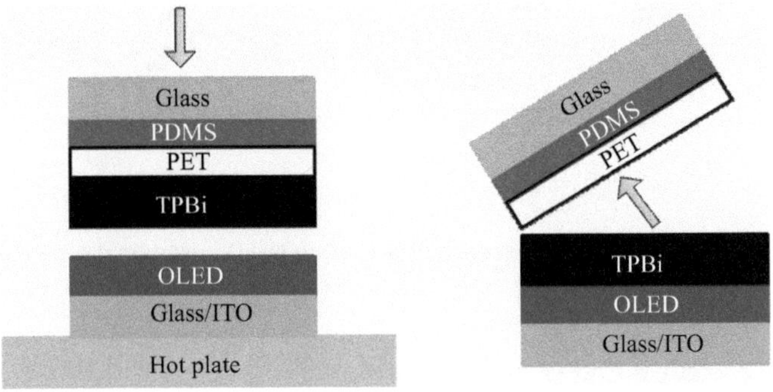

Figure 4.3 Schematic illustration of PET release film stamp printing method. Pressing load is maintained at 6 N.

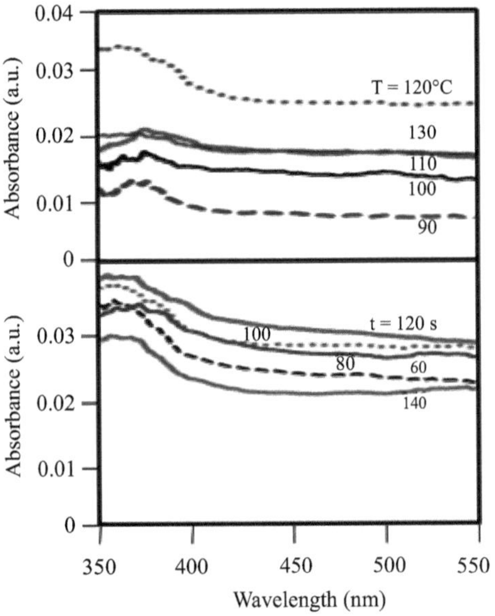

Figure 4.4 Absorbance of TPBi film at different stamping temperatures of 90–130°C (top) and stamping times of 60–140 s (bottom) at a temperature of 120°C and load of 6 N.

The absorbance of TPBi films transferred to the blue OLED is high at a UV wavelength of 390 nm originated by the π–π* transition in comparison to visible light (Fig. 4.4). According to the Beer–Lambert law, light intensity exponentially decays with TPBi film thickness. The absorption amplitude is increased when the stamping temperature and time are increased. The layer of TPBi film with 0.15 wt.% is the thickest at the stamping temperature of 120°C and stamping time of 120 s.

TPBi films from RF stamping before and after stamping are shown in Fig. 4.5. Furthermore, the TPBi film is fabricated both by spin coating and the RF-stamp printing process without any critical film fracture. According to the TPBi film on blue OLED, we introduce the following layers: PEN/ITO/spin-PEDOT:PSS/RF stamp-0.1–0.2 wt.% TPBi/spin-EML/LiF/Al. Although current density is inversely proportional to TPBi concentration, partly due to the charge balance on OLED, further high concentration of TPBi (i.e., 0.2 wt.%) is suggested to obtain high current efficiency (Table 4.5).

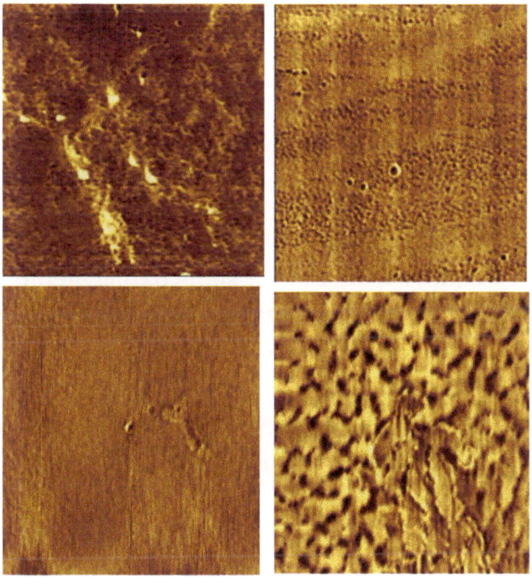

Figure 4.5 AFM images in the tapping mode of blue OLED films (5×5 μm^2), (top left) PEDOT:PSS, (top right) PEDOT:PSS/spin-TPBi, (bottom left) RF/spin-TPBi, and (bottom right) PEDOT:PSS/RF stamping-TPBi. Surface roughness values in root mean square of PEDOT:PSS, PEDOT:PSS/spin-TPBi, RF/spin-TPBi, and PEDOT:PSS/RF stamping-TPBi are 10.85, 0.56, 0.72, and 3.64 nm, respectively.

Table 4.5 Current density, luminance, and current performances with respect to the applied voltage of blue OLED

Current density, mA/cm^2	Applied voltage, V										
	1	2	3	4	5	6	7	8	9	10	11
0.1%	0	0	0	5	7	9	10	15	20	30	40
0.15%	0	0	0	3	4	6	8	10	15	20	30
0.20%	0	0	0	3	4	4	5	8	16	18	25
Luminance, Cd/m^2											
0.1%	0	0	0	1	30	42	100	500	750	900	1000
0.15%	0	0	0	2	40	50	100	550	780	952	1010
0.20%	0	0	0	5	45	100	200	600	800	1000	1100
Current efficiency, Cd/A											
0.1%	0	0	0	0	1	2	4	4	3	3	2
0.15%	0	0	0	0	2	9	6	5	5	5	4
0.20%	0	0	0	0	5	12	9	9	8	5	5

Note: TPBi concentrations are 0.10, 0.15, and 0.20% for blue OLED A, B, and C, respectively.

Problems

4.1 Show the block diagram of absorption measurement.

4.2 Draw a simple picture of the spin-coating method.

4.3 Determine the amplitude of electromagnetic field when OLED is operated under the frequency of 2.4 GHz and the power intensity of 2.0 W/m^2.

4.4 Find a distance from 10 mW OLED as the excitation source to obtain the amplitude of the electric field of 0.01 V/m.

Chapter 5

Pb–Fe-Shielded Material

5.1 Introduction

This chapter deals with the optimal design and thickness of a lead–iron double-layer tank to safely keep radioactive waste low when the input photon energy, the initial radiation intensity, and the attenuation of radiation intensity are fed to MATLAB software. The optimization model composed of logarithmic and bi-logarithmic interpolations is applicable to find out the mass attenuation coefficient and exposure buildup factor, whereas the contour-plotting method is applicable to achieve the optimal thickness of the radioactive waste tank so that the tank weight and cost finally come out.

5.2 Gamma–Matter Interaction

Nuclear generations are growing in capacity owing to their potential in energy applications worldwide (Table 5.1). Nuclear interactions, as an example of nuclear technology, always come along with radioactive waste, which is concerned with human health and environment, especially for occupational radiation workers who always work around a nuclear reactor. The most important

Linear and Nonlinear Optics: Materials, Properties, and Applications
Kitsakorn Locharoenrat

ISBN 978-981-4877-15-2 (Hardcover), 978-1-003-05588-4 (eBook)
www.jennystanford.com

phenomena once gamma rays (e.g., Cs-137, Co-60, and Eu-152) are released by the radioactive source and hit matter are as follows (Fig. 5.1):

Table 5.1 Forecast of global nuclear energy generating capacity

Year	Power capacity value, GWe
2011–15	40.6
2016–20	101.2
2021–25	176.4
2026–30	225.3

Source: World Nuclear Association

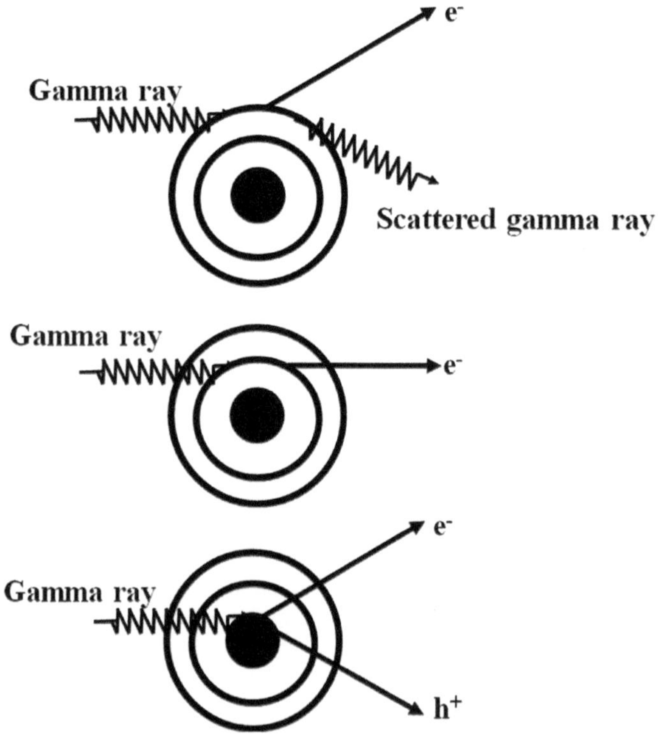

Figure 5.1 Three interactions of gamma photon with matter (Compton scattering, photoelectric absorption, and pair production).

5.2.1 Compton Scattering

Compton scattering originates from the gamma photon partial collision and energy transfer to recoiling electron. Electrons are hence excited to the conduction band.

5.2.2 Photoelectric Absorption

Photoelectric absorption occurs between the gamma photon at a low energy and matter such as metals in which the incident photon is completely absorbed, offering up all its energy by displacing and accelerating the inner shell electron.

5.2.3 Pair Production

Pair production originates from the collision of gamma photon having energy higher than 1.02 MeV with the nucleus, hence producing electron/hole pairs.

There are three factors that result in gamma ray attenuation (Fig. 5.2):

1. **Attenuation coefficient:** The attenuation coefficient μ is the sum of the cross section of photoelectric (σ_{ph}), scattering (σ_{c}), and pair production (σ_{pp}) as

$$\mu(E) = N[\sigma_{\mathrm{ph}}(E) + \sigma_{\mathrm{c}}(E) + \sigma_{\mathrm{pp}}(E)] \tag{5.1}$$

$$N = \rho N_{\mathrm{a}}/A \tag{5.2}$$

$$\lambda = \frac{\int_0^\infty x e^{-\mu x} dx}{\int_0^\infty e^{-\mu x} dx} = \frac{1}{\mu} \tag{5.3}$$

where N_{a} is Avogadro's number, A is mass number, ρ is medium density, and w is weight fraction.

$$\frac{\mu}{\rho} = \frac{N_{\mathrm{a}}}{A}[\sigma_{\mathrm{ph}}(E) + \sigma_{\mathrm{c}}(E) + \sigma_{\mathrm{pp}}(E)] = \frac{\mu_{\mathrm{ph}}}{\rho} + \frac{\mu_{\mathrm{c}}}{\rho} + \frac{\mu_{\mathrm{pp}}}{\rho} \tag{5.4}$$

$$\left(\frac{\mu}{\rho}\right)_{\mathrm{c}} = \sum_i w_i \left(\frac{\mu}{\rho}\right)_i \tag{5.5}$$

2. **Mass absorption thickness:** The absorber mass thickness ρt exists in the radiation attenuation formula as

$$\frac{I}{I_0} = e^{-(\mu/\rho)\rho t} \tag{5.6}$$

3. **Buildup factor:** The buildup factor indicates how the gamma photon deviates from the ideal collimation beam as

$$\frac{I}{I_0} = B(t, E_\gamma)e^{-\mu t} \tag{5.7}$$

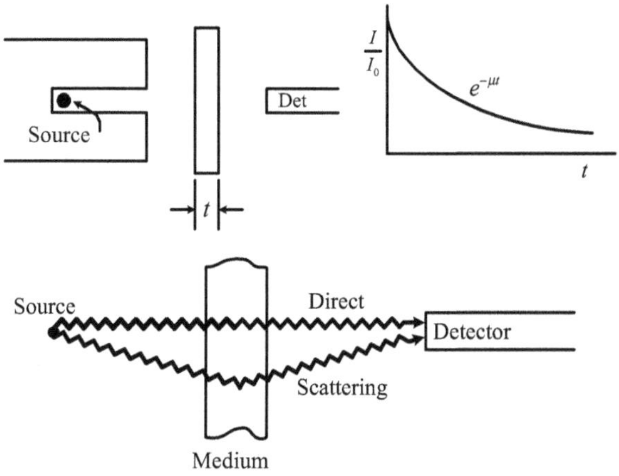

Figure 5.2 Geometrical diagram of ideal (top) and real (bottom) radiation attenuations of gamma photons into the designed medium with thickness of *t*. Det is detector. Examples of radiation beams are Cs-137, Co-60, and Eu-152.

5.3 Computational Method

As informed by the International Commission on Radiological Protection (ICRP), a radiation intensity beyond 20 mSv/year is considered risky. To deal with radioactive waste, an appropriate design of the storage tank, including material thickness and types, is necessary for radiation safety. This process is carried out with the MATLAB software due to the following benefits:

- A variety of commands are used in the for-loop and if-clause processes serving as the necessary parts of the optimization model (Fig. 5.3).

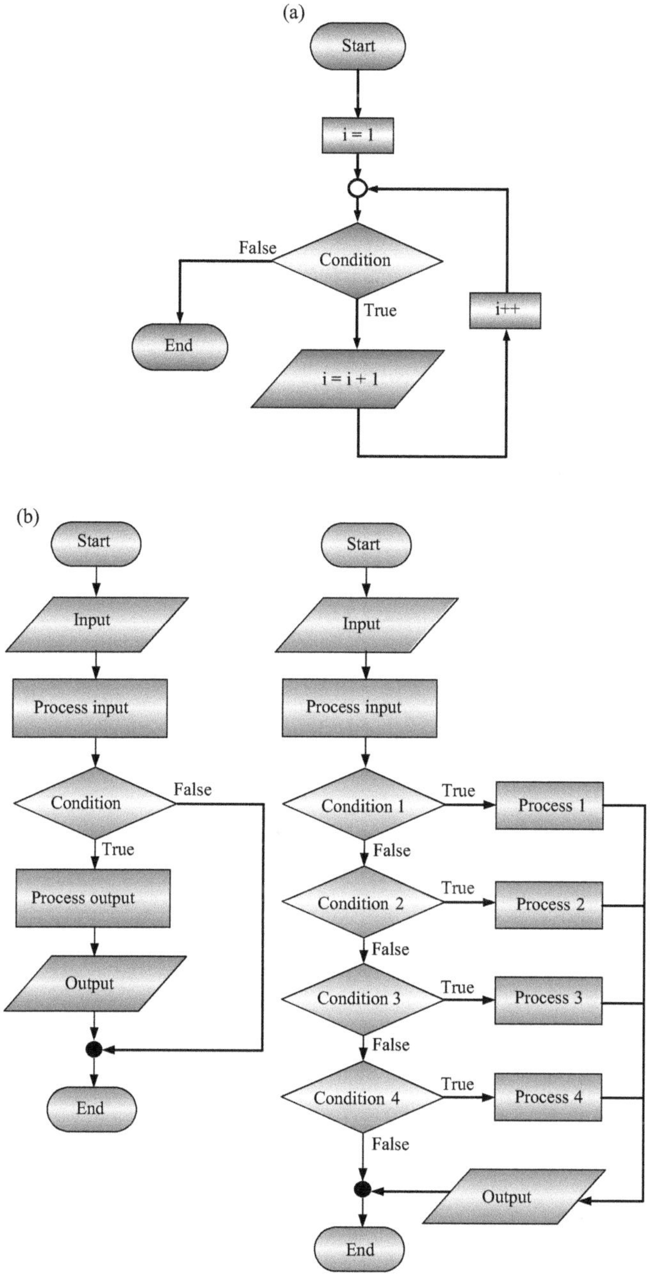

Figure 5.3 Flow chart of for-loop (a) and flowchart of if-condition and if-else condition (b).

- The contour plot method is used to determine the optimal solution, such as the optimal thickness of materials.
- The output data are simply kept in the matrix format upon mathematical processing.
- Command patterns in mathematical processing data are easy to write.
- The statistical processing time is shorter than the conventional simulation like Monte Carlo approach due to the simplex geometry design.

5.3.1 Interpolation Method

There are two interpolation techniques concerned with the MATLAB software as follows (Fig. 5.4):

1. **Linear interpolation:** Linear interpolation is a straight line given from the equation of slopes between two known data points to produce a new data point. The formula of linear interpolation is written as

$$\frac{y_2 - y_1}{x_2 - x_1} = \frac{y - y_1}{x - x_1} \tag{5.8}$$

$$y = y_1 + (y_2 - y_1)\left(\frac{x - x_1}{x_2 - x_1}\right) \tag{5.9}$$

$$y = \frac{(y_1)(x_2 - x) + (y_2)(x - x_1)}{x_2 - x_1} \tag{5.10}$$

2. **Bilinear interpolation:** Bilinear interpolation is modified from the linear interpolation using a linear interpolation in the x-direction and hence in the y-direction to produce the new two variables on the rectilinear 2D grid written as

$$\frac{A_{21} - A_{11}}{x_2 - x_1} = \frac{B_1 - A_{11}}{x - x_1} \tag{5.11}$$

$$B_1 = A_{11} + (A_{21} - A_{11})\left(\frac{x - x_1}{x_2 - x_1}\right) \tag{5.12}$$

$$B_1 = \frac{(A_{11})(x_2 - x) + (A_{21})(x - x_1)}{x_2 - x_1} \tag{5.13}$$

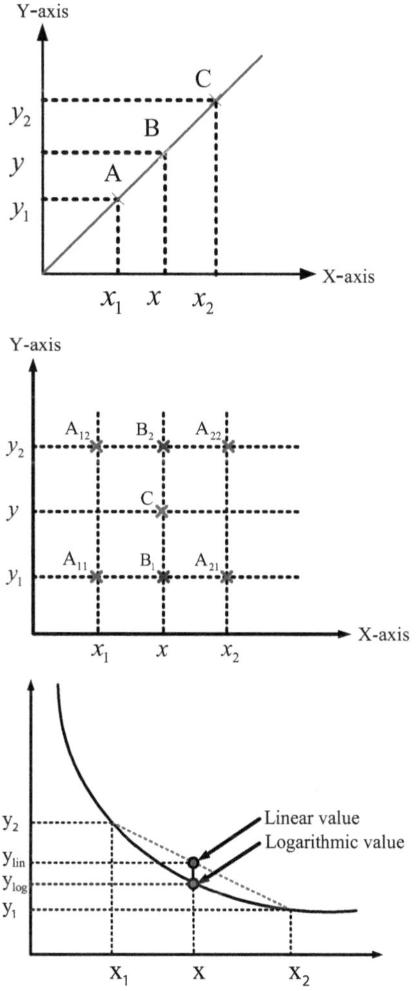

Figure 5.4 Data approximation by linear interpolation (top), bilinear interpolation (middle), and logarithm (bottom).

$$\frac{A_{22} - A_{12}}{x_2 - x_1} = \frac{B_2 - A_{12}}{x - x_1} \tag{5.14}$$

$$B_2 = A_{12} + (A_{22} - A_{12})\left(\frac{x - x_1}{x_2 - x_1}\right) \tag{5.15}$$

$$B_2 = \frac{(A_{12})(x_2 - x) + (A_{22})(x - x_1)}{x_2 - x_1} \tag{5.16}$$

3. **Logarithm interpolation:** The situation is not straightforward if the axis is a logarithmic scale, meaning that values are plotted at a position proportional to the logarithm of their value written as

$$y_{\log} = \frac{(y_1)(\log x_2 - \log x) + (y_2)(\log x - \log x_1)}{x_2 - x_1} \tag{5.17}$$

$$B_{1(\log)} = \frac{(A_{11})(\log x_2 - \log x) + (A_{21})(\log x - \log x_1)}{\log x_2 - \log x_1} \tag{5.18}$$

$$B_{2(\log)} = \frac{(A_{12})(\log x_2 - \log x) + (A_{22})(\log x - \log x_1)}{\log x_2 - \log x_1} \tag{5.19}$$

$$C_{\log} = \frac{(B_1)(\log y_2 - \log y) + (B_2)(\log y - \log y_1)}{\log y_2 - \log y_1} \tag{5.20}$$

5.3.2 Optimization Model

We define the gamma ray attenuation of ideal and real beams into the radioactive waste tank materials (i.e., lead–iron double layer) through the objective function when the radiation intensity must be less than the standard value of 0.1 mSv/hr as follows (Fig. 5.5):

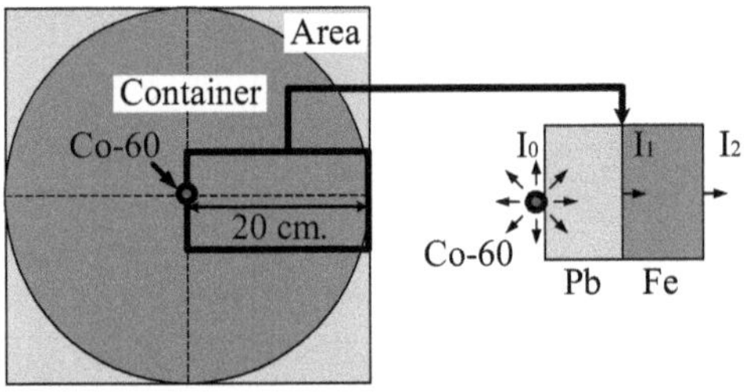

Figure 5.5 Geometrical diagram of the radiation attenuations of gamma photons passing into the designed tank materials.

$$I_1 = I_0 B_1 e^{-\mu_1 x_1} \tag{5.21}$$

$$I_2 = I_1 B_2 e^{-\mu_2 x_2} \tag{5.22}$$

$$f(x_1, x_2) = I = I_0 B_1 B_2 e^{-\mu_1 x_1 - \mu_2 x_2} \le 0.1 \begin{cases} B_1 B_2 = 1 & \text{for coherent beam} \\ B_1 B_2 > 1 & \text{for incoherent beam} \end{cases} \tag{5.23}$$

where I_0 is the initial radiation intensity (mSv/h), I is the transmission radiation intensity (mSv/h), μ_1 is the linear attenuation coefficient of lead (cm^{-1}), μ_2 is the linear attenuation coefficient of iron (cm^{-1}), B_1 is the exposure buildup factor of lead, and B_2 is the exposure buildup factor of iron.

Based on logarithmic interpolation (LI) and bi-logarithmic interpolation (BI) methods, mass attenuation coefficient μ/ρ, exposure buildup factor B, and penetration depth PD are written in the order as follows (Fig. 5.6):

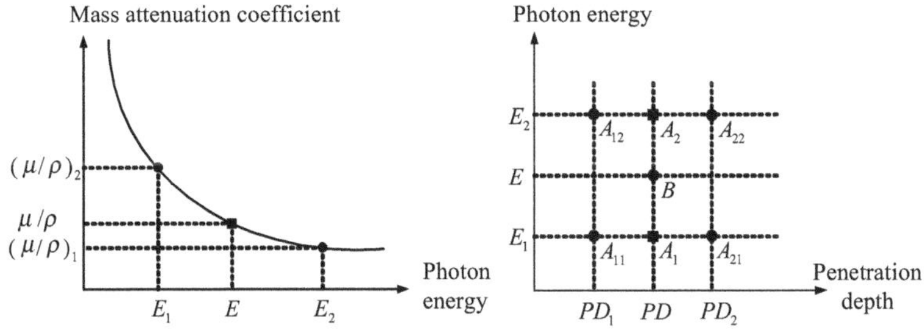

Figure 5.6 Logarithmic interpolation method (left) and bi-logarithmic interpolation method (right) to determine the mass attenuation coefficient μ/ρ, exposure buildup factor B, and penetration depth PD.

$$\mu / \rho = \frac{(\mu / \rho)_1 (\log E_2 - \log E) + (\mu / \rho)_2 (\log E - \log E_1)}{\log E_2 - \log E_1} \tag{5.24}$$

$$B = \frac{(A_1)(\log E_2 - \log E) + (A_2)(\log E - \log E_1)}{\log E_2 - \log E_1} \tag{5.25}$$

where

$$A_1 = \frac{(A_{11})(\log PD_2 - \log PD) + (A_{21})(\log PD - \log PD_1)}{\log PD_2 - \log PD_1}$$

$$A_2 = \frac{(A_{12})(\log PD_2 - \log PD) + (A_{22})(\log PD - \log PD_1)}{\log PD_2 - \log PD_1}$$

$$PD_1 = \mu_1 \cdot x_1 \text{ and } PD_2 = \mu_2 \cdot x_2 \tag{5.26}$$

We define tank thickness of less than 30 cm through the constrained function as follows (Fig. 5.7):

$$1 \le x_1 \le 30 \tag{5.27}$$

$$1 \le x_2 \le 30 \tag{5.28}$$

$$g_1(x_1, x_2): c_{\text{air}} + x_1 + x_2 \le 3 \tag{5.29}$$

We define tank weight of less than 1000 kg through the constrained function as follows (Fig. 5.7):

$$r_1 = c_{\text{air}} \tag{5.30}$$

$$h_1 = 2c_{\text{air}} \tag{5.31}$$

$$r_2 = x_1 \tag{5.32}$$

$$h_2 = 2x_1 + 2c_{\text{air}} \tag{5.33}$$

$$r_3 = x_2 \tag{5.34}$$

$$h_3 = 2x_1 + 2x_2 + 2c_{\text{air}} \tag{5.35}$$

$$v_{\text{air}} = V_1 = \pi r_1^2 h_1 = 2\pi c_{\text{air}}^3 \tag{5.36}$$

$$v_{\text{lead}} = V_2 - V_1 = \pi(r_1 + r_2)^2 h_2 - \pi r_1^2 h_1 \tag{5.37}$$

$$v_{\text{lead}} = \left(\pi(x_1 + c_{\text{air}})^2(2x_1 + 2c_{\text{air}})\right) - \left(\pi c_{\text{air}}^2(2c_{\text{air}})\right) \tag{5.38}$$

$$v_{\text{lead}} = 2\pi\left((x_1 + c_{\text{air}})^3 - c_{\text{air}}^3\right) \tag{5.39}$$

$$v_{\text{iron}} = V_3 - V_2 = \pi(r_1 + r_2 + r_3)^2 h_3 - \pi(r_1 + r_2)^2 h_2 \tag{5.40}$$

$$\begin{aligned} v_{\text{iron}} &= \left(\pi(x_1 + x_2 + c_{\text{air}})^2(2x_1 + 2x_2 + 2c_{\text{air}})\right) \\ &\quad - \left(\pi(x_1 + c_{\text{air}})^2(2x_1 + 2c_{\text{air}})\right) \end{aligned} \tag{5.41}$$

$$v_{\text{iron}} = 2\pi\left((x_1 + x_2 + c_{\text{air}})^3 - (x_1 + c_{\text{air}})^3\right) \tag{5.42}$$

$$m_{\text{total}} = m_{\text{air}} + m_{\text{lead}} + m_{\text{iron}} = \rho_{\text{air}} v_{\text{air}} + \rho_{\text{lead}} v_{\text{lead}} + \rho_{\text{iron}} v_{\text{iron}} \tag{5.43}$$

$$m_{\text{total}} = \frac{2\pi}{1000}\left\{\begin{array}{l} \rho_{\text{air}} c_{\text{air}}^3 + \rho_{\text{lead}}[(x_1 + c_{\text{air}})^3 - c_{\text{air}}^3] \\ +\rho_{\text{iron}}[(x_1 + x_2 + c_{\text{air}})^3 - (x_1 + c_{\text{air}})^3] \end{array}\right\} \tag{5.44}$$

$$g_2(x_1, x_2): \frac{2\pi}{1000}\left\{\begin{array}{l} \rho_{\text{a}} c_{\text{air}}^3 + \rho_{\text{l}}[(x_1 + c_{\text{air}}^3)^3 - c_{\text{air}}^3] \\ +\rho_{\text{i}}[(x_1 + x_2 + c_{\text{air}})^3 - (x_1 + c_{\text{air}})^3] \end{array}\right\} \le 1000 \tag{5.45}$$

We define tank cost of less than \$800 through the constrained function as follows (Fig. 5.7):

$$g_3(x_1,x_2): \frac{4.4\pi}{1000}\left\{\begin{matrix} A\rho_l[(x_1+c_{\text{air}})^3 - c_{\text{air}}^3] \\ +B'\rho_i[(x_1+x_2+c)^3 - (x_1+c_{\text{air}})^3] \end{matrix}\right\} \le 800 \tag{5.46}$$

where ρ_a is the density of air (0.001225 g/cm^3), ρ_l is the density of lead (11.35 g/cm^3), ρ_i is the density of iron (7.87 g/cm^3), c_{air} is the radius of tank space (cm), A is the cost of lead (\$0.97/lb), B' is the cost of iron (\$0.028/lb).

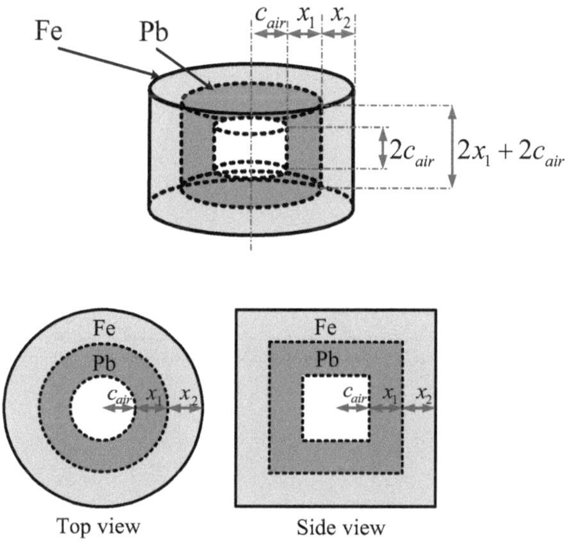

Figure 5.7 Three-dimensional model of double-layer thickness cylindrical tank (top) and 2D model of the cylindrical tank at top view (bottom left) and side view (bottom right).

5.3.3 Contour Plot Method

We import the input variables in the MATLAB software as follows (Fig. 5.8 and Table 5.2):

- Photon energy
- Initial radiation intensity
- Material 1, x_1 = lead
- Material 2, x_2 = iron

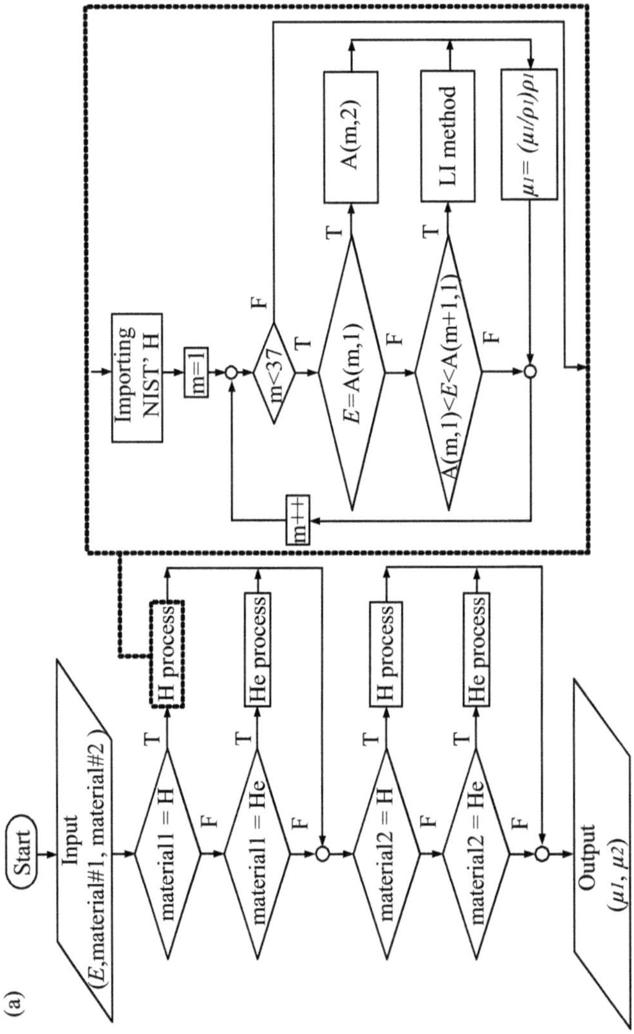
(a)
Start
Input
(E,material#1, material#2)
material1 = H
material1 = He
material2 = H
material2 = He
H process
He process
H process
He process
Output
(μ1, μ2)
Importing
NIST' H
m=1
m<37
E=A(m,1)
A(m,1)<E<A(m+1,1)
A(m,2)
LI method
μ1= (μ1/ρ1)ρ1
m++
T
F

(b)

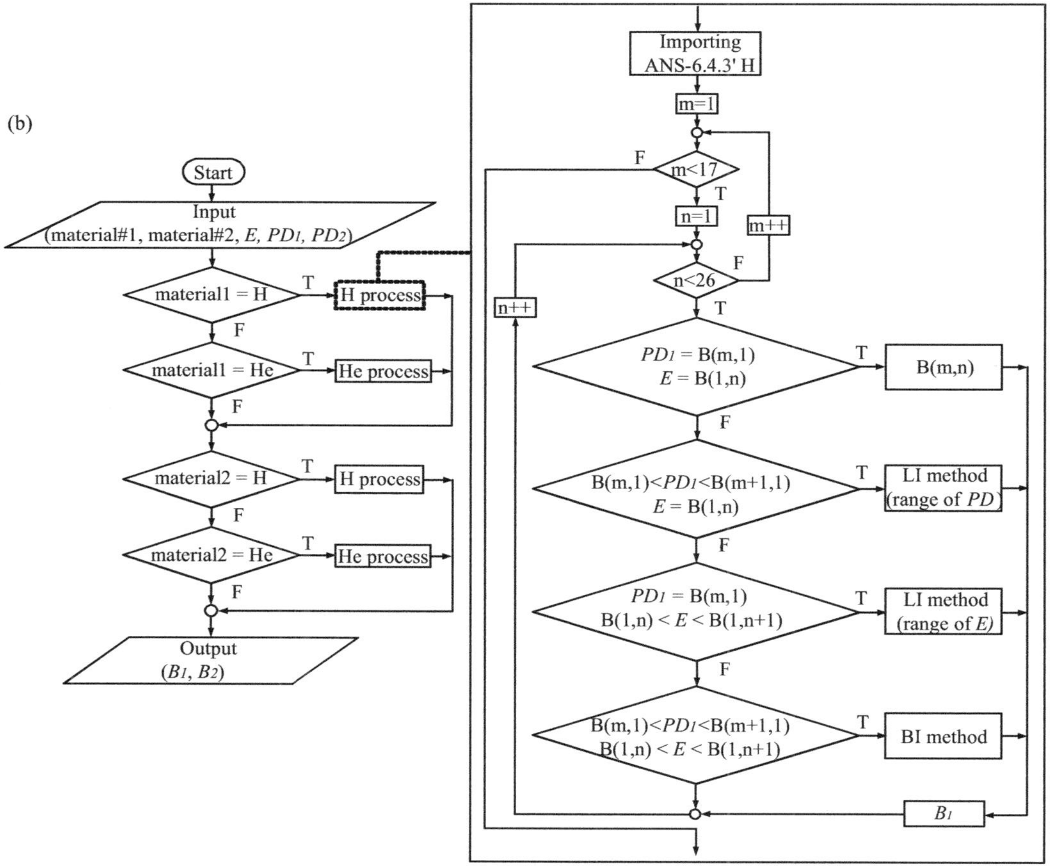

Figure 5.8 Flow chart simulation of mass attenuation coefficient (a) and exposure buildup factor (b) in order to generate a contour graph.

- Material 1 thickness = 21 cm
- Material 2 thickness = 2 cm

We thus set the matrix coordination. Next the mass attenuation coefficient, penetration depth, and exposure buildup factor are computed via the defined optimization model to generate a contour graph.

Table 5.2 (a) Representative example of the photon energy and mass attenuation coefficient of lead reported by the National Institute of Standard and Technology (NIST)

E	u/p
⋮	⋮
1.00000	0.07102
1.25000	0.05876
1.50000	0.05222
⋮	⋮

36 rows

E = Photon energy (MeV)
u/p = Mass attenuation coefficient (cm^2/g)

Table 5.2 (b) Representative example of the photon energy and mean free path of lead reported by the American National Standard (ANS-6.4.3)

MFP \ E	0.015	0.02	0.03		8	10	15
0.5				· · ·			
1.0				· · ·			
3.0				· · ·			
⋮	⋮	⋮	⋮	⋮ · ⋮	⋮	⋮	
30.0				· · ·			
35.0				· · ·			
40.0							

16 rows

25 columns

E = Photon energy (MeV)
MFP = Mean free path (-)

5.4 Implementation (see PC Code of Monte Carlo Simulation in Appendix A)

Validation of MATLAB software is first confirmed in comparison to the standard reference databases as follows:

- When the input energy is 0.15 MeV, we get the mass attenuation coefficient of 0.143600 cm^2/g. This value is similar to a number reported by the NIST.
- When the penetration depth is 5 mfp and the input energy is 20 MeV, we get the exposure buildup factor of 5.22. This value is similar to a number reported by ANS-6.4.3 and offers a better result than the invariant embedding method (IEM).

Suppose that the detailed information of gamma radiation in the radioactive waste tank is as follows:

- Photon energy is 1.3325 MeV.
- Initial gamma radiation intensity is 100 mSv/h.
- Transmission radiation intensity is less than 0.1 mSv/h.
- Material type: x_1 is lead and x_2 is iron.
- Material thickness is less than 30 cm.
- Material weight is less than 1000 kg.
- Material cost is less than $800.
- Diameter of tank space is 14 cm.

The mass attenuation coefficient and exposure buildup factor of the transmission radiation intensity are computed and summarized (Table 5.3). Then we get the computational results of tank weight and the cost of the transmission radiation intensity (Table 5.4).

Suppose that c_{air} = 7 cm, which is suitable to cover all radioactive wastes. The optimal solution of cylindrical tank thickness identified as the intersection between one objective function and three constrained functions is achieved via the contour-plotting graph (Figs. 5.9–5.11). Coordination point on the identified line is, hence, chosen to calculate each tank weight and cost (Table 5.5). It is found that the optimal thicknesses of lead and iron are 3.2 and 17.0 cm together with the tank weight and cost of 994.30 kg and $167.30, respectively.

Table 5.3 (a) Computational results of buildup factor on transmission radiation intensity

C_{air} (cm)	$g_1(x_1, x_2)$ (cm)	x_1 (cm)	x_2 (cm)	B_1	B_2	$B_1 B_2$	I (mSv/h)
7	30	1	22	1.2431	11.9342	14.8354	0.0978
		2	21	1.4763	11.2700	16.6379	0.0868
		3	20	1.6766	10.5734	17.72736	0.0732
		4	19	1.8819	9.9385	18.70326	0.0611
		5	18	2.0621	9.3386	19.25713	0.0498
		6	17	2.2348	8.7208	19.48924	0.0398
		7	16	2.4113	8.1675	19.69429	0.0318
		8	15	2.5801	7.5786	19.55355	0.0250
		9	14	2.7542	7.0394	19.38792	0.0196
		10	13	2.9272	6.5020	19.03265	0.0152
		11	12	3.0946	5.9567	18.4336	0.0117
		12	11	3.2683	5.4780	17.90375	0.0089
		13	10	3.4404	4.9537	17.04271	0.0067
		14	9	3.6095	4.5072	16.26874	0.0051
		15	8	3.7669	4.0445	15.23523	0.0037
		16	7	3.9322	3.5879	14.10834	0.0027
		17	6	4.1104	3.1991	13.14958	0.0020
		18	5	4.2785	2.7393	11.7201	0.0014
		19	4	4.4374	2.4240	10.75626	0.0010
		20	3	4.5882	2.0501	9.406269	0.0007
		21	**2**	4.7316	1.6681	7.892782	**0.0004**
		22	1	4.8684	—	—	—

Table 5.3 (b) Calculated mass attenuation coefficient, exposure buildup factor, and penetration depth under input energy of 1.3325 MeV according to C-60 source and input intensity of 100 mSv/h

Layer	Density (ρ), gm/cm^3	Mass attenuation coefficient (μ/ρ), cm^2/g	Linear attenuation coefficient (μ), cm^{-1}	Thickness (x), cm	Penetration depth (PD)	Exposure buildup factor (B)
1, Pb	11.35	0.0566	0.6424	21	13.4904	4.7316
2, Fe	7.874	0.0518	0.4083	2	0.8166	1.6681

Table 5.4 Computational results of tank weight and cost on transmission radiation intensity

C_{air} = 7 cm			Square tank		Cylindrical tank		
$g_1(x_1, x_2)$ (cm)	x_1 (cm)	x_2 (cm)	Weight (kg)	Cost (USD)	Weight (kg)	Cost (USD)	I (mSv/h)
30	1	22	1683	5	1322	106	0.0978
	2	21	1689	6	1327	138	0.0868
	3	20	1697	8	1333	179	0.0732
	4	19	1706	10	1340	228	0.0611
	5	18	1717	12	1349	287	0.0498
	6	17	1730	15	1359	357	0.0398
	7	16	1745	18	1371	439	0.0318
	8	15	1763	23	1385	533	0.0250
	9	14	1783	28	1400	640	0.0196
	10	13	1806	32	1418	762	0.0152
	11	12	1831	38	1438	899	0.0117
	12	11	1860	44	1461	1053	0.0089
	13	10	1892	52	1486	1223	0.0067
	14	9	1927	59	1513	1411	0.0051
	15	8	1965	67	1543	1618	0.0037
	16	7	2007	78	1577	1844	0.0027
	17	6	2054	89	1613	2091	0.0020
	18	5	2104	100	1652	2360	0.0014
	19	4	2158	112	1695	2651	0.0010
	20	3	2216	126	1741	2965	0.0007
	21	**2**	**2280**	140	**1790**	3303	**0.0004**
	22	1	2347	156	1843	3667	—

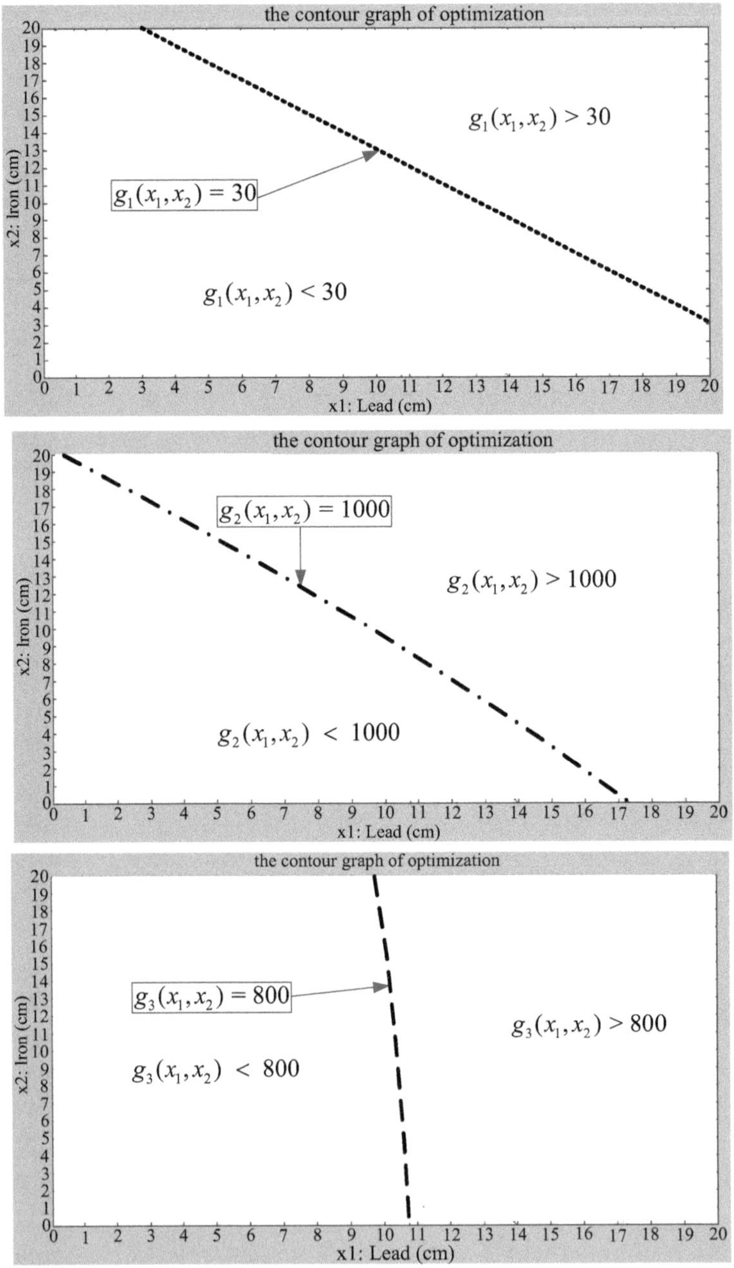

Figure 5.9 Contour-plotting graph with corresponding boundary condition of three constrained functions.

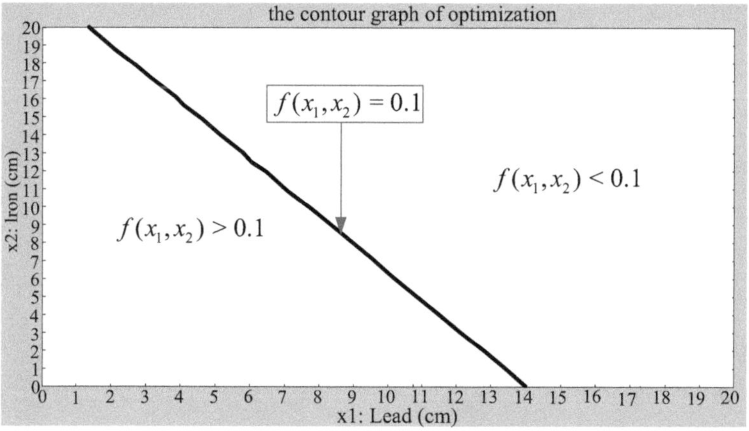

Figure 5.10 Contour-plotting graph with corresponding boundary condition of one objective function.

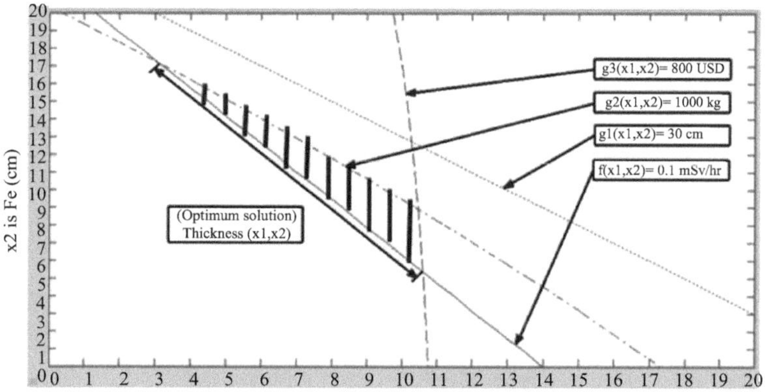

Figure 5.11 Contour-plotting graph of optimal cylindrical tank thickness through the identified line.

Table 5.5 Representative coordination points on the identified line offering the tank weight and cost (c_{air} = 7 cm)

Lead thickness (x_1), cm	Iron thickness (x_2), cm	Tank weight, kg	Tank cost, USD
7.0	11.0	808.4	404.6
6.4	12.0	838.8	356.5
5.7	13.0	860.0	305.0
5.1	14.0	893.8	266.1
4.5	15.0	929.4	231.3
3.7	16.0	943.9	188.5
3.2	17.0	994.3	167.3

Problems

5.1 Explain some radioisotopic source that produces gamma rays.

5.2 Show the equation of free electrons existing in photoelectric absorption.

5.3 Write the energy transfer and scattering angle in terms of energy and momentum conservation.

5.4 Write the condition of pair production.

5.5 Draw the relationship between photoelectric, Compton, and pair production processes.

Chapter 6

Probe Sensors

6.1 Introduction

Sensing technology has been the subject of many studies due to a variety of potential applications in biophotonics and biomedical devices. This chapter deals with probe sensors using light attached to some tools or device to probe the interactions between sample and sensing materials to quantify the physical properties of analytes.

6.2 Principle

6.2.1 Turbid Tissue

The structure of turbid tissue, including human tissue, mainly consists of three main components (Fig. 6.1):

1. Skin is the outer layer covering the human body and is composed of epidermis, dermis, and subcutis. It protects the body against pathogens and excessive water loss.
2. Adipose tissue is a loose connective tissue composed mostly of fat cells. It is located beneath the human skin. Its key function is to store the energy in lipid form.

Linear and Nonlinear Optics: Materials, Properties, and Applications
Kitsakorn Locharoenrat

ISBN 978-981-4877-15-2 (Hardcover), 978-1-003-05588-4 (eBook)
www.jennystanford.com

3. Muscles tissue is composed of three sub-tissues that are capable of contraction and relaxation to produce the movement of human body parts.
   a. Skeletal muscle tissue is a form of striated muscle tissue, which is voluntarily controlled by the somatic nervous system. Most of them are connected to the bones by tendons.
   b. Cardiac muscle tissue is a form of striated muscle tissue, which is the main tissue of walls of the heart.
   c. Smooth muscle tissue is a form of non-striated muscle tissue mostly found in the stomach, intestines, urinary bladder, and uterus.

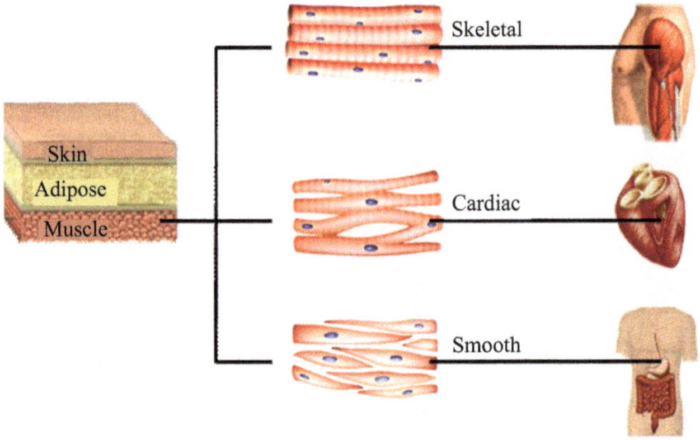

Figure 6.1 Human tissue structures composed of skin, adipose tissue, and muscle tissues.

6.2.2 Reflection Probe

Reflection probe is applicable to design and production in tissue diagnostic devices because it is very useful to observe photon behaviors in biological tissues (Fig. 6.2).

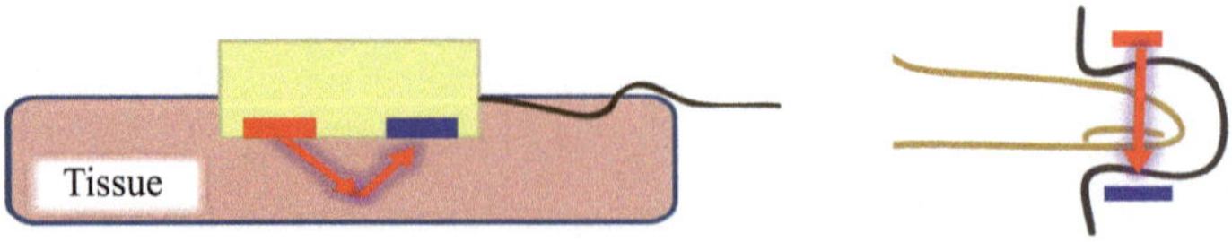

Figure 6.2 Reflection probe in comparison to transmission probe on human skin.

Since the reflection of photon intensity is dependent on light absorption and scattering, as well as tissue components (Table 6.1), it is difficult to calibrate such probe precisely. Herein we aim to explain the design of the probe via the simulation model of light–human tissue interaction using the Monte Carlo simulation method due to the following reasons:

- Photon reflection intensities are simulated for the more complex tissue structures composed of skin, fat, and skeletal muscle (Fig. 6.3).
- Photon distribution and penetration depth around the skeletal muscle are precisely calculated.
- It is used as a guideline to set up the position of point source and photodetector on the human skin.
- It is used as a precise tool to understand the human muscle conditions, such as contraction and relaxation.

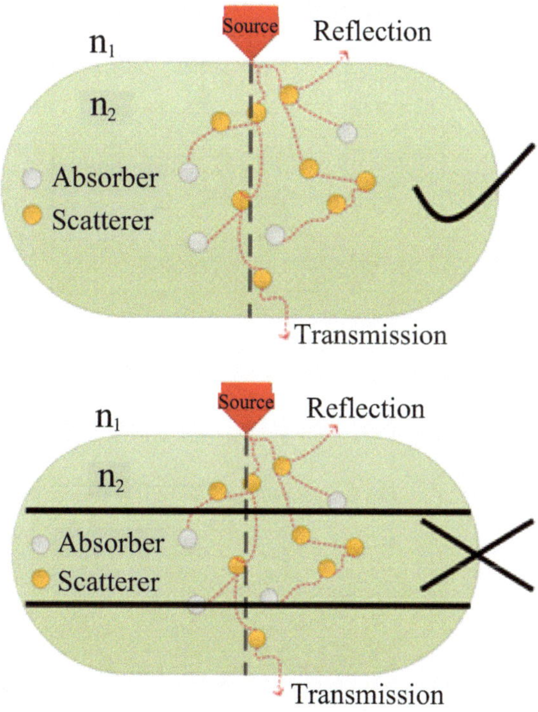

Figure 6.3 Photon transport in the simple tissue (top) and biological tissue compartments (bottom).

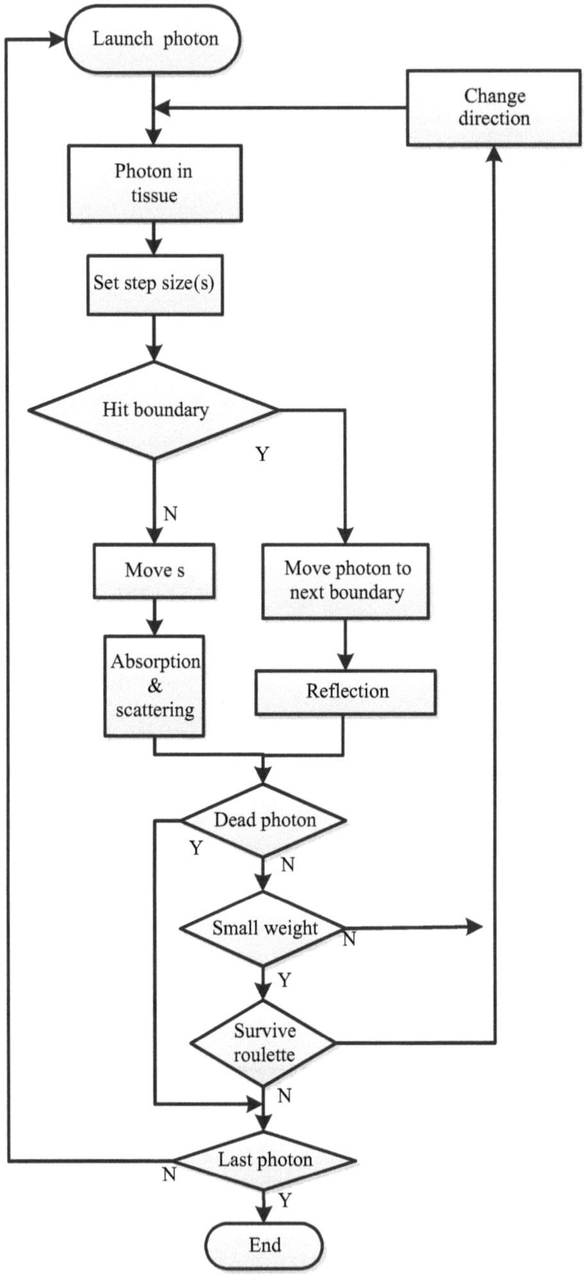

Figure 6.4 Monte Carlo simulation of photon–human tissue interaction.

Table 6.1 Geometrical parameters of each layer in the human tissue model

Layer	Thickness, d_z	Refractive index, n	Anisotropy factor, γ	Absorption coefficient, μ_a	Scattering coefficient, μ_s
1, skin	0.12 cm	$n_1 = 1.39$	0.81	2.70 1/cm	187.00 1/cm
2, fat	0.38 cm	$n_2 = 1.44$	0.85	3.60 1/cm	171.00 1/cm
3, skeletal muscle	∞	$n_3 = 1.37$	0.90	11.20 1/cm	530.00 1/cm

Photon-tissue interaction is calculated by the PC code of Monte Carlo simulation with the following steps (Fig. 6.4):

- Initial photon is located at $(x = 0, y = 0, z = 0)$ coordinates.
- Initial photon direction is set at $(\mu_x = 0, \mu_y = 0, \mu_z = 1)$.
- When the photon packet is propagated in each medium, specular reflectance R_{sp} at the interface is defined as

$$R_{sp} = \frac{(n_{\text{incidence}} - n_{\text{transmission}})^2}{(n_{\text{incidence}} + n_{\text{transmission}})^2} \tag{6.1}$$

- Probability density function $p(\sigma)$ is defined as the distribution of random variable σ over interval (0, 1).
- The sampling probability distribution of free path of photon, s $(0 \leq s < \infty)$ reveals the step size of photon packet and is defined as

$$s = -\ln\sigma/\mu_t \tag{6.2}$$

where $\mu_t = \mu_a + \mu_s$.

- Once the step size s goes forward, the updated position of photon packet is defined as

$$x \rightarrow x + \mu_x \cdot s$$

$$y \rightarrow y + \mu_y \cdot s$$

$$z \rightarrow z + \mu_z \cdot s$$

- The fraction of photon weight decreased by the medium absorption is defined as

$$\Delta W = (\mu_a / \mu_t) W \tag{6.3}$$

- Updated photon weight is defined as

$$W \rightarrow W - \Delta W$$

Photon weight W is initially set at unity, and it is then updated as

$$W = 1 - R_{sp} \tag{6.4}$$

- Deflection angle ($0 \le \theta < \pi$) and azimuthal angle ($0 < \delta < 2\pi$) of the scattered photon are defined as

$$\cos\theta = \frac{1}{2\gamma}\left[1+\gamma^2 - \left(\frac{1-\gamma^2}{1-\gamma+2\gamma\sigma}\right)^2\right] \tag{6.5}$$

$$\delta = 2\pi\sigma \tag{6.6}$$

- Under light incident at a normal angle to tissue, the photon direction due to the scattering process is simply modified as

$$\mu_x' = \frac{\sin\theta}{\sqrt{1-\mu_z^2}}(\mu_x\mu_z\cos\delta - \mu_y\sin\delta) + \mu_x\cos\theta \simeq \sin\theta\cos\delta \tag{6.7}$$

$$\mu_y' = \frac{\sin\theta}{\sqrt{1-\mu_z^2}}(\mu_y\mu_z\cos\delta - \mu_x\sin\delta) + \mu_y\cos\theta \simeq \sin\theta\sin\delta \tag{6.8}$$

$$\mu_z' = \left(-\sin\theta\cos\delta\sqrt{1-\mu_z^2}\right) + \mu_z\cos\theta \simeq \left(\mu_z / |\mu_z|\right)\cos\theta \tag{6.9}$$

- Photon crosses over to the next boundary once the step size s hits the first boundary, and hence the probability of the internal reflection is defined as

$$R_i = \frac{1}{2}\left(\frac{\sin^2(\lambda_{\text{incidence}} - \lambda_{\text{transmission}})}{\sin^2(\lambda_{\text{incidence}} + \lambda_{\text{transmission}})} + \frac{\tan^2(\lambda_{\text{incidence}} - \lambda_{\text{transmission}})}{\tan^2(\lambda_{\text{incidence}} + \lambda_{\text{transmission}})}\right) \tag{6.10}$$

where

$$\lambda_i = \cos^{-1}\mu_z$$

$$n_{\text{incidence}}\sin(\lambda_{\text{incidence}}) = n_t\sin(\lambda_{\text{transmission}})$$

- Photon termination is dependent on the photon weight W as

$$W = \begin{cases} mW & \text{if } \sigma \le 1/m \\ 0 & \text{if } \sigma > 1/m \end{cases} \tag{6.11}$$

where m is the surviving chance (i.e., $m = 100$) for the photon weight of mW.

- The photon packet is completely terminated when the photon weight becomes zero.
- Photon absorption and scattering are still executed if the photon weight is higher than the defined threshold inside the tissue (i.e., $W = 0.0001$).
- Computational results such as photon termination point and photon distribution on tissue surface by a single photon propagation to the ideal tissue model are finally calculated.

Exercise 6.1

Show the example of PC code of Monte Carlo simulation due to photon–tissue interaction.

```
%%%%%%%%%%%%%%%%%%%%%%%%%%%%%%%%Head
Code%%%%%%%%%%%%%%%%%%%%%%%%%%%
N = 1000; % number of photons
m = 100; %1 chance in m of photon packet surviving
%%%%%%%%%%%%%%%%%%%%&&%%%%%%%%%%%%%%%%%%%%%%%
%%%%%%%%%%%%%%%%%%%%%%%%%
%%%%%%%%%%%%%%%%%%%%%%%%%%%%%%%%%%Initialize
%%%%%%%%%%%%%%%%%%%%%%%%%%%
x = 0; % initialize position at x
y = 0; % initialize position at y
z = 0; % initialize position at z
mu_x = 0; % initialize direction at x
mu_y = 0; % initialize direction at y
mu_z = 1; % initialize direction at z
W = 1; % initialize photon weight
%%&&%%%%%%%%%%%%%%%%%%%%%%%%%%%%%%%%%%%%%%%%
%%%%%%%%%&&%%%%%%%%%%%%%%
%%%%%%%%%%%%%Boundary
conditions%%%%%%%%%%%%%%%%%%%%%%%%%%%%%%%%%%%%%%
if (0 <= z)&&(z < 0.12)        % boundary of skin
mu_a = 2.7; % absorption coefficient
mu_s = 187; % scattering coefficient
g = 0.81; % expected value of cos theta %
ni = 1;
nt = 1.39;
```

```
%%&&%%%%%%%%%%%%%%%%%%%%%%%%%%%%%%%%%%%
%%%%%%%%&&%%%%%%%%%%%%
%%&&%%%%%%%%%%%%%%%%%Equation%%%%%%%%%%%%%
%%%%%%%%%%%%%%%%%%%%%
mu_t = mu_a + mu_s;
Rsp = (ni – nt)^2/(ni + nt)^2;
theta=acos((1/(2*g))*(1+(g^2)-((1-g^2)/(1-g+2*g*rand))^2));
%%&&%%%%%%%%%%%%%%%%%%%%%%%%%%%%%%%%%%%
%%%%%%%%&&%%%%%%%%%%%%
%%&&%%%%%%%%%%%%%%%%%Plotting (Line 3-D
Plot)%%&&%%%%%%%%%%%%%%%%%
figure(1)
xlabel('wideness(cm)');
ylabel('Skin Depth (cm)');
plot(Xo,Zo,'ro'); % marks origin
hold on
plot(Xo,Zo,'r*'); % marks origin
hold on
plot(Xvec, -Zvec,'b.') % plots path
hold on
plot([-3 3],[0 0.0001],'r')
hold on
plot([-3 3],[-0.12 -0.1200001],'r')
hold on
plot([-3 3],[-0.5 -0.500001],'r')
axis([-xlimxlim-zlimzlim]);
hold on
%%&&%%%%%%%%%%%%%%%%%%%%%%%%%%%%%%%%%%%
%%%%%%%%&&%%%%%%%%%%%%
```

6.2.3 Barcode Scanner

Printed barcodes have grown in demand as they play a key role in our daily lives to take control of any stocks in the markets and industries. Barcodes are read by a barcode scanner using the principle of light reflection from their black and white bands. There are three types of barcode scanners:

1. **Pen scanners:** A photodiode measures the intensity of the light beam reflected on the barcode. It is manually scanned at a proper angle under a specific speed via humans.
2. **Charger coupled device (CCD) scanners:** A photodiode measures the intensity of LED beam reflected on the barcode.
3. **Laser scanners:** A photodiode measures the intensity of the laser beam reflected on the barcode. It is automatically scanned at a proper angle under a flexible speed via oscillating mirrors. The angle of view (θ) of the laser scanner (Fig. 6.5) is written as

$$\theta = 2\tan^{-1}\left(\frac{L}{2F}\right) \tag{6.12}$$

where F is the focal length from the laser scanner (i.e., F = 50 mm), L is the scanning length of the barcode (i.e., L = 12 mm), and W is the scanning width of the barcode.

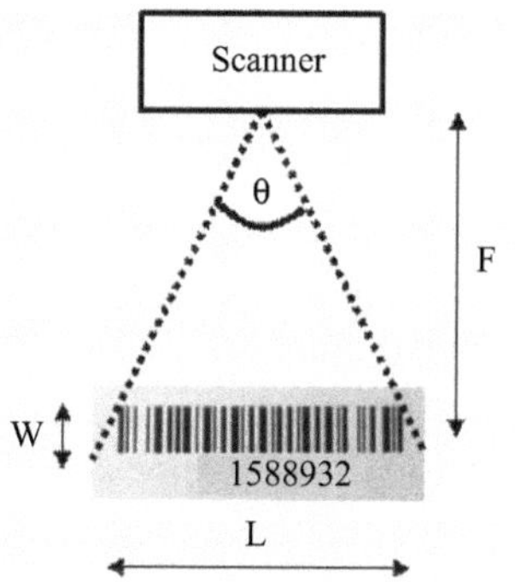

Figure 6.5 Configuration to obtain the optimal area of the barcode from the laser scanner.

Herein we introduce a portable 1D barcode laser scanner controlled by an Arduino board due to the following reasons (Fig. 6.6):

- It is a simple and compact device.
- An infrared-based sensor is used as the automatic sensing switch.
- Information of the barcode is displayed on an LCD screen.
- It is a nondestructive technique to identify the small scanning area of the barcode under scratch, abrasive, or dusty atmospheres.

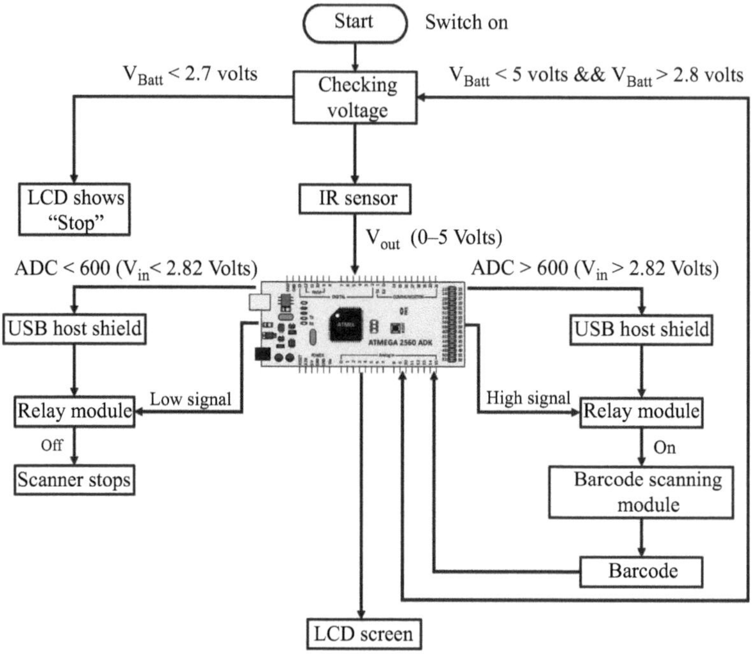

Figure 6.6 Block diagram of 1D barcode read by a laser scanner.

- It is less time consuming.
- An Arduino board converts the analog signal of "low battery' status to the digital form before sending this information to show "low battery" status on the LCD screen based on

$$V_{\text{Batt}} = \frac{5}{1024} \times \text{analog read ()} \tag{6.13}$$

  where V_{Batt} is the voltage of battery and analog read () is analog-to-digital converter.
- Libraries are downloaded for connecting the Arduino program to the Arduino board, LCD screen, and USB host shield as follows:
  - Standard library "string.h" to define constant values, mathematical functions.
  - Libraries "hidboot.h" and "usbhub.h" to establish connection between the USB host shield and the Arduino board.

- Libraries "Wire.h" and "LiquidCrystal_I2C.h" to send/receive data over the LCD screen and the Arduino board.
- Libraries "spi4teensy3.h" and "SPI.h" to communicate with Serial Peripheral Interface devices.
- Barcodes are selected and programmed into the Arduino board.

- One-dimensional barcode is easily read by a laser scanner using PC code of C++ program.

6.2.4 Smart Cane

A smart cane using an infrared-based sensor as a tracker is useful to visually impaired people due to the following reasons:

- An infrared sensor offers high accuracy of the detection system even at a longer distance.
- It can operate in a narrow environment under parallel and narrow beams.
- Pulse width modulation can measure the intensity of the beam reflected on the target. The output distance (*D* in cm) between the sensor and the target is hence written as

$$D = \frac{\text{Pulse width}}{10\ \mu\text{s/cm}} \tag{6.14}$$

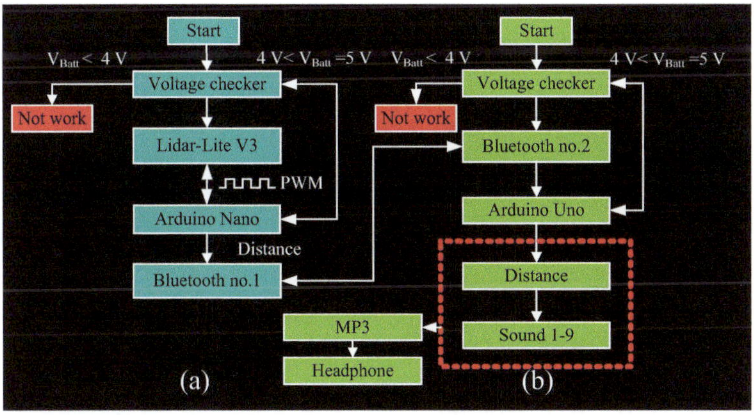

Figure 6.7 Block diagram of distance sensor (a) and alert sensor (b).

Herein we introduce a Lidar sensor controlled by an Arduino board due to the following reasons (Figs. 6.7 and 6.8):

- A lidar sensor operates under a serial communication mode: Bluetooth of headset, cell phone, and PC.
- The reflected light from the target to the sensor offers fast and precise data transmission.
- The target is detectable up to 30 m from the user.
- The user is notified by the headphone.

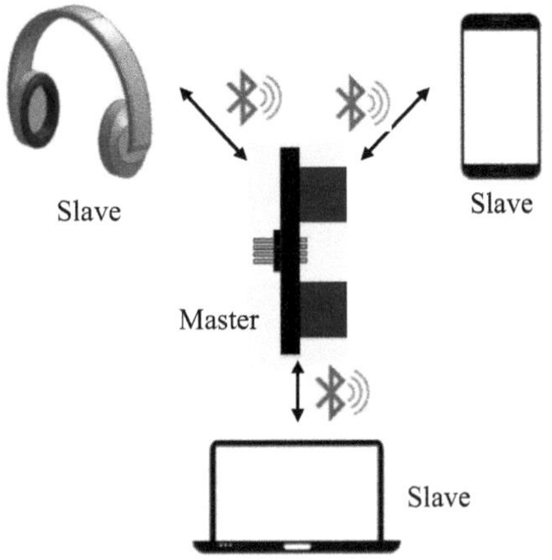

Figure 6.8 A serial communication mode through Bluetooth.

6.3 Implementations

6.3.1 Reflection Probe (see PC Code of Monte Carlo Simulation in Appendix B)

- Human tissue model is mainly composed of (Fig. 6.9) skin, fat, and skeletal muscle.
- Geometrical parameters of each tissue are defined.
- The incident plane of light is defined in the *xy* plane.
- The light propagation direction is defined in the *z*-axis.
- Visible light with the number of photons N = 1, 100, 200, ..., 100,000 is defined as a point source.

- Detection point is located on the human skin.
- Computational results are calculated.

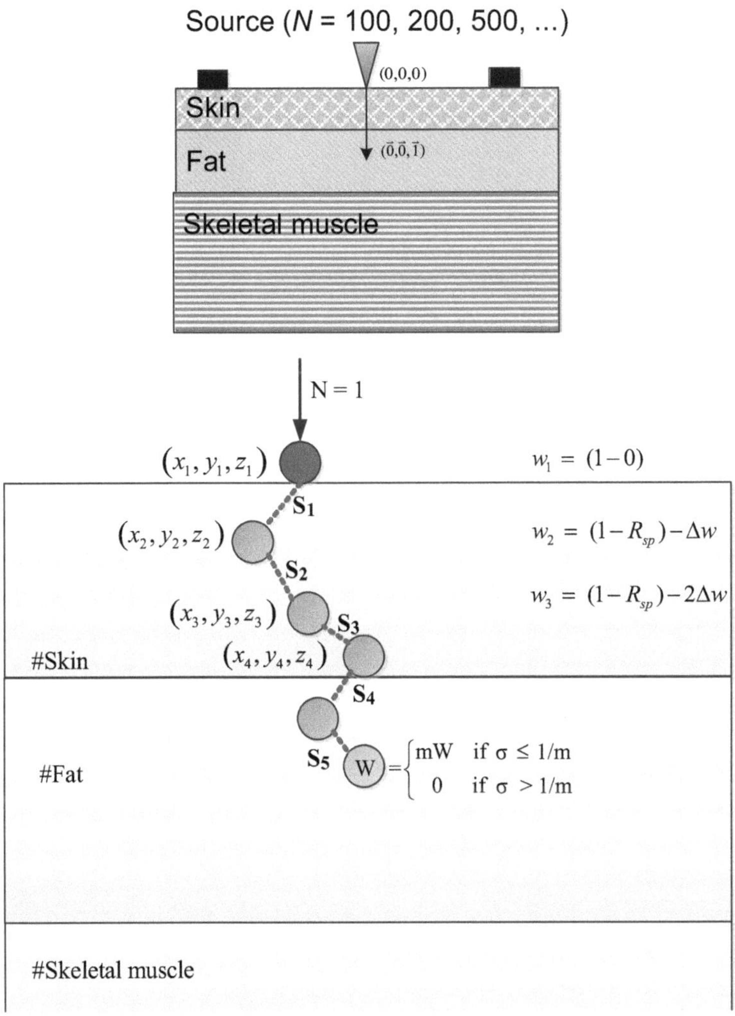

Figure 6.9 Cartesian coordinate system of the photon–human tissue (top) and representative simulation process (bottom).

Validation of the PC code of Monte Carlo simulation is confirmed because of the following reasons:

- A single photon shows the random propagation (Fig. 6.10).

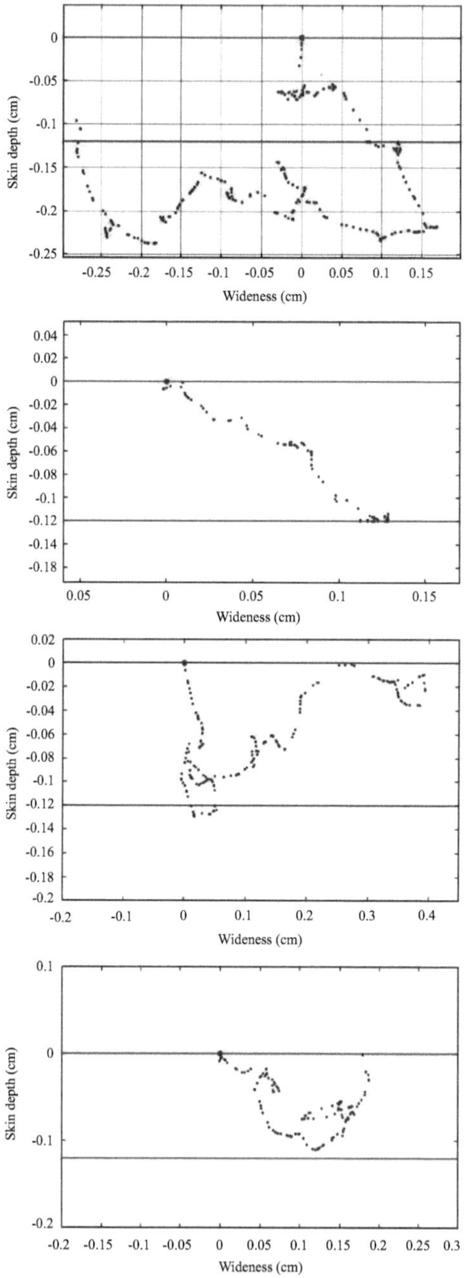

Figure 6.10 Representative single photon propagations in human tissue model depending on the probability density function and the sampling probability distribution of the free path of the photon.

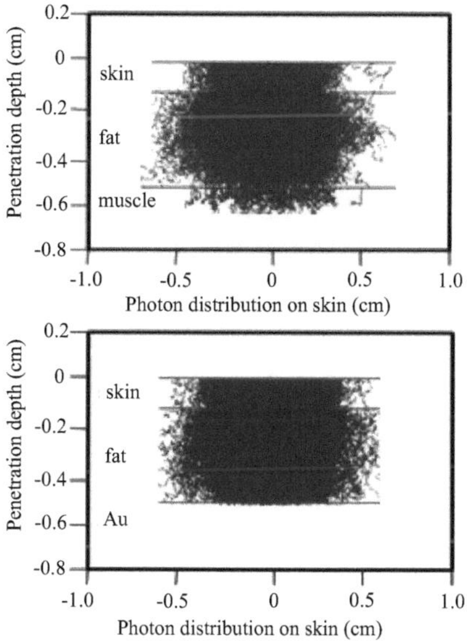

Figure 6.11 Photon distributions in human tissue (top) and human tissue with Au at the third layer at 0.5 cm in depth (bottom). Number of photons is N = 10,000.

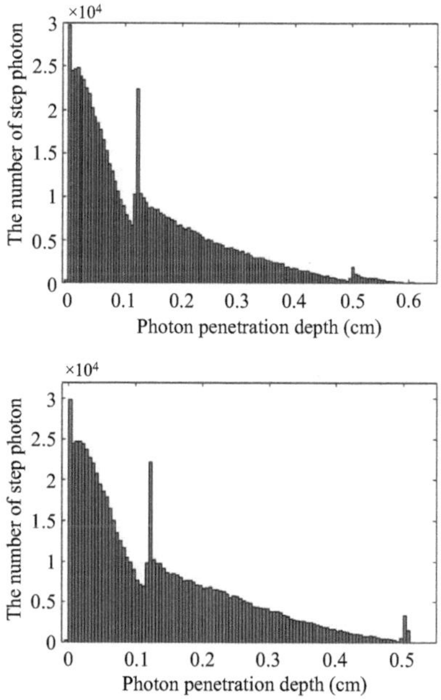

Figure 6.12 Photon penetration depth in human tissue without gold (top) and with gold layer at the third layer at 0.5 cm in depth (bottom). Number of photons is N = 10,000.

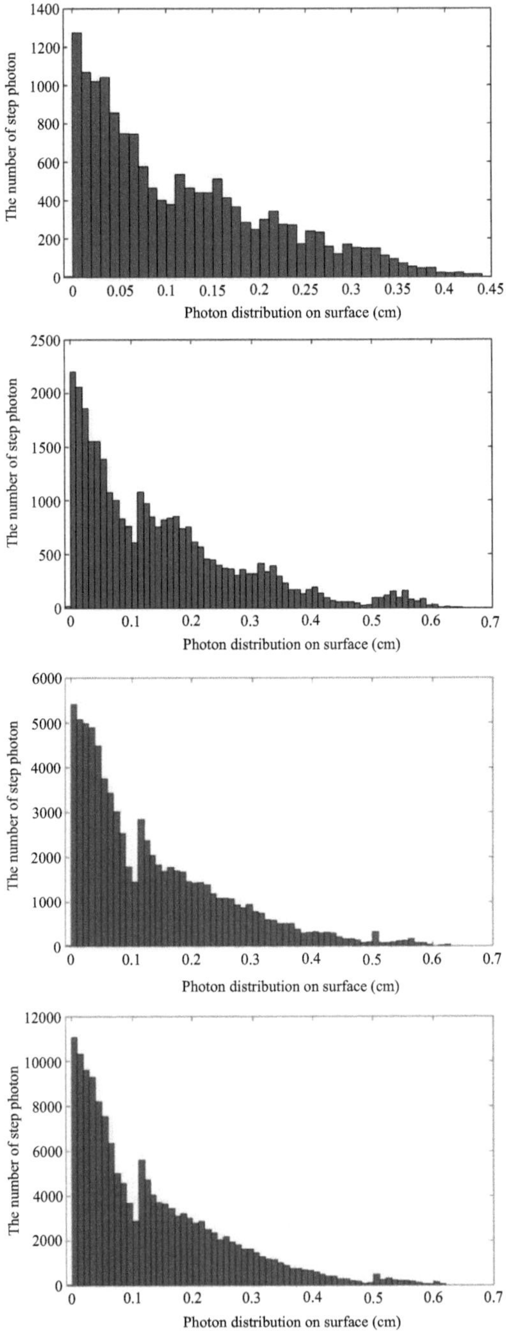
The number of step photon
Photon distribution on surface (cm)
The number of step photon
Photon distribution on surface (cm)
The number of step photon
Photon distribution on surface (cm)
The number of step photon
Photon distribution on surface (cm)

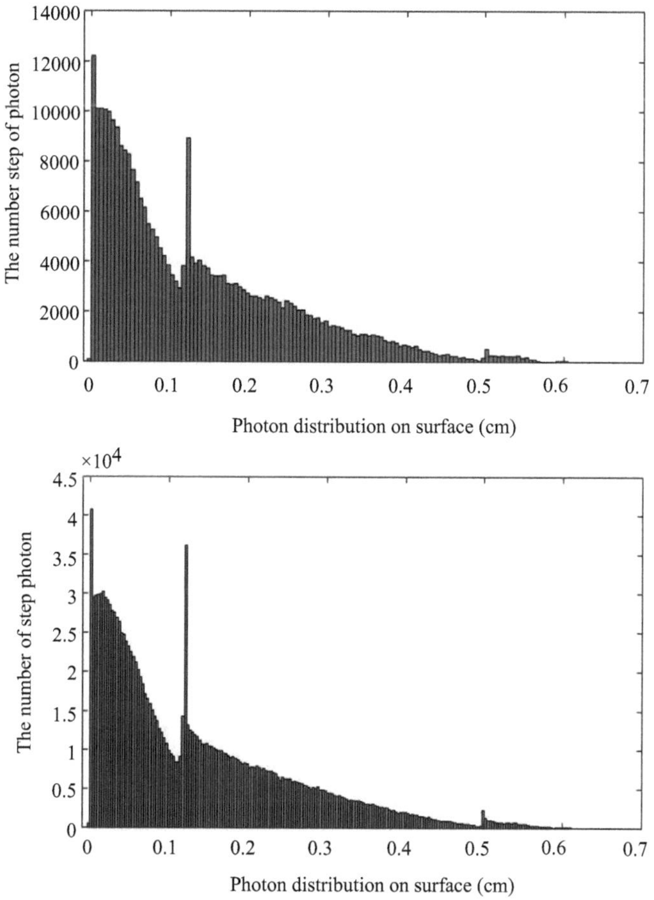

Figure 6.13 Representative photon penetration depth with respect to numbers of photons (N = 100, 200, 500, 1000, 2000, and 10000, respectively).

- The photon propagation differs from the individual layer due to the unique refractive index of each tissue (Figs. 6.11 and 6.12). Photon distributions are unable to pass through the Au film.
- The photon penetration depth is proportional to the number of input photon energy (Fig. 6.13).
- The photon distribution is proportional to the number of input photon energy (Figs. 6.14 and 6.15).

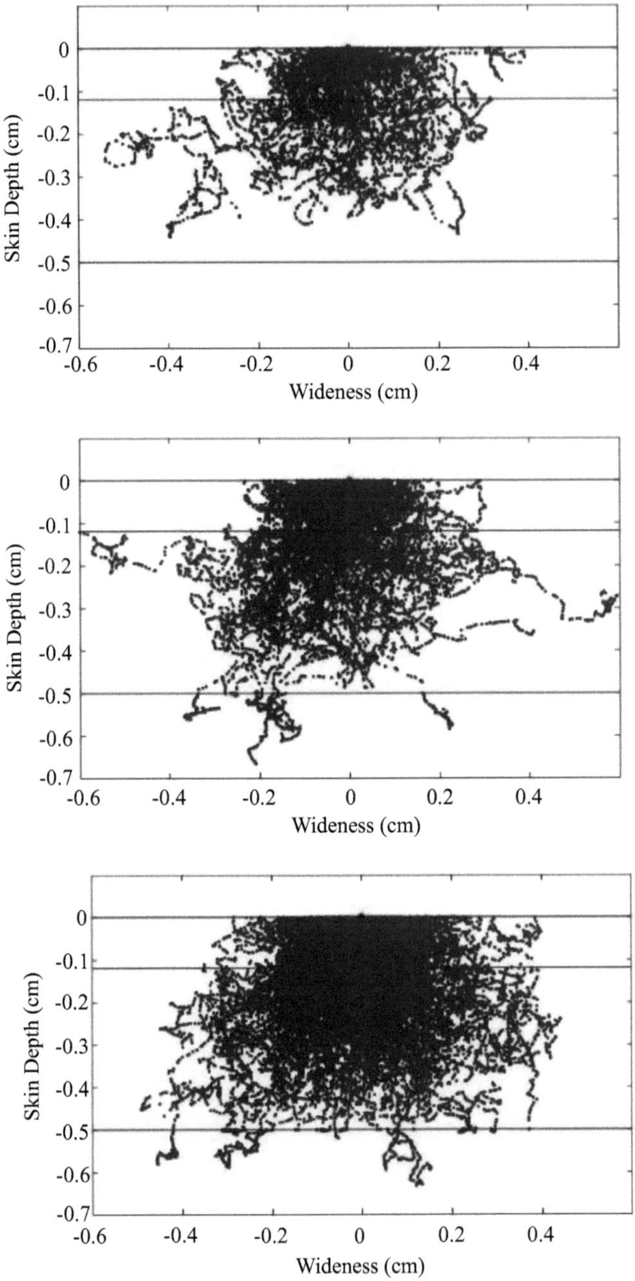
Skin Depth (cm)
0
-0.1
-0.2
-0.3
-0.4
-0.5
-0.6
-0.7
-0.6
-0.4
-0.2
0
0.2
0.4
Wideness (cm)
Skin Depth (cm)
0
-0.1
-0.2
-0.3
-0.4
-0.5
-0.6
-0.7
-0.6
-0.4
-0.2
0
0.2
0.4
Wideness (cm)
Skin Depth (cm)
0
-0.1
-0.2
-0.3
-0.4
-0.5
-0.6
-0.7
-0.6
-0.4
-0.2
0
0.2
0.4
Wideness (cm)

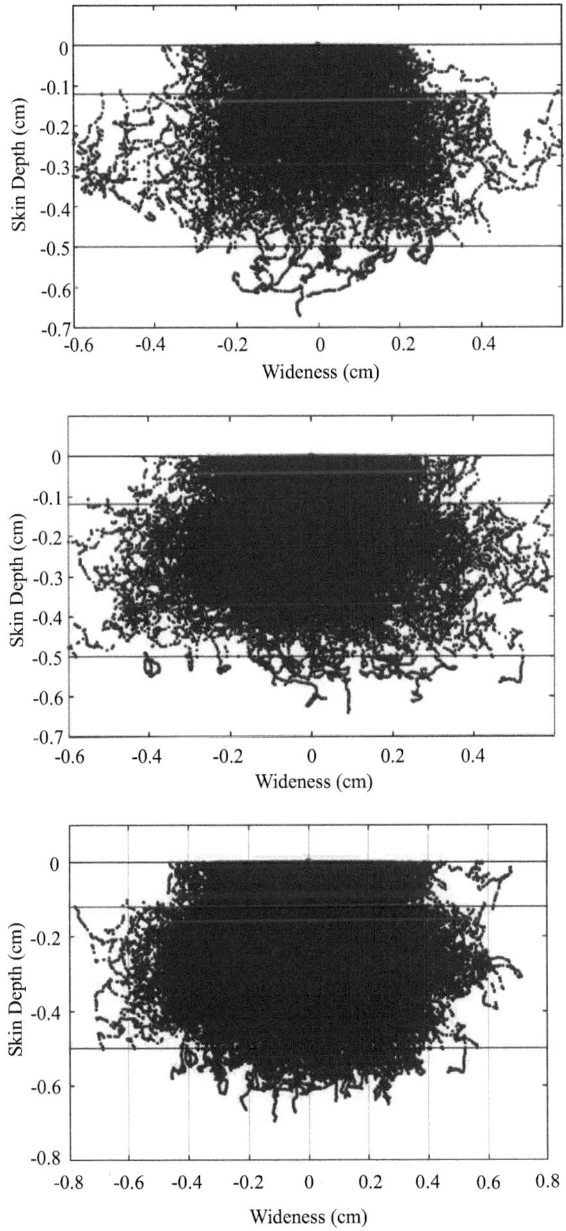

Figure 6.14 Representative photon distribution with respect to numbers of photons (N = 100, 200, 500, 1000, 2000, and 10,000, respectively).

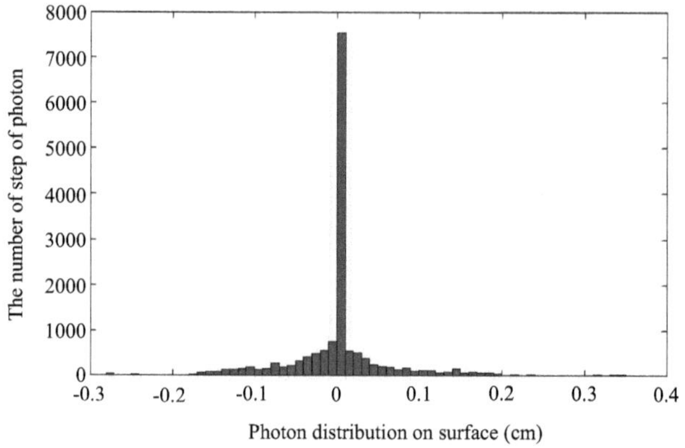
8000
7000
6000
5000
4000
3000
2000
1000
0
The number of step of photon
-0.3
-0.2
-0.1
0
0.1
0.2
0.3
0.4
Photon distribution on surface (cm)

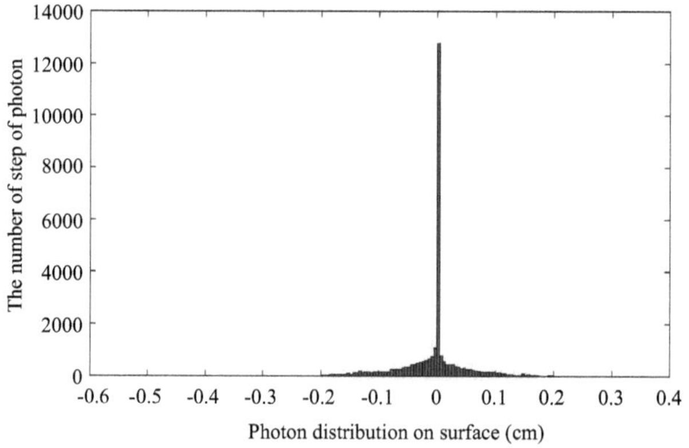
14000
12000
10000
8000
6000
4000
2000
0
The number of step of photon
-0.6
-0.5
-0.4
-0.3
-0.2
-0.1
0
0.1
0.2
0.3
0.4
Photon distribution on surface (cm)

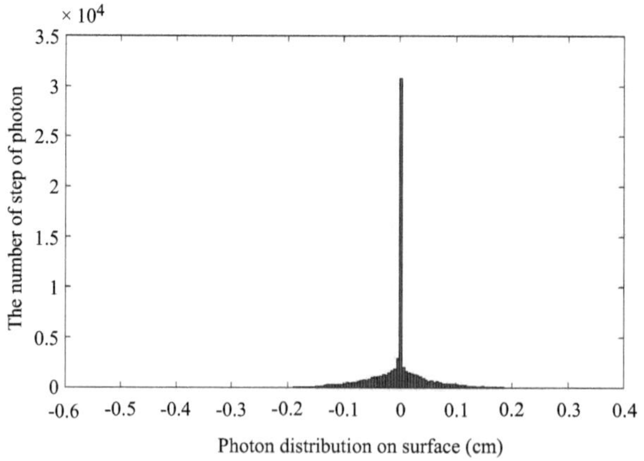
× 10^4
3.5
3
2.5
2
1.5
1
0.5
0
The number of step of photon
-0.6
-0.5
-0.4
-0.3
-0.2
-0.1
0
0.1
0.2
0.3
0.4
Photon distribution on surface (cm)

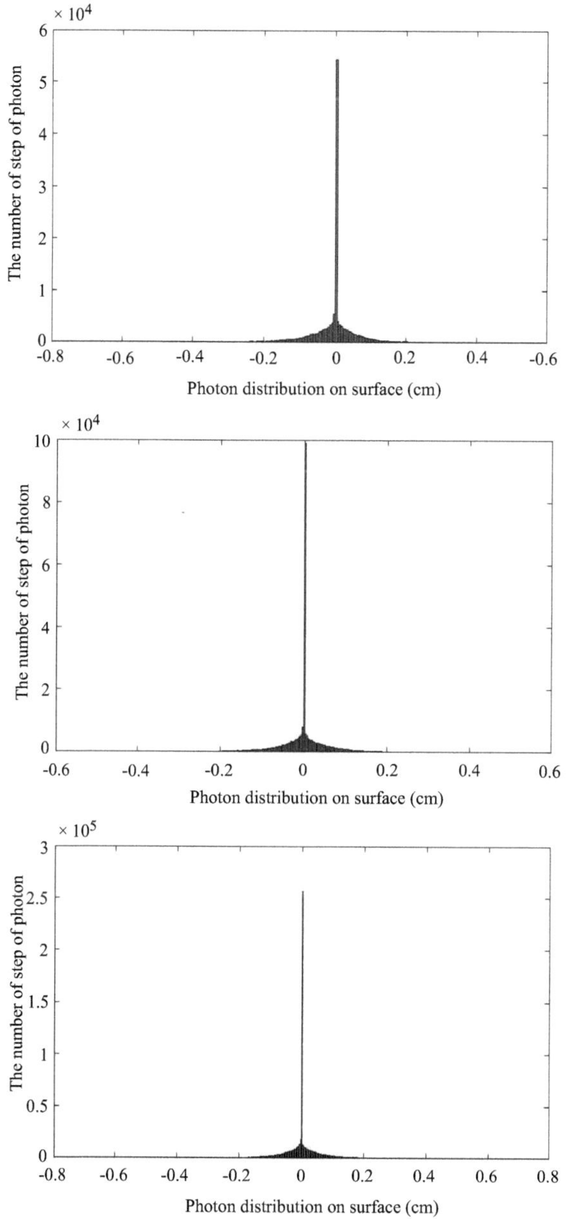

Figure 6.15 Representative photon distribution on human skin with respect to numbers of photons (N = 100, 200, 500, 1000, 2000, and 10,000, respectively).

Based on the probability function, the mathematical model of photon distribution on the human skin is finally computed as (Table 6.2 and Fig. 6.16)

$$y = 0.0825\ \ln(x) + 0.1439$$

Table 6.2 Optimal reflection probe positions of a point source and two photodetector points on the human skin

Number of photons	Z-coordinate −1.5	Z-coordinate 0.0	Z-coordinate 1.5	Distance, cm	Standard deviation	Mean value
100	−0.238	0.060	0.358	0.597	0.199	0.060
200	−0.254	0.025	0.304	0.558	0.186	0.025
500	−0.287	0.032	0.352	0.640	0.213	0.032
1000	−0.328	0.020	0.368	0.697	0.232	0.020
2000	−0.376	−0.006	0.364	0.740	0.246	−0.006
5000	−0.407	−0.006	0.395	0.802	0.267	−0.006
10000	−0.503	−0.000	0.494	0.998	0.338	−0.000
20000	−0.371	0.102	0.618	0.990	0.341	0.089
50000	−0.435	0.080	0.595	1.031	0.343	0.080
100000	−0.532	0.030	0.593	1.125	0.375	0.030

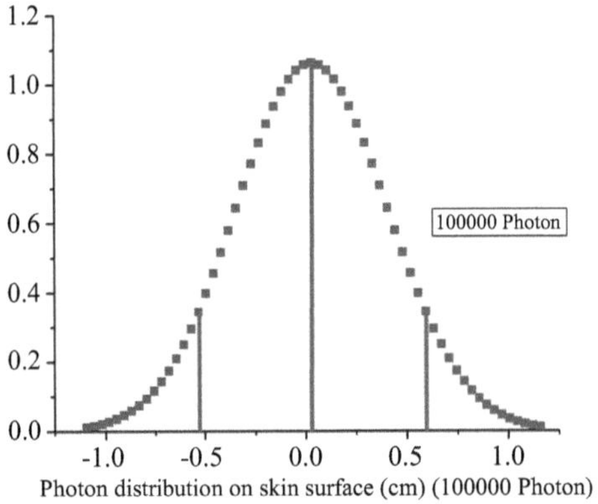

Figure 6.16 Probability distribution of photons on the human skin.

The light source practically offers more photons than the simulation model. For instance, when $N = 3.32 \times 10^{15}$ photons at the wavelength of 660 nm are applied to the mathematical model, the precise positions of the point source and detectors are achieved by setting two photodetectors at –1.0570 cm and +1.0570 cm in width in between the point source on the human skin so that the photon packet penetrates the skeletal muscle at 0.3050 cm in depth.

6.3.2 Barcode Scanner (see PC Code of Monte Carlo Simulation in Appendix C)

- The main parts of 1D laser barcode scanner are detailed in Table 6.3.

Table 6.3 Material list of 1D laser barcode scanner (Fig. 6.17)

Materials	**Function/Information**				
Infrared sensor	650 nm laser diode				
Arduino board	Open-source hardware and software as well as a central processing unit				
USB host shield	Connects the input signal to the laser barcode scanning module				
Relay module	Automatic switch				
Mindeo board	Laser barcode scanning module				
Barcode No.	**Pet**	**Name**	**Gender**	**Age (years)**	**Price (THB)**
1588932	Cat	Kit	Female	4	3000
1852698	Dog	Rex	Male	2	2000
1487526	Bird	Ben	Female	2	900
1589657	Fish	Mike	Female	2	600
"20×4" LCD module	Read and write 80 characters on LCD screen				
Charger module	Supply 5 V of battery to regulator module				

- The scanner is located at a normal plane underneath the barcode.
- The scanning length (L) is defined at 12 mm.
- The focal length (F) is defined at 10–50 mm.
- The scanning width (W) is defined at 2–20 mm.

The optimal area for reading a barcode using a laser barcode scanner is 12 × 4 mm^2 under the focal length of 50 mm. The angle of view (θ) from our laser barcode scanner is 13.7°. The scanning speed detected by the oscilloscope is 400 ms at 2.82 V. The resolution of the laser barcode scanner detected by the ImageJ program is good as the gray value range is 150, whereas the resolution is poor as the gray value is less than 80.

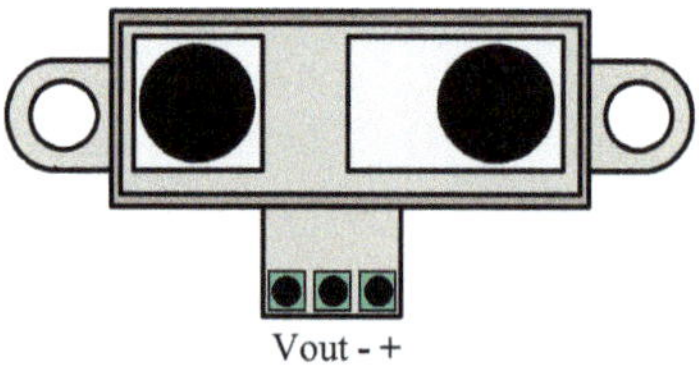

(a)

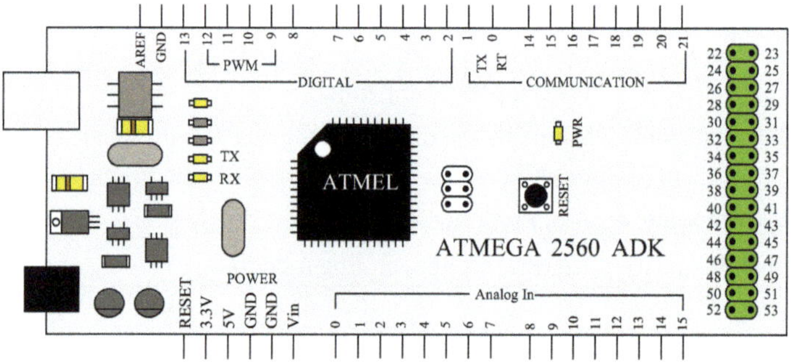

(b)

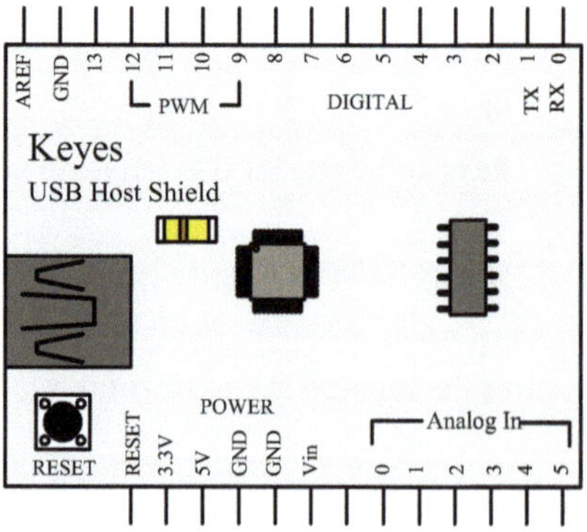

(c)

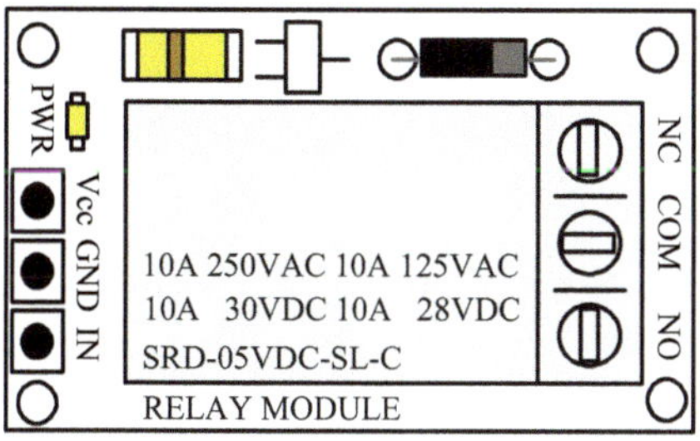
PWR
Vcc GND IN
10A 250VAC 10A 125VAC
10A 30VDC 10A 28VDC
SRD-05VDC-SL-C
RELAY MODULE
NC COM NO

(d)

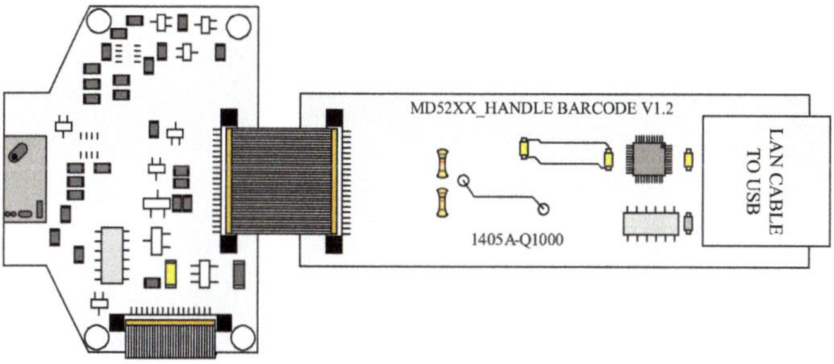
MD52XX_HANDLE BARCODE V1.2
1405A-Q1000
LAN CABLE TO USB

(e)

(f)

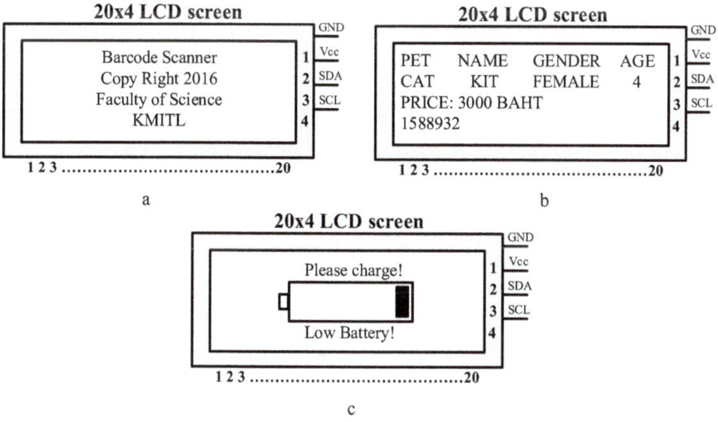
20x4 LCD screen
Barcode Scanner
Copy Right 2016
Faculty of Science
KMITL
GND
1 Vcc
2 SDA
3 SCL
4
1 2 3 20
a
20x4 LCD screen
PET NAME GENDER AGE
CAT KIT FEMALE 4
PRICE: 3000 BAHT
1588932
GND
1 Vcc
2 SDA
3 SCL
4
1 2 3 20
b
20x4 LCD screen
Please charge!
Low Battery!
GND
1 Vcc
2 SDA
3 SCL
4
1 2 3 20
c

(g)

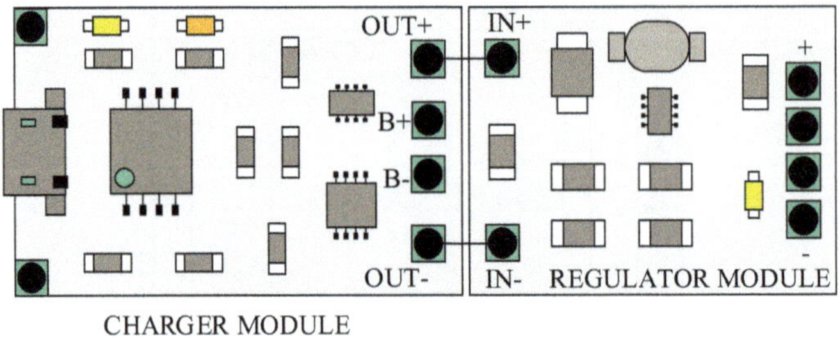

(h)

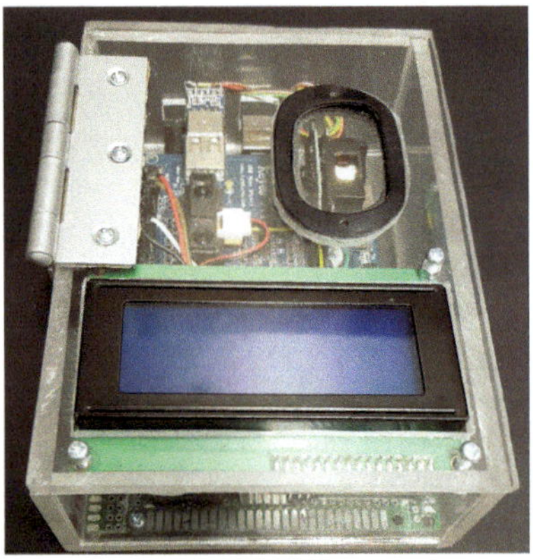

(i)

Figure 6.17 (a) Infrared sensor (Model Sharp GP2Y0A41SK). (b) Microcontroller board (Arduino Model ATMEGA 2560 ADK). (c) USB host shield (Model Keyes). (d) Relay module (Model Songle 1c5v). (e) Laser barcode scanning module (Minde Model MD52XX). (f) Representative barcode 1588932 of (a) good image and (b) poor image. (g) LCD module (Model 2004A) composed of initial screen, final screen, and "low power supply" screen. (h) Charger module (Model Arduino TP4056 18650 Li+ battery charger) and regulator module (Model Arduino USB charger 0.9–5 V to 5 V 500 mA). (i) Photograph of 1D laser barcode scanner.

6.3.3 Smart Cane

- The main components of a smart cane are detailed in Table 6.4.

Table 6.4 Material list of a smart cane (Fig. 6.19)

Materials	Function/Information
Light detection and ranging module	905 nm laser diode
Arduino board	Microcontroller
MP3 module	Voice alert player
Bluetooth module	Wireless communicator
Headphone	Voice alert player
White cane	ϕ = 21.2 mm, 1223 mm length

- Targets are human, concrete wall, and wooden table as shown in Fig. 6.18.

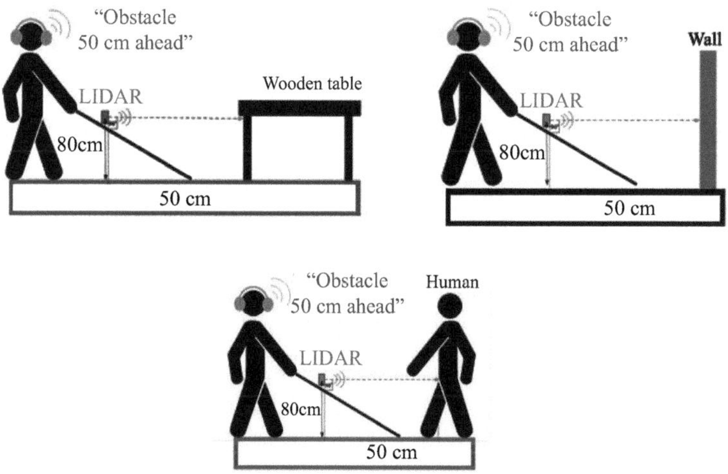

Figure 6.18 Test of a smart cane by the user on different targets.

- The design and actual distance are calibrated so that the laser sensor is very reliable.
- The distance from the user to the target (human, wall, and wooden table) is 1–3000 cm.
- The refractive indices of human, concrete wall, and wooden table are 1.36, 1.55, and 1.60, respectively.

A smart cane offers high accuracy and sensitivity to the target. Data collection and transmission from the laser sensor to the voice alert sensor conform to the response time of 51.12 ms. The span of human (10 cm) is the highest in comparison to the concrete wall (7.54 cm) and the wooden table (7.18 cm), due to the scattering property of the turbid medium.

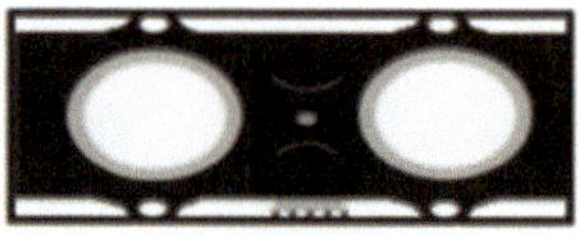

(a)

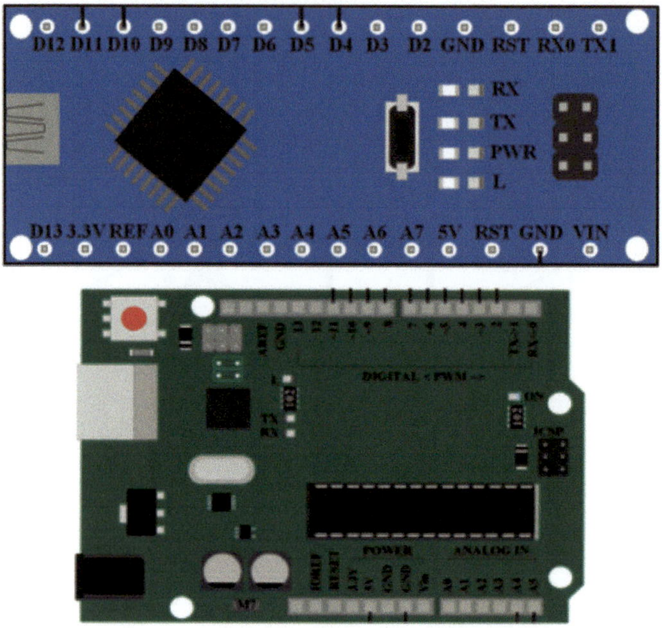

(b)

(c)

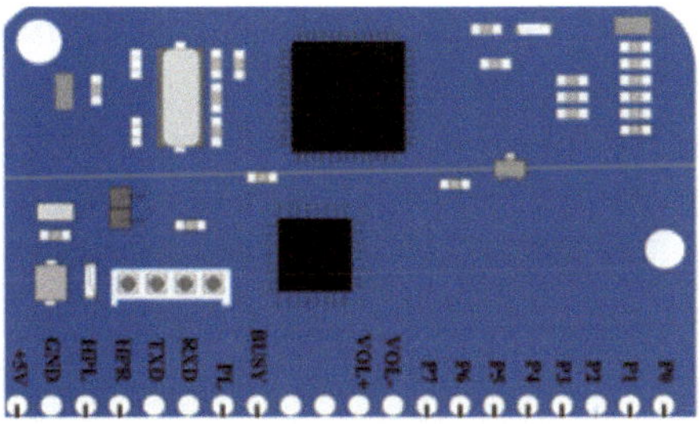

(d)

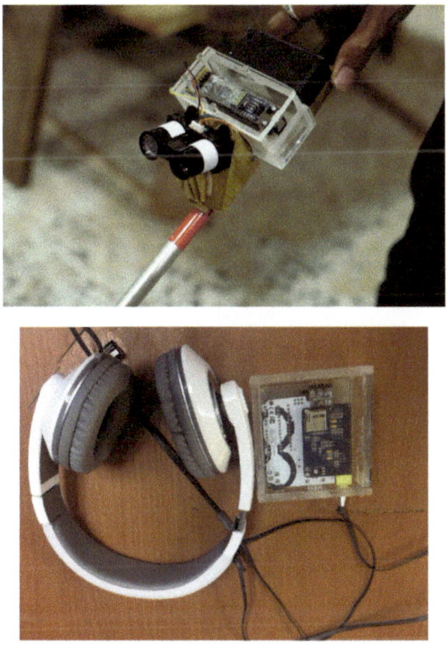

(e)

Figure 6.19 (a) Infrared sensor (Model Lidar-Lite V3). (b) Microcontroller boards (Arduino Nano V3 CH340G, Arduino Uno R3). (c) Bluetooth module (Model HC-05). (d) MP3 module (Model ETTEE043). (e) Photograph of distance and alert sensor on a smart cane.

Problems

6.1 Provide examples of point sources and detectors for the refraction probe.

6.2 Show the step-by-step process of photon propagation in each layer of human tissue.
6.3 Find the number of photons coming out of a laser diode having the power of 1 mW and the wavelength of 670 nm.
6.4 Derive the step size of a photon packet from the probability density function.

Part II
Nonlinear Optics

Chapter 7

Nonlinear Medium

7.1 Introduction

This chapter deals with the anisotropic crystals offering an additional degree of freedom to achieve the efficiency interaction of a nonlinear effect. This particular property is used to construct polarization devices such as a quarter wave plate and half wave plate.

7.2 Component of Electric Fields

Since the electromagnetic wave in a nonmagnetic media ($\mu = \mu_0$) is originally related to Maxwell's equations as

$$\vec{\nabla} \times \vec{E} = -\dot{B} \tag{7.1}$$

$$\vec{\nabla} \times \vec{H} = \vec{j} + \dot{D} \tag{7.2}$$

$$\vec{\nabla} \cdot \vec{D} = \rho_{\text{TRUE}} \tag{7.3}$$

$$\vec{\nabla} \cdot \vec{B} = 0 \tag{7.4}$$

where

$$\vec{D} = \varepsilon_0 \vec{E} + \vec{P} \tag{7.5}$$

$$\vec{B} = \mu_0 \vec{H} \tag{7.6}$$

Linear and Nonlinear Optics: Materials, Properties, and Applications
Kitsakorn Locharoenrat

ISBN 978-981-4877-15-2 (Hardcover), 978-1-003-05588-4 (eBook)
www.jennystanford.com

Maxwell's equations are modified when a plane wave propagates through a dielectric medium as

$$\vec{\nabla}\times\vec{E}=-\dot{\vec{B}} \tag{7.7}$$

$$\vec{\nabla}\times\vec{H}=\dot{\vec{D}} \tag{7.8}$$

$$\vec{\nabla}\cdot\vec{D}=0 \tag{7.9}$$

$$\vec{\nabla}\cdot\vec{B}=0 \tag{7.10}$$

Substituting the polarization term (Eq. 7.11) into the electric displacement (Eq. 7.12) of the propagating wave so that it is modified as

$$\vec{P}^{L}=\varepsilon_0\chi\vec{E} \tag{7.11}$$

$$\vec{D}=\varepsilon_0\vec{E}+\vec{P}^{L} \tag{7.12}$$

$$\vec{D}=\varepsilon_0(1+\chi)\vec{E}=\varepsilon\vec{E} \tag{7.13}$$

where χ is the electric susceptibility, ε is the electric permittivity in a medium, and ε_0 is the electric permittivity in free space.

If the electric field of the propagating wave (Eq. 7.14) and a constant phase are defined as

$$\vec{E}=E_0e^{i(\vec{k}\cdot\vec{r}-\omega t)} \tag{7.14}$$

$$\vec{k}\cdot\vec{r}=k_xx+k_yy+k_zz \tag{7.15}$$

the electric field must satisfy Maxwell's equation according to

$$\vec{\nabla}\cdot\vec{D}=0 \tag{7.16}$$

$$\vec{D}=\varepsilon\vec{E} \tag{7.17}$$

It is, therefore, implied that

$$\vec{\nabla}\cdot\vec{E}=0 \tag{7.18}$$

$$\vec{\nabla}\cdot\left[E_0e^{i(\vec{k}\cdot\vec{r}-\omega t)}\right]=0 \tag{7.19}$$

$$i\vec{k}\cdot\vec{E}=0 \tag{7.20}$$

This point reveals that the electric fields ($\vec{E}$) are perpendicular to the wavefront propagation ($\vec{k}$), as shown in Fig. 7.1.

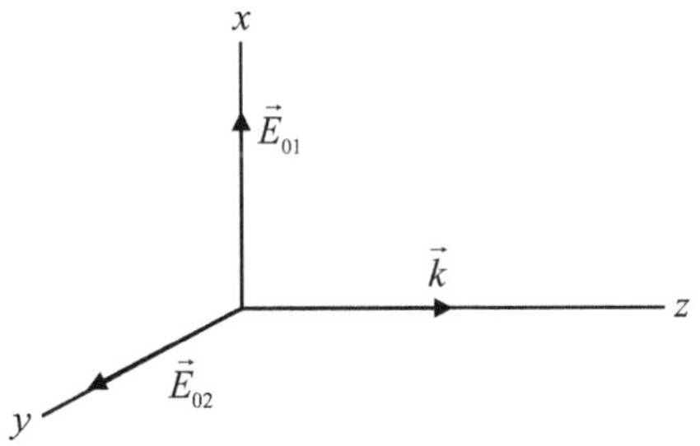

Figure 7.1 Two electric fields are perpendicular to the wavefront propagation in a dielectric medium.

Two possible electric fields are then written as

$$\vec{E}_1 = \hat{x}E_{01}e^{i(kz-\omega t)} \tag{7.21}$$

$$\vec{E}_2 = \hat{y}E_{02}e^{i(kz-\omega t)} \tag{7.22}$$

where

$$\vec{k} = \hat{z}k \tag{7.23}$$

7.3 Component of Magnetic Fields

If the magnetic field of the propagating wave (Eq. 7.24) is defined as

$$\vec{B} = B_0 e^{i(\vec{k}\cdot\vec{r}-\omega t)} \tag{7.24}$$

the magnetic field must satisfy Maxwell's equation according to

$$\vec{\nabla}\times\vec{E} = -\frac{\partial\vec{B}}{\partial t} \tag{7.25}$$

$$i\vec{k}\times\vec{E} = i\omega\vec{B} \tag{7.26}$$

Note that

$$\nabla\times \Rightarrow i\vec{k}\times \tag{7.27}$$

$$\frac{\partial}{\partial t} \Rightarrow -i\omega \tag{7.28}$$

It is, therefore, implied that

$$\vec{k}\times\vec{E} = \omega\vec{B} \tag{7.29}$$

$$\vec{E} = c\vec{B} \tag{7.30}$$

This point reveals that the magnetic and electric fields are perpendicular to each other, as shown in Fig. 7.2.

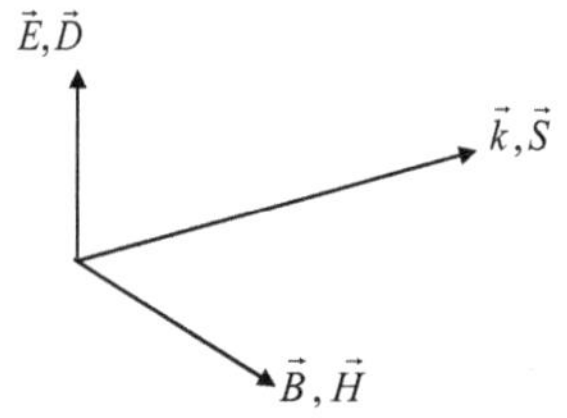

Figure 7.2 Electromagnetic fields are perpendicular to the wavefront propagation and Poynting vector in a dielectric medium.

The Poynting vector (in W/m^2) is introduced as a directional energy flux of the electromagnetic fields as

$$\vec{S}=\vec{E}\times\vec{H} \tag{7.31}$$

$$\vec{S}=\frac{1}{\mu_0}\vec{E}\times\vec{B} \tag{7.32}$$

$$S=\frac{1}{\mu_0}EB \tag{7.33}$$

$$S=\frac{1}{\mu_0(c/n)}E^2=\frac{(c/n)}{\mu_0}B^2 \tag{7.34}$$

Substituting the wavefront propagation term (Eq. (7.35)) and the speed of light (Eq. (7.36)) into Eq. (7.34), the intensity of the electromagnetic fields defined as the energy flow per unit time or the time average of Poynting vector is expressed as

$$k=\frac{\omega}{c}n \tag{7.35}$$

$$c^2=\frac{1}{\varepsilon_0\mu_0} \tag{7.36}$$

$$I=\frac{(c/n)}{\mu_0}\left\langle B^2\right\rangle=\frac{(c/n)}{2\mu_0}\left|B_0\right|^2 \tag{7.37}$$

$$I=\frac{1}{2\mu_0(c/n)}\left|E_0\right|^2 \tag{7.38}$$

$$I = \frac{n}{2c\mu_0}|E_0|^2 = \frac{nc\varepsilon_0}{2}|E_0|^2 = \frac{k}{2\mu_0\omega}|E_0|^2 \tag{7.39}$$

The incident power of the electromagnetic fields with the incident area of A is, hence, calculated as

$$P = IA \tag{7.40}$$

7.4 Classification of Nonlinear Medium

Since the intensity of electromagnetic fields in a nonlinear medium in comparison to a dielectric medium becomes weak due to a large diffraction of the output waves, the nonlinear optical properties of solids require a certain condition of effectively less diffractive interaction using fiber optics and nonlinear media, for instance. The electric displacement is, therefore, modified in a nonlinear medium as

$$\vec{D} = \varepsilon_0\vec{E} + \vec{P} = \varepsilon\vec{E} + \vec{P}^{\text{N}} \tag{7.41}$$

This equation implies that the polarization term ($\vec{P}$) is no longer proportional to the electric field ($\vec{E}$).

If possibilities of the electric fields are applied into the anisotropic crystal served as a nonlinear medium as

$$\vec{E} = \hat{x}E_x;\quad \vec{E} = \hat{y}E_y;\quad \vec{E} = \hat{z}E_z \tag{7.42}$$

the components of each electric displacement with respect to its electric field are composed of

$$D_x = \varepsilon_{xx}E_x;\ D_y = \varepsilon_{yx}E_x;\ D_z = \varepsilon_{zx}E_x \tag{7.43}$$

$$D_x = \varepsilon_{xy}E_y;\ D_y = \varepsilon_{yy}E_y;\ D_z = \varepsilon_{zy}E_y \tag{7.44}$$

$$D_x = \varepsilon_{xz}E_z;\ D_y = \varepsilon_{yz}E_z;\ D_z = \varepsilon_{zz}E_z \tag{7.45}$$

The summations of electric displacements in different directions are generally written as

$$D_x = \varepsilon_{xx}E_x + \varepsilon_{xy}E_y + \varepsilon_{xz}E_z \tag{7.46}$$

$$D_y = \varepsilon_{yx}E_x + \varepsilon_{yy}E_y + \varepsilon_{yz}E_z \tag{7.47}$$

$$D_z = \varepsilon_{zx}E_x + \varepsilon_{zy}E_y + \varepsilon_{zz}E_z \tag{7.48}$$

The summations of electric displacements in different directions in the matrix form are written as

$$\begin{pmatrix} D_x \\ D_y \\ D_z \end{pmatrix} = \begin{pmatrix} \varepsilon_{xx} & \varepsilon_{xy} & \varepsilon_{xz} \\ \varepsilon_{yx} & \varepsilon_{yy} & \varepsilon_{yz} \\ \varepsilon_{zx} & \varepsilon_{zy} & \varepsilon_{zz} \end{pmatrix} \begin{pmatrix} E_x \\ E_y \\ E_z \end{pmatrix} \tag{7.49}$$

The summation of electric displacements in different directions, in a simple form, is also written as

$$\vec{D} = \overline{\overline{\varepsilon}}\vec{E} \tag{7.50}$$

where $\varepsilon_{xy} = \varepsilon_{yx}; \varepsilon_{xz} = \varepsilon_{zx}; \varepsilon_{yz} = \varepsilon_{zy}$ are symmetric tensors satisfying energy conservation and $\varepsilon_{xx}; \varepsilon_{yy}; \varepsilon_{zz}$ are principal axis system.

Based on Eq. (7.49), nonlinear media are divided into three categories as follows:

1. Isotropic media as (i.e., glass)

$$\varepsilon_x = \varepsilon_y = \varepsilon_z$$

2. Anisotropic–biaxial media as (i.e., mica)

$$\varepsilon_x \neq \varepsilon_y \neq \varepsilon_z$$

3. Anisotropic–uniaxial media as

$$\varepsilon_x = \varepsilon_y \neq \varepsilon_z$$

Let us introduce further parameters as follows:

- Principal axes of the electric permittivity in the anisotropic media are

$$\varepsilon_x; \varepsilon_y; \varepsilon_z$$

- Principal axes of the dielectric constant in the anisotropic media are

$$K_x = \frac{\varepsilon_x}{\varepsilon_0}, K_y = \frac{\varepsilon_y}{\varepsilon_0}, K_z = \frac{\varepsilon_z}{\varepsilon_0}$$

- Principal axes of the refractive index in the anisotropic media are

$$n_x = \sqrt{K_x}, n_y = \sqrt{K_y}, n_z = \sqrt{K_z}$$

For the uniaxial media, we therefore get

$$K_x = K_y \neq K_z$$

$$n_x = n_y = n_o\ ;\ n_z = n_e$$

where velocity is

$$v = \frac{c}{n} = \frac{c}{\sqrt{K}}$$

Those terms reveal that light propagation changes the polarization in two directions: the so-called ordinary (o) and extraordinary (e) waves. Hence, the uniaxial media are divided into two subcategories as follows:

a. **Positive uniaxial crystal ($n_o < n_e$):** The ordinary wave always travels faster than the extraordinary wave, as shown in Fig. 7.3.

   Materials: Quartz SiO_2, sellaite MgF_2, rutile TiO_2

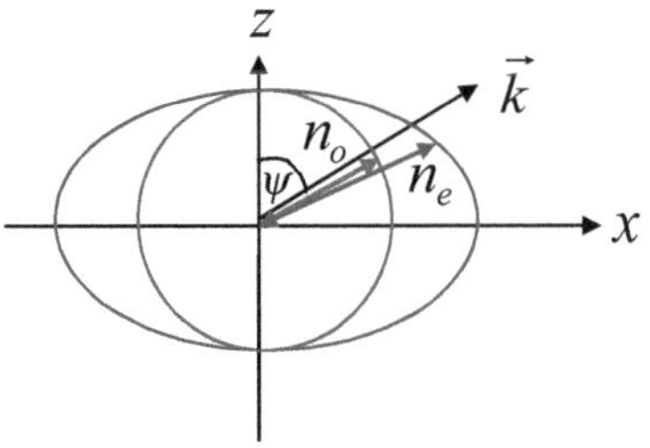

Figure 7.3 Refractive index surfaces of the positive uniaxial crystal representing a refractive index variation between ordinary and extraordinary waves.

b. **Negative uniaxial crystal ($n_o > n_e$):** The ordinary wave always travels slower than the extraordinary wave, as shown in Fig. 7.4.

   Materials: Ruby Al_2O_3, calcite $CaCO_3$, potassium dihydrogen phosphate KDP, lithium niobate $LiNbO_3$

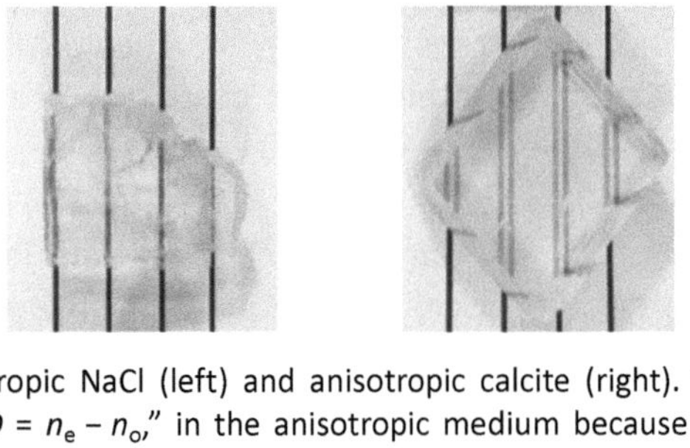

Figure 7.4 Isotropic NaCl (left) and anisotropic calcite (right). We can see "birefringence, $D = n_e - n_o$," in the anisotropic medium because unpolarized light wave is split into two polarization light waves according to two refractive indices.

Problems

7.1 Show the difference of Maxwell's equations of electric fields in dielectric and conductive media.

7.2 Explain the polarization states applicable to optic devices.

7.3 Draw the elliptic polarization state with respect to the circular polarization state in arbitrary direction if the net energy propagates along the optic axis.

7.4 Determine the intensity of a plane wave from the following relation:

$$I = \langle \vec{S} \rangle = \tfrac{1}{2} \mathrm{Re}\left[\vec{E} \times \vec{H}^* \right]$$

7.5 Draw the quadratic and cubic polarizations as a function of the external electric field.

7.6 Prove the relationship between refractive index and dielectric constant. Assume that the permeability of the medium is equivalent to 1.

7.7 Prove that the speed of light is 3×10^8 m/s.

7.8 Determine the light intensity travelling in all directions at a distance of 20 km apart from the wave source having the incident power of 100 kW.

7.9 Find the time domain expression of the electric field when the magnetic field of a plane wave propagating in a nonmagnetic medium is given by

$$\vec{H} = 0.50\, e^{i(10z - 6\pi \times 10^8 t)} \ \mathrm{A/m}$$

7.10 Determine the B-vector and E-vector when the plane wave $\vec{E} = E_0\, e^{i(kx - \omega t)}$ with the amplitude of 42 V/m passes through a UHV chamber in the x-direction.

7.11 Determine the root mean square value of the magnetic field at the user when this user is located 1.8 m apart from a light source with the power of 250 W.

7.12 A quartz crystal has $n_o(\omega) = 1.541$, $n_e(\omega) = 1.550$, $n_o(2\omega) = 1.566$, and $n_e(2\omega) = 1.577$. Draw the index surfaces of the negative as compared to the positive uniaxial crystals applicable to birefringence phase matching.

Chapter 8

Wave Propagation

8.1 Introduction

This chapter deals with basic Maxwell's equations applicable to the optical nonlinear medium. Different points of view of plane waves propagated at normal and tilt angles with respect to the optic axis will be explained.

8.2 Polarization

In addition to the linear polarization where the electric field is limited to the direction of propagation, herein we present circular and elliptical polarization as follows (Fig. 8.1). If two components of the electric fields and their amplitudes are equivalent, they generate the right and left circular polarization states in the order as

$$\vec{E}_0 = (\hat{x} + i\hat{y})E_0 \tag{8.1}$$

$$\vec{E}_0 = (\hat{x} - i\hat{y})E_0 \tag{8.2}$$

$$\vec{E} = E_0(\hat{x} + i\hat{y})e^{i(kz-\omega t)} \tag{8.3}$$

$$\vec{E} = E_0(\hat{x} - i\hat{y})e^{i(kz-\omega t)} \tag{8.4}$$

$$\vec{E}(z,t) = E_0\hat{x}\cos(kz-\omega t) + E_0 i\hat{y}\sin(kz-\omega t) \tag{8.5}$$

Linear and Nonlinear Optics: Materials, Properties, and Applications
Kitsakorn Locharoenrat

ISBN 978-981-4877-15-2 (Hardcover), 978-1-003-05588-4 (eBook)
www.jennystanford.com

$$\vec{E}(z,t) = E_0\hat{x}\cos(kz-\omega t) - E_0 i\hat{y}\sin(kz-\omega t) \quad (8.6)$$

However, when two components of the electric fields and their amplitudes are not equivalent, they generate the right and left elliptical polarization states in the order as

$$\vec{E}_0 = (\hat{x} + i\beta\hat{y})E_0 \quad (8.7)$$

$$\vec{E}_0 = (\hat{x} - i\beta\hat{y})E_0 \quad (8.8)$$

$$\vec{E} = E_0(\hat{x} + i\beta\hat{y})e^{i(kz-\omega t)} \quad (8.9)$$

$$\vec{E} = E_0(\hat{x} - i\beta\hat{y})e^{i(kz-\omega t)} \quad (8.10)$$

$$\vec{E}(z,t) = E_0\hat{x}\cos(kz-\omega t) + E_0 i\beta\hat{y}\sin(kz-\omega t) \quad (8.11)$$

$$\vec{E}(z,t) = E_0\hat{x}\cos(kz-\omega t) - E_0 i\beta\hat{y}\sin(kz-\omega t) \quad (8.12)$$

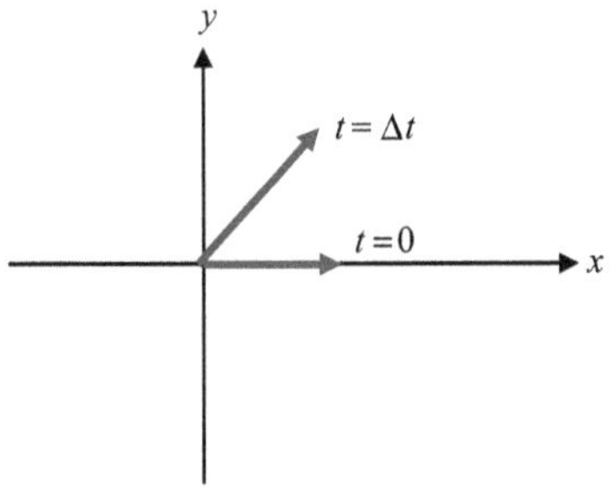

Figure 8.1 Representative polar plot of the right circular polarization state.

8.3 Wave Propagation in Nonlinear Medium

According to two Maxwell's equations as

$$\vec{\nabla} \times \vec{E} = -\frac{\partial \vec{B}}{\partial t} = -\mu_0 \frac{\partial \vec{H}}{\partial t} \quad (8.13)$$

$$\vec{\nabla} \times \vec{H} = \frac{\partial \vec{D}}{\partial t} \quad (8.14)$$

Substituting the electromagnetic fields into Eqs. (8.13) and (8.14), we get

$$\vec{E} = E_0 e^{i(\vec{k}\cdot\vec{r}-\omega t)} \quad (8.15)$$

$$\vec{H} = H_0 e^{i(\vec{k}\cdot\vec{r}-\omega t)} \tag{8.16}$$

$$i\vec{k} \times \vec{E} = i\omega\mu_0 \vec{H} \tag{8.17}$$

$$\vec{k} \times \vec{E} = \omega\vec{B} = \omega\mu_0 \vec{H} \tag{8.18}$$

$$i\vec{k} \times \vec{H} = -\,i\omega\vec{D} \tag{8.19}$$

$$\vec{k} \times \vec{H} = -\,\omega\vec{D} \tag{8.20}$$

Equations (8.18) and (8.20) imply that the wave front propagation $\vec{k}$ possibly might not be parallel to $\vec{S}$ in anisotropic crystals, as shown in Fig. 8.2.

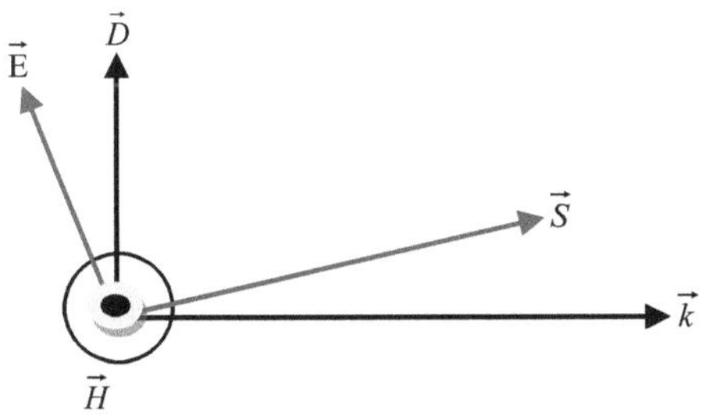

Figure 8.2 The *k*-vector is not parallel to the *S*-vector (i.e., acoustic waves propagating into the anisotropic medium).

Substituting the displacement field into those two equations, we have

$$\vec{D} = \overline{\overline{\varepsilon}}\vec{E} = \varepsilon_0 \overline{\overline{K}}\vec{E} \tag{8.21}$$

$$\vec{k} \times \vec{H} = -\omega\varepsilon_0 \overline{\overline{K}}\vec{E} \tag{8.22}$$

$$\vec{k} \times \left(\vec{k} \times \vec{E}\right) = \omega\mu_0 \left(\vec{k} \times \vec{H}\right) = -\omega^2 \varepsilon_0 \mu_0 \overline{\overline{K}}\vec{E} \tag{8.23}$$

$$\vec{k}\left(\vec{k} \cdot \vec{E}\right) - k^2\vec{E} = -\frac{\omega^2}{c^2}\overline{\overline{K}}\vec{E} \tag{8.24}$$

Equation (8.24) reveals that a component of the electric field in the *x*-direction is calculated as

$$\left(k_x E_x + k_y E_y + k_z E_z\right)k_x - k^2 E_x = -\frac{\omega^2}{c^2} K_x E_x \tag{8.25}$$

$$\frac{\omega^2}{c^2} n^2 \left(\kappa_x E_x + \kappa_y E_y + \kappa_z E_z\right)\kappa_x - \frac{\omega^2}{c^2} n^2 E_x = -\frac{\omega^2}{c^2} K_x E_x \tag{8.26}$$

$$\left(\kappa_x E_x + \kappa_y E_y + \kappa_z E_z\right)\kappa_x - E_x = -\frac{K_x E_x}{n^2} \tag{8.27}$$

$$E_x\left(\kappa_x^2 - 1 + \frac{K_x}{n^2}\right) + \kappa_x\kappa_y E_y + \kappa_x\kappa_z E_z = 0 \tag{8.28}$$

If a unit vector is defined as $\hat{\kappa}$, where $\kappa_x^2 + \kappa_y^2 + \kappa_z^2 = 1$, Eq. (8.28) is rewritten as

$$E_x\left(\frac{K_x}{n^2} - \kappa_y^2 - \kappa_z^2\right) + \kappa_x\kappa_y E_y + \kappa_x\kappa_z E_z = 0 \tag{8.29}$$

Other two components of electric fields in the x- and y-directions are similarly calculated as

$$\kappa_y\kappa_x E_x + \left(\frac{K_y}{n^2} - \kappa_x^2 - \kappa_z^2\right)E_y + \kappa_y\kappa_z E_z = 0 \tag{8.30}$$

$$\kappa_z\kappa_x E_x + \kappa_z\kappa_y E_y + \left(\frac{K_z}{n^2} - \kappa_x^2 - \kappa_y^2\right)E_z = 0 \tag{8.31}$$

Exercise 8.1

Prove the following equation:

$$\kappa_z\kappa_x E_x + \kappa_z\kappa_y E_y + \left(\frac{K_z}{n^2} - \kappa_x^2 - \kappa_y^2\right)E_z = 0$$

Solution

Substituting the displacement field into those two equations, we have

$$\vec{D} = \overline{\overline{\varepsilon}}\vec{E} = \varepsilon_0\overline{\overline{K}}\vec{E}$$

$$\vec{k} \times \vec{H} = -\omega\varepsilon_0\overline{\overline{K}}\vec{E}$$

$$\vec{k} \times \left(\vec{k} \times \vec{E}\right) = \omega\mu_0\left(\vec{k} \times \vec{H}\right) = -\omega^2\varepsilon_0\mu_0\overline{\overline{K}}\vec{E}$$

$$\vec{k}\left(\vec{k} \cdot \vec{E}\right) - k^2\vec{E} = -\frac{\omega^2}{c^2}\overline{\overline{K}}\vec{E}$$

The component of the electric field in the z-direction is calculated as

$$\left(k_x E_x + k_y E_y + k_z E_z\right)k_z - k^2 E_z = -\frac{\omega^2}{c^2}K_z E_z$$

$$\frac{\omega^2}{c^2}n^2\left(\kappa_x E_x + \kappa_y E_y + \kappa_z E_z\right)\kappa_z - \frac{\omega^2}{c^2}n^2 E_z = -\frac{\omega^2}{c^2}K_z E_z$$

$$\left(\kappa_x E_x + \kappa_y E_y + \kappa_z E_z\right)\kappa_z - E_z = -\frac{K_z E_z}{n^2}$$

$$\kappa_x \kappa_z E_x + \kappa_y \kappa_z E_y + \left(\kappa_z^2 - 1 + \frac{K_z}{n^2}\right) E_z = 0$$

$$\kappa_z \kappa_x E_x + \kappa_z \kappa_y E_y + \left(\frac{K_z}{n^2} - \kappa_x^2 - \kappa_y^2\right) E_z = 0$$

8.4 Implementations

8.4.1 Plane Wave Propagation in Nonlinear Medium at Normal Angle

If a plane wave generally propagates along the *z*-direction in the anisotropic medium, we have the unit vectors as

$$\kappa_x = 0,\ \kappa_y = 0,\ \kappa_z = 1$$

The plane wave Eq. (8.31) is modified as

$$\left(\frac{K_z}{n^2}\right) E_z = 0 \tag{8.32}$$

This point confirms that the electric field in the *z*-direction is zero.

On the other hand, the plane waves Eqs. (8.29) and (8.30) are modified as

$$E_x\left(\frac{K_x}{n^2} - 1\right) = 0 \tag{8.33}$$

$$E_y\left(\frac{K_y}{n^2} - 1\right) = 0 \tag{8.34}$$

The first and the second plane wave solutions in anisotropic medium in the order are

$$n = \sqrt{K_x},\ E_x \neq 0,\ E_y = 0,\ E_z = 0$$

$$n = \sqrt{K_y},\ E_x = 0,\ E_y \neq 0,\ E_z = 0$$

These solutions imply that the electric fields move along the *x*- and *y*-directions, respectively (Fig. 8.3). When these electric fields propagate along the principal axis, the corresponding displacement fields tend to move at the same direction.

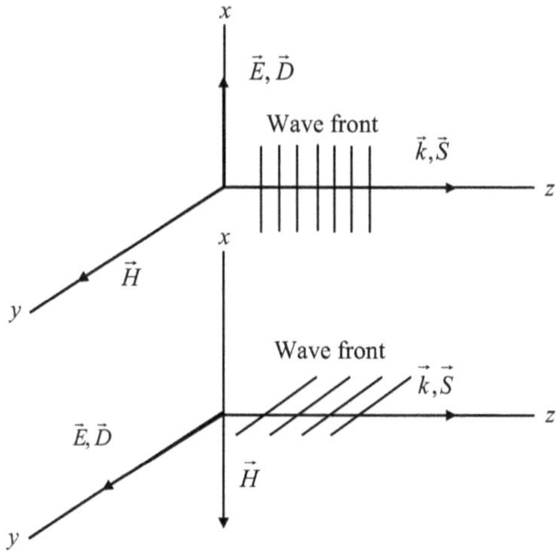

Figure 8.3 Possible electric and displacement fields move along the x-direction (top) and the y-direction (bottom) when a plane wave propagates in the z-direction. They also move perpendicular to the paralleled $\vec{k}, \vec{S}$.

8.4.2 Plane Wave Propagation in Nonlinear Medium at a Tile Angle

When a plane wave particularly propagates at a certain angle, ψ above the z-direction, we have the unit vectors as

$$\kappa_x = \sin\psi,\ \kappa_y = 0,\ \kappa_z = \cos\psi$$

The plane wave Eq. (8.30) is modified as

$$E_y\left(\frac{K_y}{n^2} - 1\right) = 0 \tag{8.35}$$

The first plane wave solution is

$$n = n_o = \sqrt{K_y},\ E_x = E_z = 0,\ E_y \neq 0$$

This solution implies that the electric field moves along the y-direction (Fig. 8.4). When this electric field propagates along the principal axis, the corresponding displacement field tends to move at the same direction. Velocity of this (ordinary) wave is independent of the certain angle ψ.

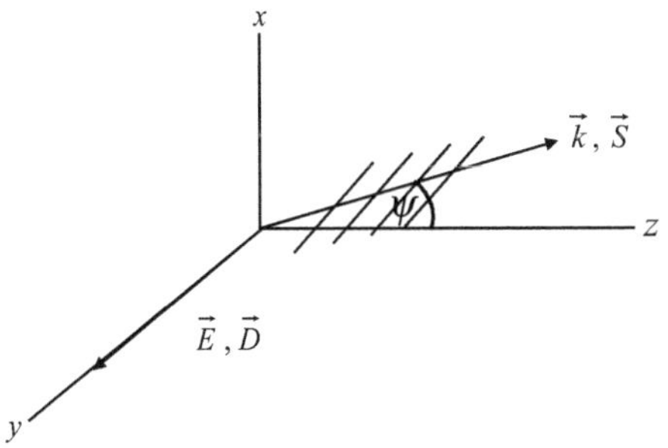

Figure 8.4 Possible electric and displacement fields move along the *y*-axis when a plane wave propagates at a certain angle ψ with respect to the *z*-direction. They also move perpendicular to the paralleled $\vec{k}, \vec{S}$.

One the other hand, the plane waves Eqs. (8.29) and (8.31) are modified as

$$E_x\left(\frac{K_x}{n^2}-\cos^2\psi\right)+\sin\psi\cos\psi E_z=0 \tag{8.36}$$

$$\sin\psi\cos\psi E_x+\left(\frac{K_z}{n^2}-\sin^2\psi\right)E_z=0 \tag{8.37}$$

They are rewritten in the matrix form as

$$\begin{pmatrix} \frac{K_x}{n^2}-\cos^2\psi & \sin\psi\cos\psi \\ \sin\psi\cos\psi & \frac{K_z}{n^2}-\sin^2\psi \end{pmatrix}\begin{pmatrix} E_x \\ E_z \end{pmatrix}=\begin{pmatrix} 0 \\ 0 \end{pmatrix} \tag{8.38}$$

$$\left(\frac{K_x}{n^2}-\cos^2\psi\right)\left(\frac{K_z}{n^2}-\sin^2\psi\right)-\sin^2\psi\cos^2\psi=0 \tag{8.39}$$

$$\frac{K_xK_z}{n^4}-\frac{1}{n^2}\left(K_x\sin^2\psi+K_z\cos^2\psi\right)=0 \tag{8.40}$$

$$\frac{1}{n^2}=\frac{\cos^2\psi}{K_x}+\frac{\sin^2\psi}{K_z} \tag{8.41}$$

$$\frac{1}{n_e^2(\psi)}=\frac{\cos^2\psi}{n_o^2}+\frac{\sin^2\psi}{n_e^2} \tag{8.42}$$

The second plane wave solution is

$$n_o = \sqrt{K_x},\ n_e = \sqrt{K_z}$$

$$E_y = 0,\ E_x \neq 0,\ E_z \neq 0$$

This solution implies that the velocity of the extraordinary wave is dependent on a certain angle ψ in which $n_e(\psi)$ is in between n_o and n_e. The maximum difference of velocity between ordinary and extraordinary waves will appear when a plane wave propagates in the xz-direction and we have E_x (with n_o) and E_z (with n_e) components.

Arranging Eq. (8.41) and hence substituting into Eq. (8.36), we have

$$\frac{K_x}{n^2} - \cos^2\psi = \frac{K_x}{K_z}\sin^2\psi \tag{8.43}$$

$$\frac{K_x}{K_z}\sin^2\psi E_x + \sin\psi\cos\psi E_z = 0 \tag{8.44}$$

$$\frac{K_x}{K_z}\sin\psi E_x = -\cos\psi E_z \tag{8.45}$$

$$\frac{K_x}{K_z}\tan\psi E_x = -E_z \tag{8.46}$$

$$\frac{K_z E_z}{K_x E_x} = -\tan\psi \tag{8.47}$$

$$\frac{K_z E_z}{K_x E_x} = \frac{\varepsilon_z / \varepsilon_0}{\varepsilon_x / \varepsilon_0}\frac{E_z}{E_x} \tag{8.48}$$

$$\frac{K_z E_z}{K_x E_x} = \frac{D_z}{D_x} = -\tan\psi \tag{8.49}$$

$$\frac{E_z}{E_x} = -\frac{K_x}{K_z}\tan\psi \tag{8.50}$$

These points imply that $\vec{D}$ is perpendicular to $\vec{k}$ in which $\vec{D}$ and $\vec{E}$ are not parallel to each other. Also $\vec{S}$ and $\vec{k}$ are not parallel to each other (Fig. 8.5).

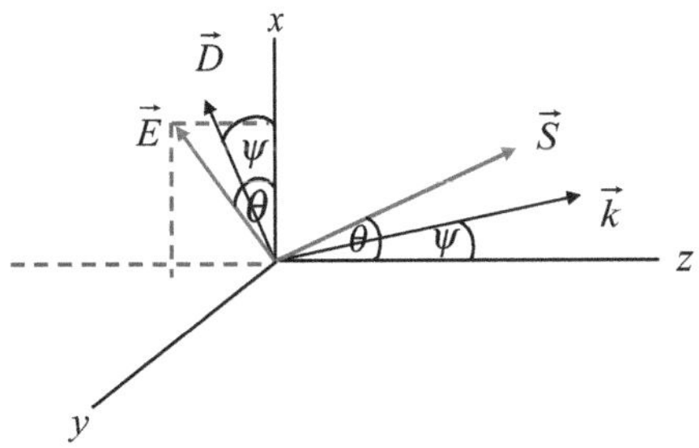

Figure 8.5 Possible electric field and displacement field move along the *xz*-direction when a plane wave propagates at a certain angle ψ with respect to the *z*-direction. They also move perpendicular to the paralleled $\vec{k}, \vec{S}$.

Problems

8.1 Prove the following equations:

$$\kappa_y \kappa_x E_x + \left(\frac{K_y}{n^2} - \kappa_x^2 - \kappa_z^2 \right) E_y + \kappa_y \kappa_z E_z = 0$$

8.2 Show how the electric field propagates in the *yz*-plane if a plane wave propagates at a certain angle ψ with respect to the optic axis.

8.3 A plane wave propagates along the direction given by the unit vector $\frac{\sqrt{3}}{2}\hat{x} + \frac{1}{2}\hat{z}$ into the uniaxial medium with $n_0 = 2.3$ and $n_e = 2.2$. Determine the angle made by $\vec{k}$ of the ordinary wave as compared with the optic axis.

8.4 If we select the arbitrary direction of the electric vector and a net energy propagates along the z-direction, show the details of electric fields in *x*- and *y*-axes.

Chapter 9

Second Harmonic Generation

9.1 Introduction

This chapter deals with second harmonic generation (SHG), a process in which one photon at a fundamental frequency and another photon at the same frequency merge to form a photon at double frequency. The SHG power conversion efficiency is generally dependent on a sine function; however, the maximum of SHG power conversion efficiency grows along the crystal length under birefringence phase matching condition. Furthermore, the maximum of SHG power conversion efficiency relies on a periodic length and crystal length under quasi-phase matching (QPM) condition.

9.2 Second Harmonic Generation

The linear approximations of polarization in a vector form and a component form are as follows:

$$\vec{P} = \varepsilon_0 \chi \vec{E} \tag{9.1}$$

$$P_i = \varepsilon_0 \sum_{j=1}^{3} \chi_{ij} E_j \tag{9.2}$$

Based on this relationship, the applied electric dipole of displacement is assumed to be very small, meaning that the

---

Linear and Nonlinear Optics: Materials, Properties, and Applications
Kitsakorn Locharoenrat

ISBN 978-981-4877-15-2 (Hardcover), 978-1-003-05588-4 (eBook)
www.jennystanford.com

oscillation is harmonic. This condition is correct if the electric field is small. If the electric field is strong comparable to the interatomic electric field in the medium, this approximation, therefore, fails, meaning that the oscillation is no more harmonic. This is a concept of nonlinear optic that needs to modify the above-mentioned polarization approximation as

$$P_i = P^{(1)} + P^{(2)} + P^{(3)} + \cdots \tag{9.3}$$

$$P_i = \varepsilon_0 \left[\sum_j \chi_{ij}^{(1)} E_j + \sum_{jk} \chi_{ijk}^{(2)} E_j E_k + \sum_{jkl} \chi_{ijkl}^{(3)} E_j E_k E_l + \cdots \right] \tag{9.4}$$

where $\chi_{ij}^{(1)}$ is linear susceptibility, $\chi_{ijk}^{(2)}$ is the second-order nonlinear susceptibility, $\chi_{ijkl}^{(3)}$ is the third-order nonlinear susceptibility, j = 1, 2, 3 $\rightarrow x, y, z$; k = 1, 2, 3 $\rightarrow x, y, z$; l = 1, 2, 3 $\rightarrow x, y, z$.

Nonlinear polarization in the scalar form is written as

$$P = P^{(1)} + P^{(NL)} \tag{9.5}$$

$$P = \varepsilon_0 \left[\chi^{(1)} E + \chi^{(2)} E^2 \right] \tag{9.6}$$

The second term on the right-hand side of Eq. (9.6) is proportional to two power electric fields, leading to the second-order nonlinear effect, including the SHG by launching light at the fundamental frequency (i.e., red) into the nonlinear optic (NLO) crystal in order to generate a constant polarization at the double frequency out of this medium (i.e., green laser). Other terms generating new frequencies are unable to generate the electromagnetic waves or they are negligible.

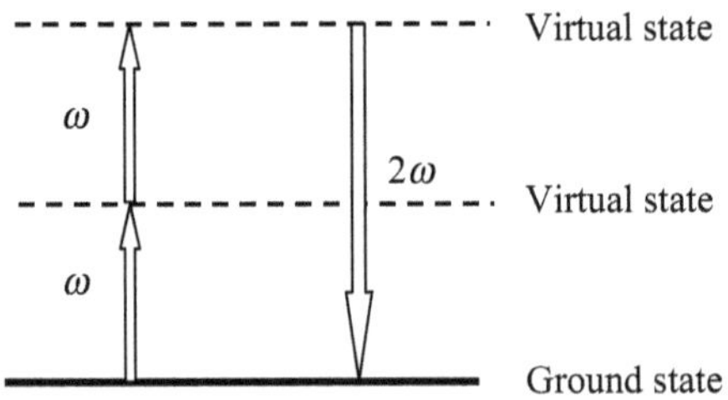

Figure 9.1 Energy level diagram of the electric field at the fundamental frequency propagates in the nonlinear optical medium generating an electric field at double frequency.

When the SHG takes place, we need to satisfy two conditions so that the efficiency of the generation of double frequency is not little as follows (Fig. 9.1):

- Photon energy conservation

$$\hbar\omega_1 + \hbar\omega_1 = \hbar\omega_2 \Rightarrow \omega_2 = 2\omega_1 \tag{9.7}$$

- Photon momentum conservation (phase matching)

$$\hbar k_1 + \hbar k_1 = \hbar k_2 \Rightarrow k_2 = 2k_1 \tag{9.8}$$

Suppose that the nonlinear optical medium of interest has no center of inversion symmetry to generate the second-order process, as shown in Fig. 9.2. Upon launching the input electric field at the fundamental frequency into this medium (no free current and electric charge), the total displacement is modified as

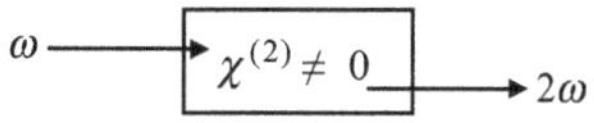

Figure 9.2 Nonlinear optical medium having no center of inversion symmetry generates the electromagnetic wave at double frequency along the optic axis, *z*.

$$\vec{D} = \varepsilon_0 \vec{E} + \vec{P} \tag{9.9}$$

$$\vec{D} = \varepsilon_0 \vec{E} + \vec{P}^{\mathrm{L}} + \vec{P}^{\mathrm{NL}} \tag{9.10}$$

$$\vec{D} = \varepsilon_0 \vec{E} + \varepsilon_0 \chi \vec{E} + \vec{P}^{\mathrm{NL}} \tag{9.11}$$

$$\vec{D} = \varepsilon_0 (1 + \chi) \vec{E} + \vec{P}^{\mathrm{NL}} \tag{9.12}$$

$$\vec{D} = \varepsilon \vec{E} + \vec{P}^{\mathrm{NL}} \tag{9.13}$$

According to the two Maxwell's equations,

$$\vec{\nabla} \times \vec{E} = -\frac{\partial \vec{B}}{\partial t} = -\mu_0 \frac{\partial \vec{H}}{\partial t} \tag{9.14}$$

$$\vec{\nabla} \times \vec{H} = \frac{\partial \vec{D}}{\partial t} \tag{9.15}$$

we get

$$\vec{\nabla} \times \left(\vec{\nabla} \times \vec{E} \right) = -\mu_0 \frac{\partial (\vec{\nabla} \times \vec{H})}{\partial t} \tag{9.16}$$

$$\vec{\nabla}(\vec{\nabla} \cdot \vec{E}) - \vec{\nabla}^2 \vec{E} = -\mu_0 \frac{\partial}{\partial t} \left(\varepsilon \frac{\partial \vec{E}}{\partial t} + \frac{\partial \vec{P}^{\mathrm{NL}}}{\partial t} \right) \tag{9.17}$$

$(\vec{\nabla}\cdot\vec{E})$ is equivalent to zero in an isotropic medium. In an anisotropic medium, $\vec{D}$ is perpendicular to $\vec{k}$ and $\vec{E}$ is not, in general, perpendicular to $\vec{k}$. Therefore, an angle between $\vec{E}$ and $\vec{D}$ is very small, and $(\vec{\nabla}\cdot\vec{E})$ is not equivalent to zero. To simplify our analysis here, we ignore $(\vec{\nabla}\cdot\vec{E})$ and have

$$-\vec{\nabla}^2\vec{E} = -\mu_0\varepsilon\frac{\partial^2\vec{E}}{\partial t^2} - \mu_0\frac{\partial^2\vec{P}^{\mathrm{NL}}}{\partial t^2} \tag{9.18}$$

$$\vec{\nabla}^2\vec{E} - \mu_0\varepsilon\frac{\partial^2\vec{E}}{\partial t^2} = \mu_0\frac{\partial^2\vec{P}^{\mathrm{NL}}}{\partial t^2} \tag{9.19}$$

When the wave propagations with fundamental and double frequency pass simultaneously through the nonlinear optical medium, interaction between these two frequencies is brought about by the nonlinearity in the medium as

$$\vec{\nabla}^2\vec{E}^{(\omega)} - \mu_0\varepsilon(\omega)\frac{\partial^2\vec{E}^{(\omega)}}{\partial t^2} = \mu_0\frac{\partial^2\vec{P}^{\mathrm{NL}(\omega)}}{\partial t^2} \tag{9.20}$$

$$\vec{\nabla}^2\vec{E}^{(2\omega)} - \mu_0\varepsilon(2\omega)\frac{\partial^2\vec{E}^{(2\omega)}}{\partial t^2} = \mu_0\frac{\partial^2\vec{P}^{\mathrm{NL}(2\omega)}}{\partial t^2} \tag{9.21}$$

Let us define an SHG of a nonlinear polarization as

$$P^{\mathrm{NL}} = \varepsilon_0\chi^{(2)}E^2 = 2d\varepsilon_0E^2 \tag{9.22}$$

where d is a nonlinear coefficient ($d \approx 10^{-12}\,\mathrm{m/V}$). The optical amplification via a finite d-tensor is very similar to what we can amplify light via population inversion in lasers, meaning that we put more atoms in the excited state as compared to the ground state, and due to most emissions, the input light then gets amplified, and we make the optical amplifier.

E is the total complex electric field composed of

$$E = E^{(\omega)} + E^{(2\omega)} \tag{9.23}$$

If a complex electric field at any frequency propagating in the z-direction into the medium is defined as

$$\tilde{z} = \mathrm{Re}[z] + \mathrm{Im}[z] \tag{9.24}$$

$$\tilde{z} = \frac{1}{2}\left(z + z^*\right) + \frac{1}{2}\left(z - z^*\right) \tag{9.25}$$

$$E = E_0\left(\cos\theta + i\sin\theta\right) = E_0\,\mathrm{e}^{i\phi} \tag{9.26}$$

$$E = \mathrm{Re}\left[E_0\cos(kz-\omega t+\phi)\right] = \frac{1}{2}[E_0 e^{i(kz-\omega t)}e^{i\phi} + cc] \tag{9.27}$$

The total complex electric fields at the two corresponding frequencies and two propagation constants are rewritten as

$$E^{(\omega)} = \frac{1}{2}[E_1 e^{i(k_1 z-\omega t)} + cc] \tag{9.28}$$

$$E^{(2\omega)} = \frac{1}{2}[E_2 e^{i(k_2 z-2\omega t)} + cc] \tag{9.29}$$

where

$$k_1 = \frac{\omega}{c}\cdot n(\omega) = \frac{\omega}{c}\cdot n_1$$

$$v_{P_{\mathrm{NL}}^{(2\omega)}} = \frac{\omega}{k_1} \tag{9.30}$$

$$k_2 = \frac{2\omega}{c}\cdot n(2\omega) = \frac{2\omega}{c}\cdot n_2$$

$$v_{E^{(2\omega)}} = \frac{2\omega}{k_2} \tag{9.31}$$

A second-order nonlinear response in Eq. (9.22) is also rewritten as

$$P^{\mathrm{NL}} = 2\varepsilon_0 d\left[\frac{1}{2}[E_1 e^{i(k_1 z-\omega t)} + cc] + \frac{1}{2}[E_2 e^{i(k_2 z-2\omega t)} + cc]\right]^2 \tag{9.32}$$

$$P^{\mathrm{NL}} = 2\varepsilon_0 d\frac{1}{4}\left[E_1 e^{i(k_1 z-\omega t)} + cc + E_2 e^{i(k_2 z-2\omega t)} + cc\right]^2 \tag{9.33}$$

$$P^{\mathrm{NL}} = 2\varepsilon_0 d\frac{1}{4}\left[E_1 e^{i(k_1 z-\omega t)} + E_1 e^{-i(k_1 z-\omega t)} + E_2 e^{i(k_2 z-2\omega t)} + E_2 e^{-i(k_2 z-2\omega t)}\right]^2 \tag{9.34}$$

$$(a+b+c+d)^2 = a^2+b^2+c^2+d^2+2ab+2ac+2ad+2bc+2bd+2cd \Rightarrow$$

$$P^{\mathrm{NL}} = 2\varepsilon_0 d\frac{1}{4}\left[E_1^2 e^{2i(k_1 z-\omega t)} + E_1^2 e^{-2i(k_1 z-\omega t)} + E_2^2 e^{2i(k_2 z-2\omega t)} + E_2^2 e^{-2i(k_2 z-2\omega t)}\right]$$
$$+2\varepsilon_0 d\frac{1}{4}\left[2E_1^2 + 2E_2 E_1 e^{i(k_1 z-\omega t)+i(k_2 z-2\omega t)} + 2E_2 E_1 e^{i(k_1 z-\omega t)-i(k_2 z-2\omega t)}\right]$$
$$+2\varepsilon_0 d\frac{1}{4}\left[2E_2 E_1 e^{-i(k_1 z-\omega t)+i(k_2 z-2\omega t)} + 2E_2 E_1 e^{-i(k_1 z-\omega t)-i(k_2 z-2\omega t)} + 2E_2^2\right] \tag{9.35}$$

$$P^{\mathrm{NL}} = 2\varepsilon_0 d\frac{1}{4}\left[E_1^2 e^{2i(k_1z-\omega t)} + E_1^2 e^{-2i(k_1z-\omega t)} + E_2^2 e^{2i(k_2z-2\omega t)} + E_2^2 e^{-2i(k_2z-2\omega t)}\right]$$
$$+2\varepsilon_0 d\frac{1}{4}\left[2E_1^2 + 2E_2E_1 e^{i[(k_1+k_2)z-3\omega t]} + 2E_2E_1 e^{-[i(k_2-k_1)z-\omega t]}\right]$$
$$+2\varepsilon_0 d\frac{1}{4}\left[2E_2E_1 e^{[i(k_2-k_1)z-\omega t]} + 2E_2E_1 e^{-i[(k_1+k_2)z-3\omega t]} + 2E_2^2\right] \tag{9.36}$$

$$P^{\mathrm{NL}} = 2\varepsilon_0 d\frac{1}{4}\left[E_1^2 e^{2i(k_1z-\omega t)} + E_1^2 e^{-2i(k_1z-\omega t)} + E_2^2 e^{2i(k_2z-2\omega t)} + E_2^2 e^{-2i(k_2z-2\omega t)}\right]$$
$$+2\varepsilon_0 d\frac{1}{4}\left[2E_1^2 + 2E_2E_1 e^{i[(k_1+k_2)z-3\omega t]} + 2E_2E_1^* e^{[i(k_2-k_1)z-\omega t]}\right]$$
$$+2\varepsilon_0 d\frac{1}{4}\left[2E_2E_1^* e^{-[i(k_2-k_1)z-\omega t]} + 2E_2E_1 e^{-i[(k_1+k_2)z-3\omega t]} + 2E_2^2\right] \tag{9.37}$$

$$P^{\mathrm{NL}} = 2\varepsilon_0 d\frac{1}{4}\left[E_1^2 e^{2i(k_1z-\omega t)} + cc + 0 + 0 + 0 + 2E_2E_1^* e^{[i(k_2-k_1)z-\omega t]} + cc + 0 + 0\right] \tag{9.38}$$

Nonlinear polarization at the corresponding fundamental and double frequencies exists as

$$P_{\mathrm{NL}}^{(2\omega)} = \frac{1}{2}\left[\varepsilon_0 d E_1^2 e^{2i(k_1z-\omega t)} + cc\right] \tag{9.39}$$

$$P_{\mathrm{NL}}^{(\omega)} = \frac{1}{2}\left[\varepsilon_0 d 2E_2E_1^* e^{[i(k_2-k_1)z-\omega t]} + cc\right] \tag{9.40}$$

$$\frac{\partial P_{\mathrm{NL}}^{(2\omega)}}{\partial t} = \frac{\partial}{\partial t}\frac{1}{2}\left[\varepsilon_0 d E_1^2 e^{2i(k_1z-\omega t)} + cc\right] \tag{9.41}$$

$$\frac{\partial P_{\mathrm{NL}}^{(2\omega)}}{\partial t} = \frac{1}{2}\varepsilon_0 d\left[(-2i\omega)E_1^2 e^{2i(k_1z-\omega t)} + e^{2i(k_1z-\omega t)}\frac{\partial E_1^2}{\partial t} + cc\right] \tag{9.42}$$

$$\frac{\partial^2 P_{\mathrm{NL}}^{(2\omega)}}{\partial t^2} = \frac{1}{2}\varepsilon_0 d\left[(-2i\omega)^2 E_1^2 e^{2i(k_1z-\omega t)} + (-2i\omega)e^{2i(k_1z-\omega t)}\frac{\partial E_1^2}{\partial t}\right.$$
$$\left. + e^{2i(k_1z-\omega t)}\frac{\partial^2 E_1^2}{\partial t^2} + (-2i\omega)e^{2i(k_1z-\omega t)}\frac{\partial E_1^2}{\partial t} + cc\right] \tag{9.43}$$

$$\frac{\partial^2 P_{\text{NL}}^{(2\omega)}}{\partial t^2} = \frac{1}{2}\varepsilon_0 d\left[(-2i\omega)^2 E_1^2 e^{2i(k_1 z-\omega t)} + (-4i\omega)e^{2i(k_1 z-\omega t)}\frac{\partial E_1^2}{\partial t} + e^{2i(k_1 z-\omega t)}\frac{\partial^2 E_1^2}{\partial t^2} + cc\right] \quad (9.44)$$

$$\frac{\partial^2 P_{\text{NL}}^{(2\omega)}}{\partial t^2} = \frac{1}{2}\varepsilon_0 d\left[\frac{\partial^2 E_1^2}{\partial t^2} - 4i\omega\frac{\partial E_1^2}{\partial t} - 4\omega^2 E_1^2\right]e^{2i(k_1 z-\omega t)} + cc \quad (9.45)$$

The second derivative of the electric field at the corresponding double frequency is written as

$$\frac{\partial E^{(2\omega)}}{\partial t} = \frac{\partial}{\partial t}\frac{1}{2}[E_2 e^{i(k_2 z-2\omega t)} + cc] \quad (9.46)$$

$$\frac{\partial E^{(2\omega)}}{\partial t} = \frac{1}{2}\left[(-2i\omega)E_2 e^{i(k_2 z-2\omega t)} + e^{i(k_2 z-2\omega t)}\frac{\partial E_2}{\partial t} + cc\right] \quad (9.47)$$

$$\frac{\partial^2 E^{(2\omega)}}{\partial t^2} = \frac{1}{2}\left[(-2i\omega)^2 E_2 e^{i(k_2 z-2\omega t)} + (-2i\omega)e^{i(k_2 z-2\omega t)}\frac{\partial E_2}{\partial t} + e^{i(k_2 z-2\omega t)}\frac{\partial^2 E_2}{\partial t^2} + (-2i\omega)e^{i(k_2 z-2\omega t)}\frac{\partial E_2}{\partial t} + cc\right] \quad (9.48)$$

$$\frac{\partial^2 E^{(2\omega)}}{\partial t^2} = \frac{1}{2}\left[(-2i\omega)^2 E_2 e^{i(k_2 z-2\omega t)} + (-4i\omega)e^{i(k_2 z-2\omega t)}\frac{\partial E_2}{\partial t} + e^{i(k_2 z-2\omega t)}\frac{\partial^2 E_2}{\partial t^2} + cc\right] \quad (9.49)$$

$$\frac{\partial^2 E^{(2\omega)}}{\partial t^2} = \frac{1}{2}\left[\frac{\partial^2 E_2}{\partial t^2} - 4i\omega\frac{\partial E_2}{\partial t} - 4\omega^2 E_2\right]e^{i(k_2 z-2\omega t)} + cc \quad (9.50)$$

$$\frac{\partial E^{(2\omega)}}{\partial z} = \frac{\partial}{\partial z}\frac{1}{2}[E_2 e^{i(k_2 z-2\omega t)} + cc] \quad (9.51)$$

$$\frac{\partial E^{(2\omega)}}{\partial z}=\frac{1}{2}\left[(ik_2)E_2e^{i(k_2z-2\omega t)}+e^{i(k_2z-2\omega t)}\frac{\partial E_2}{\partial z}+cc\right] \quad (9.52)$$

$$\frac{\partial^2 E^{(2\omega)}}{\partial z^2}=\frac{1}{2}\left[(ik_2)^2E_2e^{i(k_2z-2\omega t)}+(ik_2)e^{i(k_2z-2\omega t)}\frac{\partial E_2}{\partial z}\right.$$
$$\left.+e^{i(k_2z-2\omega t)}\frac{\partial^2 E_2}{\partial z^2}+(ik_2)e^{i(k_2z-2\omega t)}\frac{\partial E_2}{\partial z}+cc\right] \quad (9.53)$$

$$\frac{\partial^2 E^{(2\omega)}}{\partial z^2}=\frac{1}{2}\left[(ik_2)^2E_2e^{i(k_2z-2\omega t)}+(2ik_2)e^{i(k_2z-2\omega t)}\frac{\partial E_2}{\partial z}\right.$$
$$\left.+e^{i(k_2z-2\omega t)}\frac{\partial^2 E_2}{\partial z^2}+cc\right] \quad (9.54)$$

$$\nabla^2 E^{(2\omega)}=\frac{1}{2}\left[\frac{d^2E_2}{dz^2}+2ik_2\frac{dE_2}{dz}-k_2^2E_2\right]e^{i(k_2z-2\omega t)}+cc \quad (9.55)$$

We neglect the second derivative terms and assume that the growth of the second harmonic is slow. Substituting Eqs. (9.50) and (9.55) into Eq. (9.56), we get

$$\vec{\nabla}^2\vec{E}^{(2\omega)}-\mu_0\varepsilon(2\omega)\frac{\partial^2\vec{E}^{(2\omega)}}{\partial t^2}=\mu_0\frac{\partial^2\vec{P}^{\mathrm{NL}(2\omega)}}{\partial t^2} \quad (9.56)$$

$$\frac{1}{2}\left[\frac{d^2E_2}{dz^2}+2ik_2\frac{dE_2}{dz}-k_2^2E_2\right]e^{i(k_2z-2\omega t)}+cc$$
$$-\mu_0\varepsilon(2\omega)\frac{1}{2}\left[\frac{\partial^2E_2}{\partial t^2}-4i\omega\frac{\partial E_2}{\partial t}-4\omega^2E_2\right]e^{i(k_2z-2\omega t)}+cc$$
$$=\mu_0\frac{1}{2}\varepsilon_0 d\left[\frac{\partial^2E_1^2}{\partial t^2}-4i\omega\frac{\partial E_1^2}{\partial t}-4\omega^2E_1^2\right]e^{2i(k_1z-\omega t)}+cc$$

$$\left(2ik_2\frac{dE_2}{dz}-k_2^2E_2\right)e^{ik_2z}-\mu_0\varepsilon(2\omega)(-4\omega^2)E_2e^{ik_2z}$$
$$=\mu_0(-4\omega^2)\varepsilon_0 d\,E_1^2e^{2ik_1z} \quad (9.57)$$

Substituting the wave propagation at the double frequency into Eq. (9.57), we have

$$k_2 = \frac{2\omega}{c} n_2 = 2\omega\sqrt{\mu_0\varepsilon_0}\, n_2 = 2\omega\sqrt{\mu_0\varepsilon_0}\sqrt{\varepsilon_r} \tag{9.58}$$

$$k_2^2 = 4\omega^2\mu_0\varepsilon \tag{9.59}$$

$$2ik_2\frac{dE_2}{dz} = -4\mu_0\varepsilon_0\omega^2 d\,E_1^2 e^{-i(k_2-2k_1)z} \tag{9.60}$$

$$\frac{dE_2}{dz} = \frac{4}{2k_2} i\mu_0\varepsilon_0\omega^2 d\,E_1^2 e^{-i(k_2-2k_1)z} \tag{9.61}$$

$$\frac{dE_2}{dz} = i\frac{\omega^2}{c^2}\frac{4}{2k_2} d\,E_1^2 e^{-i\Delta kz} \tag{9.62}$$

$$\frac{dE_2}{dz} = i\frac{\omega^2}{c^2}\frac{2}{\frac{2\omega}{c}n_2} d\,E_1^2 e^{-i\Delta kz} \tag{9.63}$$

$$\frac{dE_2}{dz} = i\frac{\omega d}{cn_2} E_1^2 e^{-i\Delta kz} \tag{9.64}$$

where

$$\Delta k = k_2 - 2k_1 \tag{9.65}$$

To sum up, the nonlinear polarization and electric field at the double frequency offer the first wave equation as

$$\nabla^2 E^{(2\omega)} - \mu_0\varepsilon(2\omega)\frac{\partial^2 E^{(2\omega)}}{\partial t^2} = \mu_0\frac{\partial^2 P^{\mathrm{NL}(2\omega)}}{\partial t^2} \tag{9.66}$$

$$P_{\mathrm{NL}}^{(2\omega)} = \frac{1}{2}\left[\varepsilon_0 dE_1^2 e^{2i(k_1 z-\omega t)} + cc\right] \tag{9.67}$$

$$E^{(2\omega)} = \frac{1}{2}[E_2(z)e^{i(k_2 z-2\omega t)} + cc] \tag{9.68}$$

$$\frac{dE_2}{dz} = i\frac{\omega d}{cn_2} E_1^2 e^{-i\Delta kz} \tag{9.69}$$

Exercise 9.1

Derive the electric field at the fundamental frequency using the same procedure according to the nonlinear polarization and electric field at double frequency.

Solution

$$P_{\mathrm{NL}}^{(\omega)}=\frac{1}{2}\left[2\varepsilon_0 dE_2E_1^*e^{i\left[(k_2-k_1)z-\omega t\right]}+cc\right]$$

$$\frac{\partial P_{\mathrm{NL}}^{(\omega)}}{\partial t}=\varepsilon_0 d\left[E_2\left(-i\omega E_1^*e^{i\left[(k_2-k_1)z-\omega t\right]}+e^{i\left[(k_2-k_1)z-\omega t\right]}\frac{\partial E_1^*}{\partial t}\right)\right.$$
$$\left.+E_1^*e^{i\left[(k_2-k_1)z-\omega t\right]}\frac{\partial E_2}{\partial t}+cc\right]$$

$$\frac{\partial P_{\mathrm{NL}}^{(\omega)}}{\partial t}=\varepsilon_0 d\left[-i\omega E_2E_1^*e^{i\left[(k_2-k_1)z-\omega t\right]}+E_2e^{i\left[(k_2-k_1)z-\omega t\right]}\frac{\partial E_1^*}{\partial t}\right.$$
$$\left.+E_1^*e^{i\left[(k_2-k_1)z-\omega t\right]}\frac{\partial E_2}{\partial t}+cc\right]$$

$$\frac{\partial^2 P_{\mathrm{NL}}^{(\omega)}}{\partial t^2}=\varepsilon_0 d\left[(-i\omega)^2E_2E_1^*e^{i\left[(k_2-k_1)z-\omega t\right]}-i\omega E_1^*e^{i\left[(k_2-k_1)z-\omega t\right]}\frac{\partial E_2}{\partial t}\right.$$
$$\left.-i\omega E_1^*e^{i\left[(k_2-k_1)z-\omega t\right]}\frac{\partial E_2}{\partial t}\right]$$
$$+\varepsilon_0 d\left[e^{i\left[(k_2-k_1)z-\omega t\right]}E_2\frac{\partial E_1^{*2}}{\partial t^2}-i\omega E_2\frac{\partial E_1^*}{\partial t}e^{i\left[(k_2-k_1)z-\omega t\right]}\right.$$
$$\left.+\frac{\partial E_2}{\partial t}\frac{\partial E_1^*}{\partial t}e^{i\left[(k_2-k_1)z-\omega t\right]}\right]$$
$$+\varepsilon_0 d\left[e^{i\left[(k_2-k_1)z-\omega t\right]}E_1^*\frac{\partial E_2^2}{\partial t^2}-i\omega E_1^*\frac{\partial E_2}{\partial t}e^{i\left[(k_2-k_1)z-\omega t\right]}\right.$$
$$\left.+\frac{\partial E_2}{\partial t}\frac{\partial E_1^*}{\partial t}e^{i\left[(k_2-k_1)z-\omega t\right]}\right]+cc$$

$$E^{(\omega)}=\frac{1}{2}[E_1(z)e^{i(k_1z-\omega t)}+cc]$$

$$\frac{\partial E^{(\omega)}}{\partial z}=\frac{1}{2}\frac{\partial}{\partial z}[E_1(z)e^{i(k_1z-\omega t)}+cc]$$

$$\frac{\partial E^{(\omega)}}{\partial z}=\frac{1}{2}[ik_1E_1e^{i(k_1z-\omega t)}+e^{i(k_1z-\omega t)}\frac{\partial E_1}{\partial z}+cc]$$

$$\frac{\partial^2 E^{(\omega)}}{\partial z^2} = \frac{1}{2}[(ik_1)^2 E_1 e^{i(k_1 z-\omega t)} + ik_1 e^{i(k_1 z-\omega t)} \frac{\partial E_1}{\partial z}$$

$$+ e^{i(k_1 z-\omega t)} \frac{\partial^2 E_1}{\partial z^2} + ike^{i(k_1 z-\omega t)} \frac{\partial E_1}{\partial z} + cc]$$

$$\frac{\partial^2 E^{(\omega)}}{\partial z^2} = \frac{1}{2}[(ik_1)^2 E_1 e^{i(k_1 z-\omega t)} + 2ik_1 e^{i(k_1 z-\omega t)} \frac{\partial E_1}{\partial z}$$

$$+ e^{i(k_1 z-\omega t)} \frac{\partial^2 E_1}{\partial z^2}] + cc$$

$$\frac{\partial E^{(\omega)}}{\partial t} = \frac{1}{2}\frac{\partial}{\partial t}[\mathrm{E}_1(\mathrm{z})\mathrm{e}^{i(k_1 z-\omega t)} + \mathrm{cc}]$$

$$\frac{\partial E^{(\omega)}}{\partial t} = \frac{1}{2}[-i\omega E_1 e^{i(k_1 z-\omega t)} + e^{i(k_1 z-\omega t)} \frac{\partial E_1}{\partial t} + cc]$$

$$\frac{\partial^2 E^{(\omega)}}{\partial t^2} = \frac{1}{2}[(-i\omega)^2 E_1 e^{i(k_1 z-\omega t)} - i\omega e^{i(k_1 z-\omega t)}$$

$$\frac{\partial E_1}{\partial t} + e^{i(k_1 z-\omega t)} \frac{\partial^2 E_1}{\partial t^2} - i\omega e^{i(k_1 z-\omega t)} \frac{\partial E_1}{\partial t} + cc]$$

$$\frac{\partial^2 E^{(\omega)}}{\partial t^2} = \frac{1}{2}[(-i\omega)^2 E_1 e^{i(k_1 z-\omega t)} - 2i\omega e^{i(k_1 z-\omega t)} \frac{\partial E_1}{\partial t}$$

$$+ e^{i(k_1 z-\omega t)} \frac{\partial^2 E_1}{\partial t^2}] + cc$$

$$\vec{\nabla}^2 \vec{E}^{(\omega)} - \mu_0 \varepsilon(\omega) \frac{\partial^2 \vec{E}^{(\omega)}}{\partial t^2} = \mu_0 \frac{\partial^2 \vec{P}^{\mathrm{NL}(\omega)}}{\partial t^2}$$

$$\frac{1}{2}[(ik_1)^2 E_1 + 2ik_1 \frac{\partial E_1}{\partial z} + \frac{\partial^2 E_1}{\partial z^2}] e^{i(k_1 z-\omega t)}$$

$$-\frac{1}{2}\mu_0 \varepsilon[(-i\omega)^2 E_1 - 2i\omega \frac{\partial E_1}{\partial t} + \frac{\partial^2 E_1}{\partial t^2}] e^{i(k_1 z-\omega t)}$$

$$= \mu_0(-i\omega)^2 E_2 E_1^* e^{i[(k_2-k_1)z-\omega t]}$$

$$[(ik_1)^2E_1+2ik_1\frac{\partial E_1}{\partial z}+\frac{\partial^2 E_1}{\partial z^2}]-\mu_0\varepsilon[(-i\omega)^2E_1-2i\omega\frac{\partial E_1}{\partial t}+\frac{\partial^2 E_1}{\partial t^2}]$$
$$=\mu_0(-i\omega)^2 2E_2E_1^{*}e^{i[(k_2-2k_1)z]}$$

$$[(ik_1)^2E_1+2ik_1\frac{\partial E_1}{\partial z}+\frac{\partial^2 E_1}{\partial z^2}]-\mu_0\varepsilon[(-i\omega)^2E_1-2i\omega\frac{\partial E_1}{\partial t}+\frac{\partial^2 E_1}{\partial t^2}]$$
$$=\mu_0(-i\omega^2)\varepsilon_0 d2E_2E_1^{*}\,e^{i[(k_2-2k_1)z]}$$

Substituting the wave propagation at the fundamental frequency, we have

$$k_1=\frac{\omega}{c}n_1=\omega\sqrt{\mu_0\varepsilon_0}\,n_1=\omega\sqrt{\mu_0\varepsilon_0}\sqrt{\varepsilon_r}$$

$$k_1^2=\omega^2\mu_0\varepsilon$$

$$[2ik_1\frac{\partial E_1}{\partial z}+\frac{\partial^2 E_1}{\partial z^2}]-2i\omega\frac{\partial E_1}{\partial t}+\frac{\partial^2 E_1}{\partial t^2}=\mu_0(-\omega^2)\varepsilon_0 d2E_2E_1^{*}e^{i[(k_2-2k_1)z]}$$

$$2ik_1\frac{\partial E_1}{\partial z}=\mu_0(-\omega^2)\varepsilon_0 d2E_2E_1^{*}e^{i[(k_2-2k_1)z]}$$

$$\frac{\partial E_1}{\partial z}=\mu_0(-\omega^2)\varepsilon_0 d\frac{2}{2ik_1}E_2E_1^{*}e^{i[(k_2-2k_1)z]}$$

$$\frac{\partial E_1}{\partial z}=i\mu_0\omega^2\varepsilon_0 d\frac{2}{2\frac{\omega}{c}n_1}E_2E_1^{*}e^{i[(k_2-2k_1)z]}$$

$$\frac{\partial E_1}{\partial z}=i\mu_0\omega\varepsilon_0 d\frac{1}{\frac{1}{c}n_1}E_2E_1^{*}e^{i[(k_2-2k_1)z]}$$

$$\frac{\partial E_1}{\partial z}=i\omega\frac{1}{c^2}d\frac{1}{\frac{1}{c}n_1}E_2E_1^{*}e^{i[(k_2-2k_1)z]}$$

$$\frac{dE_1}{dz}=i\frac{\omega d}{cn_1}E_2E_1^{*}e^{i\Delta kz}$$

To sum up, the nonlinear polarization and electric field at the fundamental frequency offer the second wave equation as

$$\vec{\nabla}^2\vec{E}^{(\omega)} - \mu_0\varepsilon(\omega)\frac{\partial^2\vec{E}^{(\omega)}}{\partial t^2} = \mu_0\frac{\partial^2\vec{P}^{\mathrm{NL}(\omega)}}{\partial t^2}$$

$$E^{(\omega)} = \frac{1}{2}[E_1(z)e^{i(k_1 z-\omega t)} + cc]$$

$$P_{\mathrm{NL}}^{(\omega)} = \frac{1}{2}\left[2\varepsilon_0 dE_2E_1^* e^{i[(k_2-k_1)z-\omega t]} + cc\right]$$

$$\frac{dE_1}{dz} = i\frac{\omega d}{cn_1}E_2E_1^* e^{i\Delta kz}$$

Equation (9.69) reveals that if we have a nonlinear optical medium with a finite second-order nonlinear coefficient (d), by launching the electric field (E_1) of the incident wave at the fundamental frequency, we can calculate the amplitude of electric field of the generated second harmonic field (E_2) at the double frequency along the optic axis in the following steps.

Assuming that the electric field (E_1) is a constant (strong pump), we get

$$\int_0^z dE_2 = i\frac{\omega d}{cn_2}E_1^2\int_0^z e^{-i\Delta kz}dz \tag{9.70}$$

$$E_2(z) - E_2(0) = i\frac{\omega d}{cn_2}E_1^2\left[\frac{e^{-i\Delta kz}}{-i\Delta k}\right]_0^z \tag{9.71}$$

$$E_2(z) = i\frac{\omega d}{cn_2}E_1^2\frac{e^{-i\Delta kz}-1}{-i\Delta k} \tag{9.72}$$

$$E_2(z) = i\frac{\omega d}{cn_2}E_1^2\frac{1-e^{-i\Delta kz}}{i\Delta k} \tag{9.73}$$

$$E_2(z) = \frac{\omega d}{cn_2}E_1^2 e^{-i\Delta kz/2}\frac{e^{i\Delta kz/2}-e^{-i\Delta kz/2}}{\Delta k} \tag{9.74}$$

$$E_2(z) = \frac{\omega d}{cn_2} E_1^2 e^{-i\Delta kz/2} \frac{2i\sin(\Delta kz/2)}{\Delta k} \tag{9.75}$$

$$E_2(z) = \frac{\omega d}{cn_2} E_1^2 e^{-i\Delta kz/2} \frac{i\sin(\Delta kz/2)}{\Delta k/2} \tag{9.76}$$

Recall that the incident power at the fundamental frequency and the output second harmonic power at double frequencies are written as

$$P_1 = \frac{n_1}{2c\mu_0} |E_1|^2 S \tag{9.77}$$

$$|E_1|^4 = \frac{4c^2\mu_0^2 P_1^2}{n_1^2 S^2} \tag{9.78}$$

$$P_2(z) = \frac{n_2}{2c\mu_0} |E_2(z)|^2 S \tag{9.79}$$

where S is the cross-sectional area of the light beam.

Substituting Eqs. (9.76) and (9.78) into Eq. (9.79), we get

$$P_2(z) = \frac{n_2}{2c\mu_0} \left| i\frac{\omega d}{cn_2} E_1^2 \frac{\sin(\Delta kz/2)}{\Delta k/2} \right|^2 S \tag{9.80}$$

$$P_2(z) = \frac{n_2}{2c\mu_0} \frac{\omega^2 d^2}{c^2 n_2^2} |E_1|^4 \frac{\sin^2(\Delta kz/2)}{(\Delta k/2)^2} S \tag{9.81}$$

$$P_2(z) = \frac{n_2}{2c\mu_0} \frac{\omega^2 d^2}{c^2 n_2^2} \frac{4c^2\mu_0^2 P_1^2}{n_1^2 S^2} \frac{\sin^2(\Delta kz/2)}{(\Delta k/2)^2} S \tag{9.82}$$

$$P_2(z) = \frac{2\mu_0\omega^2}{cn_1^2 n_2} \frac{d^2 P_1^2}{S} \frac{\sin^2(\Delta kz/2)}{(\Delta k/2)^2} \tag{9.83}$$

- If Δk is not equivalent to zero, the power conversion efficiency of the SHG will be oscillating along the crystal length z as (Fig. 9.3)

$$\eta = \frac{P_2(z)}{P_1} = \frac{2\mu_0\omega^2}{cn_1^2 n_2} \frac{d^2 P_1}{S} \frac{\sin^2(\Delta kz/2)}{(\Delta k/2)^2} \tag{9.84}$$

- If Δk is equivalent to zero, the power conversion efficiency of the SHG will be maximum along the crystal length z as (Fig. 9.3)

$$\lim_{\Delta k \to 0} \frac{\sin^2(\Delta kz/2)}{(\Delta k/2)^2} = z^2 \tag{9.85}$$

$$\eta = \frac{2\mu_0\omega^2}{cn_1^2 n_2} \frac{d^2 P_1}{S} z^2 \tag{9.86}$$

Since a phase matching condition exists, we will have a maximum conversion efficiency of the SHG at the coherence length defined as

$$\frac{\Delta kz}{2} = \frac{\pi}{2} \Rightarrow L_c = \frac{\pi}{\Delta k} \tag{9.87}$$

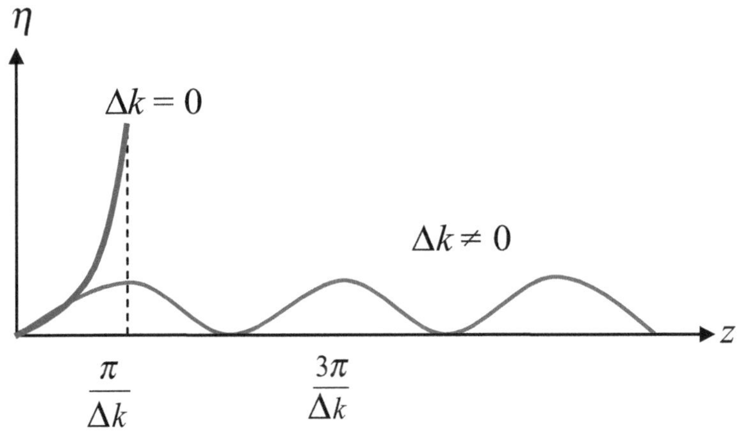

Figure 9.3 SHG power conversion efficiency with and without phase matching as a function of the optic axis *z*. The larger the Δk, the smaller the maximum conversion efficiency we can generate from the SHG. However, the power conversion efficiency of the SHG is usually very weak; a few percent.

In the case of a phase mismatching condition, the SHG amplitude first generates to build up until the coherence length number. At the coherence length, the nonlinear polarization and the electromagnetic wave get de-phased and the SHG amplitude decreases if they are not phase matched. Beyond this point, the SHG feeds energy back to the fundamental frequency. Hence, at this point, all the energy is back in the fundamental frequency, again the double frequency builds up, and it oscillates periodically. Furthermore, the SHG power efficiency drops down with some reasonable efficiency depending on the width of the coherence length, $\pm\pi/\Delta k$. The narrower the curve, the more we achieve phase matching. The longer the crystal length, the narrower the phase matching. The shorter the crystal length, the less critical the phase matching is.

Exercise 9.2

Derive that the phase difference between nonlinear polarization and electromagnetic wave is equivalent to pi.

Solution

The phase shift travelling through the optic axis suffered by the nonlinear polarization is

$$P_{\text{NL}}^{(2\omega)}=\frac{1}{2}\left[\varepsilon_0 dE_1^2 e^{2i(k_1 z-\omega t)}+cc\right]$$

$$\phi_{P_{\text{NL}}^{(2\omega)}}=2i(k_1 z-\omega t)$$

The phase shift travelling through the optic axis suffered by the electromagnetic wave is

$$E^{(2\omega)}=\frac{1}{2}[E_2(z)e^{i(k_2 z-2\omega t)}+cc]$$

$$\phi_{E^{(2\omega)}}=i(k_2 z-2\omega t)$$

The phase difference between the nonlinear polarization and the electromagnetic wave is written as

$$\Delta\phi=\phi_{E^{(2\omega)}}-\phi_{P_{\text{NL}}^{(2\omega)}}$$

$$\Delta\phi=(k_2 z-2\omega t)-2(k_1 z-\omega t)$$

$$\Delta\phi=k_2 z-2\omega t-2k_1 z+2\omega t$$

$$\Delta\phi=(k_2-2k_1)z=\Delta k\,z$$

$$\Delta\phi\big|_{z=L_c}=\Delta k\,L_c$$

$$\Delta\phi\big|_{z=L_c}=\Delta k\frac{\pi}{\Delta k}=\pi$$

9.3 Realistic SHG

The realistic analytical second harmonic solution to a maximum SHG under a perfect phase matching situation is related to amplitudes and phases of the complex electric fields as follows (Fig. 9.4):

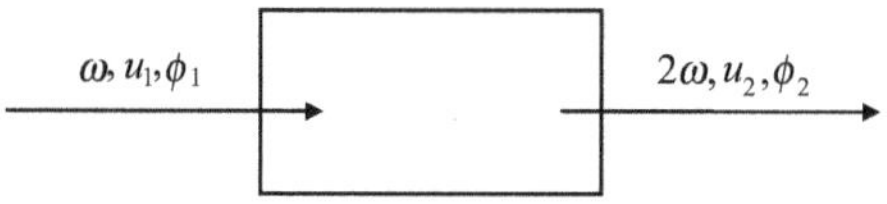

Figure 9.4 Amplitudes of the electric fields (u_1, u_2) and phases of the electric fields (ϕ_1, ϕ_2) are all real terms for the SHG process.

$$E_1(z) = u_1(z) e^{i\phi_1(z)} \quad (9.88)$$

$$E_1^*(z) = u_1(z) e^{-i\phi_1(z)} \quad (9.89)$$

$$E_2(z) = u_2(z) e^{i\phi_2(z)} \quad (9.90)$$

$$E_2^*(z) = u_2(z) e^{-i\phi_2(z)} \quad (9.91)$$

Recall the two wave equations as

$$\frac{dE_1}{dz} = i\frac{\omega d}{cn_1} E_2 E_1^* e^{i\Delta kz} \quad (9.92)$$

$$\frac{dE_2}{dz} = i\frac{\omega d}{cn_2} E_1^2 e^{-i\Delta kz} \quad (9.93)$$

Under a phase matching condition, those two equations are rewritten as

$$\frac{dE_1}{dz} = i\frac{\omega d}{cn_1} E_2 E_1^* \quad (9.94)$$

$$\frac{dE_2}{dz} = i\frac{\omega d}{cn_2} E_1^2 \quad (9.95)$$

If we write the complex electric field E_1 as the product of the amplitude and phase terms, we have the amplitude and phase equations for the SHG as

$$\frac{dE_1}{dz} = \frac{du_1}{dz} e^{i\phi_1} + i\frac{d\phi_1}{dz} u_1 e^{i\phi_1} \quad (9.96)$$

$$\frac{dE_1}{dz} = i\frac{\omega d}{cn_1} u_2 e^{i\phi_2} u_1 e^{-i\phi_1} \tag{9.97}$$

$$\frac{du_1}{dz} + i\frac{d\phi_1}{dz} u_1 = i\frac{\omega d}{cn_1} u_1 u_2 e^{i\theta(z)} \Rightarrow \theta(z) = \phi_2 - 2\phi_1 \tag{9.98}$$

$$\frac{du_1}{dz} + i\frac{d\phi_1}{dz} u_1 = i\frac{\omega d}{cn_1} u_1 u_2 \left(\cos\theta(z) + i\sin\theta(z)\right) \tag{9.99}$$

$$\frac{du_1}{dz} + i\frac{d\phi_1}{dz} u_1 = i\frac{\omega d}{cn_1} u_1 u_2 \cos\theta(z) - \frac{\omega d}{cn_1} u_1 u_2 \sin\theta(z) \tag{9.100}$$

$$\frac{du_1}{dz} = -\frac{\omega d}{cn_1} u_1 u_2 \sin\theta(z) \tag{9.101}$$

$$\frac{d\phi_1}{dz} u_1 = \frac{\omega d}{cn_1} u_1 u_2 \cos\theta(z) \tag{9.102}$$

Similarly, we write the complex electric field E_2 as the product of the amplitude and phase terms and we get other amplitude and phase equations for the SHG as

$$\frac{du_2}{dz} = \frac{\omega d}{cn_2} u_1^2 \sin\theta(z) \tag{9.103}$$

$$\frac{d\phi_2}{dz} u_2 = \frac{\omega d}{cn_2} u_1^2 \cos\theta(z) \tag{9.104}$$

We finally have four real amplitude and phase equations for the SHG as

$$\frac{du_1}{dz} = -\kappa u_1 u_2 \sin\theta \tag{9.105}$$

$$\frac{d\phi_1}{dz} u_1 = \kappa u_1 u_2 \cos\theta \tag{9.106}$$

$$\frac{du_2}{dz} = \kappa u_1^2 \sin\theta \tag{9.107}$$

$$\frac{d\phi_2}{dz} u_2 = \kappa u_1^2 \cos\theta \tag{9.108}$$

where

$$\kappa = \frac{\omega d}{cn_1} = \frac{\omega d}{cn_2} \tag{9.109}$$

Substituting the differential equation of the phase matching term from Eqs. (9.106) and (9.108), we get

$$\frac{d\theta}{dz} = \frac{d\phi_2}{dz} - 2\frac{d\phi_1}{dz} \tag{9.110}$$

$$\frac{d\theta}{dz} = \frac{\kappa u_1^2}{u_2}\cos\theta - 2\kappa u_2 \cos\theta \tag{9.111}$$

$$\frac{d\theta}{dz} = \kappa\cos\theta\left(\frac{u_1^2}{u_2} - 2u_2\right) \tag{9.112}$$

$$\frac{d\theta}{dz} = \frac{\kappa}{u_2}\cos\theta\left(u_1^2 - 2u_2^2\right) \tag{9.113}$$

Substituting Eqs. (9.105) and (9.107) into

$$\frac{d}{dz}\left(\ln u_1^2 u_2\right) = \frac{1}{u_1^2 u_2}\left(u_1^2\frac{du_2}{dz} + 2u_1u_2\frac{du_1}{dz}\right) \tag{9.114}$$

$$\frac{d}{dz}\left(\ln u_1^2 u_2\right) = \frac{1}{u_1^2 u_2}\left(\kappa u_1^4 \sin\theta - 2\kappa u_1^2 u_2^2 \sin\theta\right) \tag{9.115}$$

$$\frac{d}{dz}\left(\ln u_1^2 u_2\right) = \frac{\kappa\sin\theta}{u_2}\left(u_1^2 - 2u_2^2\right) \tag{9.116}$$

$$\begin{gathered}\frac{d}{dz}\left(\ln u_1^2 u_2\right) = \frac{\sin\theta}{\cos\theta}\frac{d\theta}{dz} = -\frac{d}{dz}\ln\cos\theta \\ \Rightarrow \frac{d\theta}{dz} = \frac{\kappa}{u_2}\cos\theta\left(u_1^2 - 2u_2^2\right)\end{gathered} \tag{9.117}$$

$$\frac{d}{dz}\left(\ln u_1^2 u_2 + \ln\cos\theta\right) = 0 \tag{9.118}$$

$$\frac{d}{dz}\left[\ln\left(u_1^2 u_2 \cos\theta\right)\right] = 0 \tag{9.119}$$

Equation (9.119) reveals that $u_1^2 u_2 \cos\theta$ is a constant of motion independent of the optic axis z.

There are two conditions of SHG after the fundamental wave incident on the nonlinear optical medium along the optic axis z as follows:

$$u_2(0) = 0 \Rightarrow u_1^2(0)u_2(0)\cos\theta(0) = 0 \tag{9.120}$$

$$u_2(z) = z \Rightarrow u_1^2(z)u_2(z)\cos\theta(z) = 0 \tag{9.121}$$

$$\theta(z) = \phi_2(z) - 2\phi_1(z) = \pm\frac{\pi}{2} \tag{9.122}$$

- When $\theta = -\pi/2$, Eqs. (9.105) and (9.107) get simplified as

$$\frac{du_1}{dz} = \kappa u_1 u_2\ ;\ \frac{du_2}{dz} = -\kappa u_1^2 \tag{9.123}$$

$$\frac{du_1}{u_1} = \kappa u_2 dz \tag{9.124}$$

$$\int_0^z \frac{du_1}{u_1} = \kappa u_2 \int_0^z dz \tag{9.125}$$

$$\left[\ln u_1\right]_0^z = \kappa u_2 z \tag{9.126}$$

$$\ln\frac{u_1(z)}{u_1(0)} = \kappa u_2 z \tag{9.127}$$

$$e^{\ln\frac{u_1(z)}{u_1(0)}} = e^{\kappa u_2 z} \tag{9.128}$$

$$u_1(z) = u_1(0)e^{\kappa u_2 z} \tag{9.129}$$

This equation reveals that although the power of the fundamental light possibly increases with the crystal length of z, this process does not exist because there is no incident field of the second harmonic light.

- When $\theta = +\pi/2$, Eqs. (9.105) and (9.107) get simplified as

$$\frac{du_1}{dz} = -\kappa u_1 u_2 \; ; \; \frac{du_2}{dz} = \kappa u_1^2 \tag{9.130}$$

$$\frac{d}{dz}\left(u_1^2 + u_2^2\right) = 2u_1 \frac{du_1}{dz} + 2u_2 \frac{du_2}{dz} \tag{9.131}$$

$$\frac{d}{dz}\left(u_1^2 + u_2^2\right) = -2u_1 \kappa u_1 u_2 + 2u_2 \kappa u_1^2 \tag{9.132}$$

$$\frac{d}{dz}\left(u_1^2 + u_2^2\right) = 0 \tag{9.133}$$

This equation satisfies the energy conservation, implying that

$$u_1^2(z) + u_2^2(z) = u_1^2(0) = u_{10}^2 \tag{9.134}$$

$$u_1^2 = u_{10}^2 - u_2^2 \tag{9.135}$$

$$\frac{du_2}{dz} = \kappa u_1^2 = \kappa\left(u_{10}^2 - u_2^2\right) \tag{9.136}$$

$$\int_0^{u_2} \frac{du_2}{u_{10}^2 - u_2^2} = \int_0^z \kappa\, dz \tag{9.137}$$

$$\frac{1}{u_{10}^2} \int_0^{u_2} \frac{du_2}{1 - \frac{u_2^2}{u_{10}^2}} = \kappa z \tag{9.138}$$

$$\frac{u_2}{u_{10}} = x \Rightarrow du_2 = u_{10} dx \tag{9.139}$$

$$\int_0^x \frac{dx}{1 - x^2} = \kappa z u_{10} \tag{9.140}$$

$$\frac{1}{2} \int_0^x \left[\frac{1}{1+x} + \frac{1}{1-x}\right] dx = \kappa z u_{10} \tag{9.141}$$

$$\frac{1}{2} \ln \frac{1+x}{1-x} = \kappa z u_{10} \tag{9.142}$$

$$\ln\frac{1+x}{1-x} = 2\kappa\, zu_{10} \tag{9.143}$$

$$\frac{1+x}{1-x} = e^{2\kappa\, zu_{10}} \tag{9.144}$$

$$1+x = e^{2\kappa\, zu_{10}}(1-x) \tag{9.145}$$

$$1+x = e^{2\kappa\, zu_{10}} - e^{2\kappa\, zu_{10}}x \tag{9.146}$$

$$x(1+e^{2\kappa\, zu_{10}}) = e^{2\kappa\, zu_{10}} - 1 \tag{9.147}$$

$$x = \frac{e^{2\kappa\, zu_{10}} - 1}{e^{2\kappa\, zu_{10}} + 1} \tag{9.148}$$

Using algebra calculation, the final solution is written as

$$\frac{e^{2\kappa\, zu_{10}} - 1}{e^{2\kappa\, zu_{10}} + 1} = \tanh(u_{10}\kappa z) \tag{9.149}$$

$$\frac{u_2}{u_{10}} = x \Rightarrow u_2(z) = u_{10}\tanh(u_{10}\kappa z) \tag{9.150}$$

$$u_1^2 = u_{10}^2 - u_2^2 = u_{10}^2\left[1-\tanh^2(u_{10}\kappa z)\right] \tag{9.151}$$

$$u_1^2 = u_{10}^2 - u_2^2 = u_{10}^2\,\text{sech}^2(u_{10}\kappa z) \tag{9.152}$$

The power of fundamental light as a function of the optic axis is finally written as

$$P_1(z) = \frac{n_1}{2c\mu_0}\left|E_1(z)\right|^2 S \tag{9.153}$$

$$P_1(z) = \frac{n_1}{2c\mu_0}{u_1}^2 S \tag{9.154}$$

$$P_1(z) = \frac{n_1}{2c\mu_0}u_{10}^2\,\text{sech}^2(u_{10}\kappa z)S \tag{9.155}$$

The power of second harmonic light as a function of the optic axis is finally written as

$$P_2(z) = \frac{n_2}{2c\mu_0}\left|E_2(z)\right|^2 S \tag{9.156}$$

$$P_2(z) = \frac{n_2}{2c\mu_0} u_2^2 S \tag{9.157}$$

$$P_2(z) = \frac{n_2}{2c\mu_0} u_{10}^2 \tanh^2(u_{10}\kappa z)S \Rightarrow \frac{u_2}{u_{10}} = x \tag{9.158}$$

The SHG goes to the maximum value and then saturates according to Eq. (9.158) (Fig. 9.5). This means that as the SHG is gaining the maximum energy out of the fundamental wave, there is no way that it can change its amplitude because the total energy is conserved. On the other hand, all energy from the fundamental wave is now converted to the second harmonic wave.

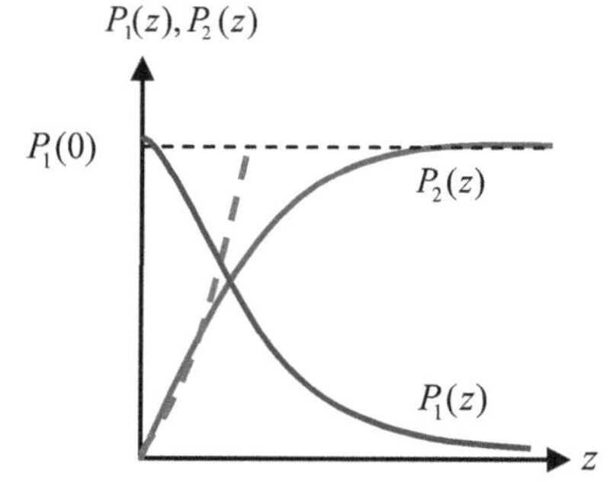

Figure 9.5 Evolution of the fundamental and second harmonic amplitudes with respect to the length of interaction z without pump depletion approximation (dash line) and with pump depletion as a realistic approach (solid line). $P_1(0)$ is the actual maximum power.

9.4 Implementations

9.4.1 Birefringence Phase Matching

There are two methods in order to achieve the maximum efficiency of the second harmonic power according to the phase matching condition (Fig. 9.6):

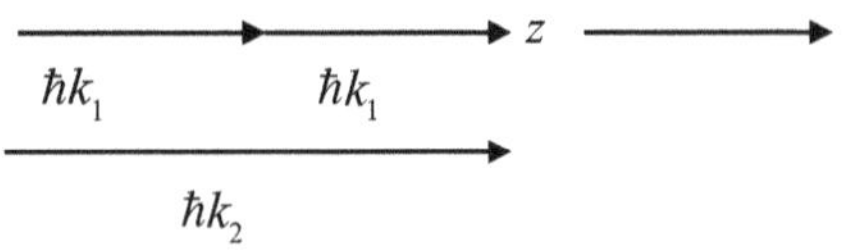

Figure 9.6 Vector diagram of momentum conservation: $\Delta k = 0 \Rightarrow k_2 = 2k_1$.

1. Birefringence phase matching (BPM)
2. Quasi-phase matching (QPM)

For BPM, we use the orientation property of the "anisotropy" of a crystal. We use the different refractive indices (n_o and n_e) of this nonlinear optical medium to acquire a phase matching between the two waves: ordinary and extraordinary waves.

For instance, the SHG of 500 nm is produced by a 1000 nm input light out of a negative uniaxial-KDP (KH_2PO_4) crystal ($n_o > n_e$) composed of

$$n_0(\omega) = 1.5050, n_e(\omega) = 1.4653$$

$$n_0(2\omega) = 1.5327, n_e(2\omega) = 1.4971$$

The SHG of 400 nm is produced by an 800 nm input light out of a negative uniaxial-KDP (KH_2PO_4) crystal ($n_o > n_e$) composed of

$$n_0(\omega) = 1.5015, n_e(\omega) = 1.4633$$

$$n_0(2\omega) = 1.5238, n_e(2\omega) = 1.4797$$

we have $\psi_m = 50.50°$ and 44.69°, respectively, due to different input wavelengths.

The ordinary and extraordinary refractive index surfaces are related to the ordinary and extraordinary waves at the fundamental and double frequencies as shown in Fig. 9.7.

$$n_0(2\omega) > n_0(\omega), \qquad n_e(2\omega) > n_e(\omega)$$

The SHG processes upon angle tuning method to satisfy the BPM and the finite effective nonlinear polarization not to be zero so that a pair of wavelengths at which the amplifier will possibly work are as follows:

- In a negative uniaxial crystal, we launch the ordinary wave at the fundamental frequency and convert to the extraordinary wave of the double frequency in the birefringence crystal with specified *d*-tensor, $n_0(\omega) = n_e(2\omega, \psi_m)$. We should have a phase matching angle as

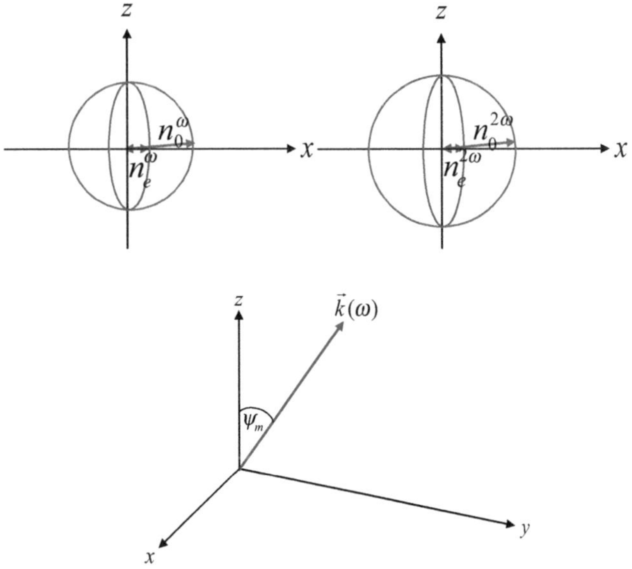

Figure 9.7 Refractive index surfaces of a negative uniaxial-KDP (KH_2PO_4) crystal corresponding to the fundamental frequency (top left) and the double frequency (top right). The maximum conversion efficiency of the second harmonic power exists at a certain angle of ψ_m = 50.5° with the optic axis OA (bottom).

$$\frac{1}{n_e^2(2\omega,\psi_m)} = \frac{\cos^2\psi_m}{n_0^2(2\omega)} + \frac{\sin^2\psi_m}{n_e^2(2\omega)} = \frac{1}{n_0^2(\omega)} \tag{9.159}$$

$$\frac{1}{n_0^2(\omega)} = \frac{1-\sin^2\psi_m}{n_0^2(2\omega)} + \frac{\sin^2\psi_m}{n_e^2(2\omega)} \tag{9.160}$$

$$\frac{1}{n_0^2(\omega)} = \frac{1}{n_0^2(2\omega)} + \frac{\sin^2\psi_m}{n_e^2(2\omega)} - \frac{\sin^2\psi_m}{n_0^2(2\omega)} \tag{9.161}$$

$$\frac{1}{n_0^2(\omega)} - \frac{1}{n_0^2(2\omega)} = \sin^2\psi_m\left[\frac{1}{n_e^2(2\omega)} - \frac{1}{n_0^2(2\omega)}\right] \tag{9.162}$$

$$\sin^2\psi_m = \frac{\left[\frac{1}{n_0^2(\omega)} - \frac{1}{n_0^2(2\omega)}\right]}{\left[\frac{1}{n_e^2(2\omega)} - \frac{1}{n_o^2(2\omega)}\right]} \tag{9.163}$$

Under a noncritical phase matching at $\psi_m = 90°$ as compared to the optic axis, $\vec{S}$ of extraordinary wave is parallel to $\bar{k}$ of the extraordinary wave. In contrast, under a critical phase matching at a certain angle, since $\vec{S}$ of the extraordinary wave is not parallel to $\bar{k}$ of the extraordinary wave, the energy of the SHG is not propagating at that angle; it is propagating at the tilt angle. The ordinary wave at the fundamental frequency goes along the same direction; however, the extraordinary wave at the double frequency is deviated from the main beam. This is called "beam walk-off," as shown in Fig. 9.8.

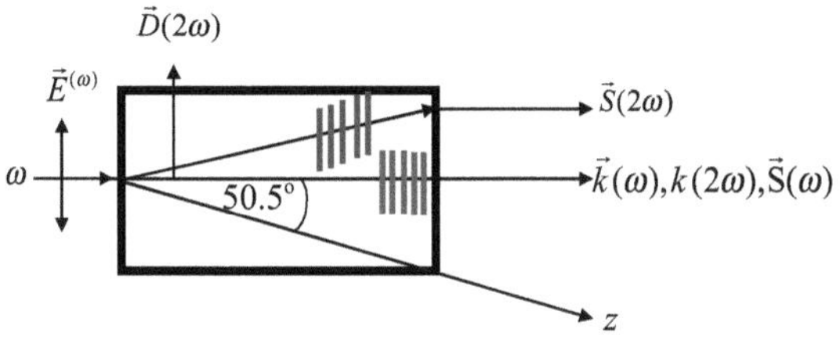

Figure 9.8 Representative beam walk-off from the horizontal polarized light into the KDP (KH_2PO_4) crystal when the ordinary wave is perpendicular to the *k*-vector.

- In a positive uniaxial crystal, we launch the extraordinary wave at the fundamental frequency to generate the ordinary wave of the double frequency via birefringence crystal, $n_e(\omega, \psi_m) = n_0(2\omega)$. We should have a phase matching angle as

$$\frac{1}{n_e^2(\omega,\psi_m)} = \frac{\cos^2\psi_m}{n_0^2(\omega)} + \frac{\sin^2\psi_m}{n_e^2(\omega)} = \frac{1}{n_0^2(2\omega)} \tag{9.164}$$

$$\sin^2\psi_m = \frac{\left[\frac{1}{n_0^2(2\omega)} - \frac{1}{n_0^2(\omega)}\right]}{\left[\frac{1}{n_e^2(\omega)} - \frac{1}{n_0^2(\omega)}\right]} \tag{9.165}$$

9.4.2 Effective Nonlinear Polarization

In order to deeply understand the orientation effect of the electric field on the SHG under a certain phase matching angle, we start a relationship between the total nonlinear polarization and the total complex electric fields at two frequencies and two propagation constants under the second-order effect as

$$P^{\mathrm{NL}} = 2\varepsilon_0 d E^2 \tag{9.166}$$

$$P_i^{\mathrm{NL}} = 2\varepsilon_0 d_{ijk} E_j E_k \tag{9.167}$$

$$E^{(\omega)} = \frac{1}{2}[E_1(z)e^{i(k_1 z-\omega t)} + cc] \tag{9.168}$$

$$E^{(2\omega)} = \frac{1}{2}[E_2(z)e^{i(k_2 z-2\omega t)} + cc] \tag{9.169}$$

$$E_j = \frac{1}{2}[E_j^{(\omega)}e^{i(k_1 z-\omega t)} + cc] + \frac{1}{2}[E_j^{(2\omega)}e^{i(k_2 z-2\omega t)} + cc] \tag{9.170}$$

$$E_k = \frac{1}{2}[E_k^{(\omega)}e^{i(k_1 z-\omega t)} + cc] + \frac{1}{2}[E_k^{(2\omega)}e^{i(k_2 z-2\omega t)} + cc] \tag{9.171}$$

Substituting these into the nonlinear polarization, we get

$$P_i^{(2\omega)} = 2\varepsilon_0 d_{ijk} \frac{1}{4}\left[E_j^{(\omega)}E_k^{(\omega)}e^{2i(k_1 z-\omega t)} + cc\right] \tag{9.172}$$

$$P_i^{(2\omega)} = \frac{1}{2}\left[\varepsilon_0 d_{ijk} E_j^{(\omega)}E_k^{(\omega)}e^{2i(k_1 z-\omega t)} + cc\right] \tag{9.173}$$

$$P_i^{(2\omega)} = \frac{1}{2}\left[\tilde{P}_i^{(2\omega)}e^{2i(k_1 z-\omega t)} + cc\right] \tag{9.174}$$

Table 9.1 Standard convection for tensor contraction

The last two indices as *jk*	Represented by
11	1
22	2
33	3
23, 32	4
13, 31	5
12, 21	6

The effective nonlinear polarization is defined as

$$\tilde{P}_i^{(2\omega)} = \varepsilon_0 d_{ijk} E_j^{(\omega)} E_k^{(\omega)} \tag{9.175}$$

where d is the nonlinear coefficient (d-tensor) representing the symmetry properties of any anisotropic crystals.

A standard convection is used to destroy the d-tensor, as shown in Table 9.1.

where

$$\begin{pmatrix} 11 & 12 & 13 \\ 21 & 22 & 23 \\ 31 & 32 & 33 \end{pmatrix} = \begin{pmatrix} 1 & 6 & 5 \\ 6 & 2 & 4 \\ 5 & 4 & 3 \end{pmatrix} \tag{9.176}$$

Due to the intrinsic permutation theory $d_{\text{ijk}} = d_{\text{ikj}}$, we have the d-tensors of 27 − 9 = 18 elements in the matrix form as

$$d = \begin{pmatrix} d_{11} & d_{12} & d_{13} & d_{14} & d_{15} & d_{16} \\ d_{21} & d_{22} & d_{23} & d_{24} & d_{25} & d_{26} \\ d_{31} & d_{32} & d_{33} & d_{34} & d_{35} & d_{36} \end{pmatrix} \tag{9.177}$$

For example, all the tensor elements belonging to a class of KDP and lithium niobate crystals are in the following order:

$$d = \begin{pmatrix} 0 & 0 & 0 & d_{14} & 0 & 0 \\ 0 & 0 & 0 & 0 & d_{25} & 0 \\ 0 & 0 & 0 & 0 & 0 & d_{36} \end{pmatrix} \tag{9.178}$$

$$d = \begin{pmatrix} 0 & 0 & 0 & 0 & d_{15} & -d_{22} \\ -d_{22} & d_{22} & 0 & d_{15} & 0 & 0 \\ d_{31} & d_{31} & d_{33} & 0 & 0 & 0 \end{pmatrix} \tag{9.179}$$

Using the contraction indices in Eq. (9.125), we get

$$\begin{aligned} \tilde{P}_x^{(2\omega)} &= \varepsilon_0 d_{xxx} E_x^{(\omega)^2} + \varepsilon_0 d_{xyy} E_y^{(\omega)^2} + \varepsilon_0 d_{xzz} E_z^{(\omega)^2} + 2\varepsilon_0 d_{xyz} E_y^{(\omega)} E_z^{(\omega)} \\ &+ 2\varepsilon_0 d_{xxz} E_x^{(\omega)} E_z^{(\omega)} + 2\varepsilon_0 d_{xxy} E_x^{(\omega)} E_y^{(\omega)} \end{aligned} \tag{9.180}$$

Three components of the nonlinear polarizations related to the complex electric field components at the fundamental frequency are, therefore, written as

$$\begin{pmatrix} \tilde{P}_x^{(2\omega)} \\ \tilde{P}_y^{(2\omega)} \\ \tilde{P}_z^{(2\omega)} \end{pmatrix} = \varepsilon_0 \begin{pmatrix} d_{11} & d_{12} & d_{13} & d_{14} & d_{15} & d_{16} \\ d_{21} & d_{22} & d_{23} & d_{24} & d_{25} & d_{26} \\ d_{31} & d_{32} & d_{33} & d_{34} & d_{35} & d_{36} \end{pmatrix} \begin{pmatrix} E_x^{(\omega)^2} \\ E_y^{(\omega)^2} \\ E_z^{(\omega)^2} \\ 2E_y^{(\omega)}E_z^{(\omega)} \\ 2E_x^{(\omega)}E_z^{(\omega)} \\ 2E_x^{(\omega)}E_y^{(\omega)} \end{pmatrix} \tag{9.181}$$

Suppose that we select one orientation of the electric field direction into the KDP crystal, as shown in Fig. 9.9. The "ordinary polarization (y-direction)" is perpendicular to the optic axis (z) and to the direction of wave propagation, $\vec{k}$.

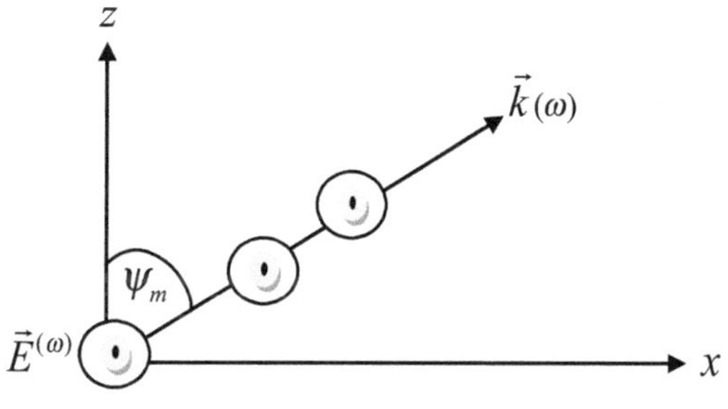

Figure 9.9 Ordinary polarization in the y-direction is perpendicular to the optic axis (z) and to $\vec{k}$.

Components of electric fields and generation of nonlinear polarization are written as

$$E_x^{(\omega)} = E_z^{(\omega)} = 0$$

$$E_y^{(\omega)} \neq 0$$

$$\begin{pmatrix} \tilde{P}_x^{(2\omega)} \\ \tilde{P}_y^{(2\omega)} \\ \tilde{P}_z^{(2\omega)} \end{pmatrix} = \varepsilon_0 \begin{pmatrix} 0 & 0 & 0 & d_{14} & 0 & 0 \\ 0 & 0 & 0 & 0 & d_{25} & 0 \\ 0 & 0 & 0 & 0 & 0 & d_{36} \end{pmatrix} \begin{pmatrix} E_x^{(\omega)^2} \\ E_y^{(\omega)^2} \\ E_z^{(\omega)^2} \\ 2E_y^{(\omega)}E_z^{(\omega)} \\ 2E_x^{(\omega)}E_z^{(\omega)} \\ 2E_x^{(\omega)}E_y^{(\omega)} \end{pmatrix} \qquad (9.182)$$

We get

$$\begin{pmatrix} \tilde{P}_x^{(2\omega)} \\ \tilde{P}_y^{(2\omega)} \\ \tilde{P}_z^{(2\omega)} \end{pmatrix} = \varepsilon_0 \begin{pmatrix} 0 & 0 & 0 & d_{14} & 0 & 0 \\ 0 & 0 & 0 & 0 & d_{25} & 0 \\ 0 & 0 & 0 & 0 & 0 & d_{36} \end{pmatrix} \begin{pmatrix} 0 \\ E_y^{(\omega)^2} \\ 0 \\ 0 \\ 0 \\ 0 \end{pmatrix} = \begin{pmatrix} 0 \\ 0 \\ 0 \end{pmatrix} \qquad (9.183)$$

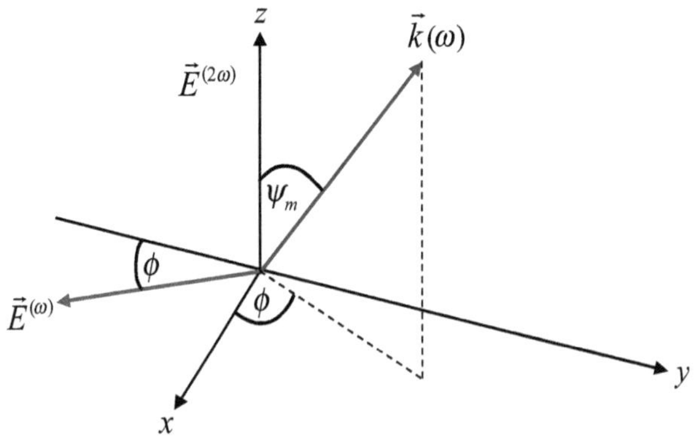

Figure 9.10 Arbitrary direction of wave propagation with the phase matching angle ψ_m of the KDP crystal. The ordinary wave in the *xy* plane at the fundamental frequency generates the extraordinary second harmonic wave at the double frequency. The high power of the 400 nm output extraordinary wave and the low power of the 800 nm input ordinary wave get amplified or attenuated depending on the phase difference.

Since all nonlinear polarization elements are zero, there is no SHG of nonlinear polarization although the phase matching is correct. However, let us modify the orientation of the electric field direction to some other arbitrary plane, as shown in Fig. 9.10. The projection of k-vector on the xy plane makes the angle ϕ with the x-axis. Then we get the electric field at the xy plane because the "ordinary wave" is always perpendicular to the optic axis z and $\vec{k}$. The components of electric fields and nonlinear polarization are written as

$$E_x^{(\omega)} = E_1 \sin\phi$$

$$E_y^{(\omega)} = -E_1 \cos\phi$$

$$E_z^{(\omega)} = 0$$

$$\begin{pmatrix} \tilde{P}_x^{(2\omega)} \\ \tilde{P}_y^{(2\omega)} \\ \tilde{P}_z^{(2\omega)} \end{pmatrix} = \varepsilon_0 \begin{pmatrix} 0 & 0 & 0 & d_{14} & 0 & 0 \\ 0 & 0 & 0 & 0 & d_{25} & 0 \\ 0 & 0 & 0 & 0 & 0 & d_{36} \end{pmatrix} \begin{pmatrix} E_x^{(\omega)^2} \\ E_y^{(\omega)^2} \\ E_z^{(\omega)^2} \\ 2E_y^{(\omega)}E_z^{(\omega)} \\ 2E_x^{(\omega)}E_z^{(\omega)} \\ 2E_x^{(\omega)}E_y^{(\omega)} \end{pmatrix} \quad (9.184)$$

We get

$$\begin{pmatrix} \tilde{P}_x^{(2\omega)} \\ \tilde{P}_y^{(2\omega)} \\ \tilde{P}_z^{(2\omega)} \end{pmatrix} = \varepsilon_0 \begin{pmatrix} 0 & 0 & 0 & d_{14} & 0 & 0 \\ 0 & 0 & 0 & 0 & d_{25} & 0 \\ 0 & 0 & 0 & 0 & 0 & d_{36} \end{pmatrix} \begin{pmatrix} E_1^2 \sin^2\phi \\ E_1^2 \cos^2\phi \\ 0 \\ 0 \\ 0 \\ -2E_1^2 \sin\phi\cos\phi \end{pmatrix} \quad (9.185)$$

$$\begin{pmatrix} \tilde{P}_x^{(2\omega)} \\ \tilde{P}_y^{(2\omega)} \\ \tilde{P}_z^{(2\omega)} \end{pmatrix} = \varepsilon_0 \begin{pmatrix} 0 \\ 0 \\ -2d_{36}E_1^2 \sin\phi\cos\phi \end{pmatrix} \quad (9.186)$$

Since the optimum value of ϕ is $\phi = \pi/4$ with the xy plane, the effective nonlinear polarization of the SHG under a certain phase matching angle is written as

$$\tilde{P}_z^{(2\omega)} = -\varepsilon_0 d_{36} E_1^2 \sin 2\phi = -\varepsilon_0 d_{36} E_1^2 \qquad (9.187)$$

Suppose that we launch the ordinary polarization at the fundamental frequency along the optic axis of the lithium niobate crystal with d-tensors where $d_{21} = -d_{22}$; $d_{32} = d_{31}$; $d_{24} = d_{15}$; $d_{16} = -d_{22}$. The components of electric fields and nonlinear polarization are written as (Fig. 9.11)

$$E_x^{(\omega)} = E_y^{(\omega)} = 0$$

$$E_z^{(\omega)} = E_1$$

$$\begin{pmatrix} \tilde{P}_x^{(2\omega)} \\ \tilde{P}_y^{(2\omega)} \\ \tilde{P}_z^{(2\omega)} \end{pmatrix} = \varepsilon_0 \begin{pmatrix} 0 & 0 & 0 & 0 & d_{15} & -d_{22} \\ -d_{22} & d_{22} & 0 & d_{15} & 0 & 0 \\ d_{31} & d_{31} & d_{33} & 0 & 0 & 0 \end{pmatrix} \begin{pmatrix} 0 \\ 0 \\ E_1^2 \\ 0 \\ 0 \\ 0 \end{pmatrix} \qquad (9.188)$$

$$\tilde{P}_x^{(2\omega)} = \tilde{P}_y^{(2\omega)} = 0 \qquad (9.189)$$

$$\tilde{P}_z^{(2\omega)} = \varepsilon_0 d_{33} E_1^2 \qquad (9.190)$$

Although the effective nonlinear polarization is strong under this orientation, we are unable to reach the phase matching $(n_0(\omega) \neq n_e(2\omega))$ generated by the different velocities of these two waves.

Next we suppose the orientation of the electric field direction at the fundamental frequency to some arbitrary plane in the lithium niobate crystal. The electric field propagating along the x-axis makes some angle θ with the optic axis in the yz plane. The components of electric fields and nonlinear polarization are written as

$$E_x^{(\omega)} = 0$$

$$E_y^{(\omega)} = E_1 \sin\theta$$

$$E_z^{(\omega)} = E_1 \cos\theta$$

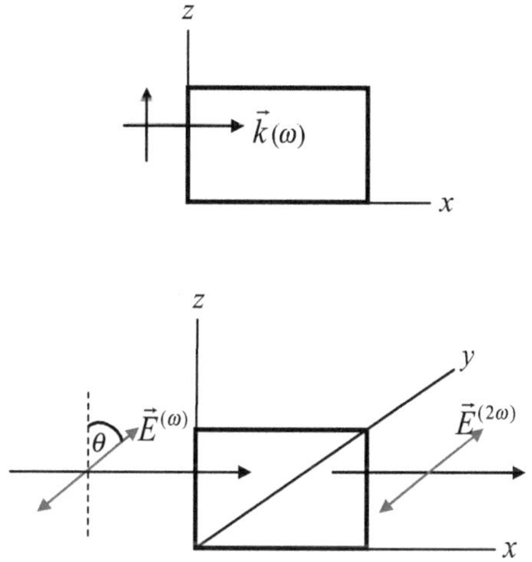

Figure 9.11 (Top) Electric field in the z-direction propagates along the optic axis of the lithium niobate crystal and perpendicular to $\vec{k}$. (Bottom) Electric field in the *yz* plane propagates along the *x*-axis of the lithium niobate crystal with an angle of θ to the optic axis.

$$\begin{pmatrix} \tilde{P}_x^{(2\omega)} \\ \tilde{P}_y^{(2\omega)} \\ \tilde{P}_z^{(2\omega)} \end{pmatrix} = \varepsilon_0 \begin{pmatrix} 0 & 0 & 0 & 0 & d_{15} & -d_{22} \\ -d_{22} & d_{22} & 0 & d_{15} & 0 & 0 \\ d_{31} & d_{31} & d_{33} & 0 & 0 & 0 \end{pmatrix} \begin{pmatrix} E_x^{(\omega)^2} \\ E_y^{(\omega)^2} \\ E_z^{(\omega)^2} \\ 2E_y^{(\omega)}E_z^{(\omega)} \\ 2E_x^{(\omega)}E_z^{(\omega)} \\ 2E_x^{(\omega)}E_y^{(\omega)} \end{pmatrix} \tag{9.191}$$

$$\begin{pmatrix} \tilde{P}_x^{(2\omega)} \\ \tilde{P}_y^{(2\omega)} \\ \tilde{P}_z^{(2\omega)} \end{pmatrix} = \varepsilon_0 \begin{pmatrix} 0 & 0 & 0 & 0 & d_{15} & -d_{22} \\ -d_{22} & d_{22} & 0 & d_{15} & 0 & 0 \\ d_{31} & d_{31} & d_{33} & 0 & 0 & 0 \end{pmatrix} \begin{pmatrix} 0 \\ E_1^2 \sin^2\theta \\ E_1^2 \cos^2\theta \\ 2E_1^2 \sin\theta\cos\theta \\ 0 \\ 0 \end{pmatrix} \tag{9.192}$$

$$\begin{pmatrix} \tilde{P}_x^{(2\omega)} \\ \tilde{P}_y^{(2\omega)} \\ \tilde{P}_z^{(2\omega)} \end{pmatrix} = \varepsilon_0 \begin{pmatrix} 0 \\ d_{22}E_1^2 \sin^2\theta + 2d_{15}E_1^2 \sin\theta\cos\theta \\ d_{31}E_1^2 \sin^2\theta + d_{33}E_1^2 \cos^2\theta \end{pmatrix} \quad (9.193)$$

Effective nonlinear polarizations are still unable to generate the maximum efficiency of second harmonic of the extraordinary wave because the BPM condition is unable to merge the ordinary photon and the extraordinary photon at the fundamental frequencies to form the extraordinary photon at the double frequency (Fig. 9.12). Therefore, the QPM condition is introduced to overcome the problem.

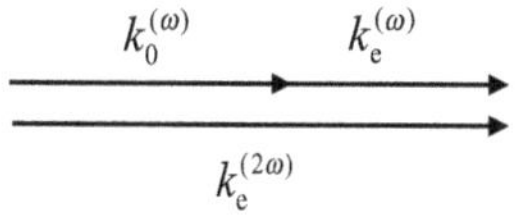

Figure 9.12 Photon picture of the lithium niobite crystal. $k_0^{(\omega)}$ is bigger than $k_e^{(\omega)}$ because $n_0 > n_e$.

9.4.3 Quasi-Phase Matching

Previously there was one technique used to achieve phase matching, the so-called birefringence phase matching in which we use the "anisotropy" of the crystal: one is the ordinary wave and another is the extraordinary wave. We use the n_0 and n_e of this medium to achieve phase matching between these two waves. However, there are so many crystals, such as GaAs, GaN, etc., that they are the cubic crystals in which are "isotropic" but are nonlinear crystals. Therefore, another technique, the so-called QPM, has been introduced to achieve the phase matching condition. According to the second harmonic power efficiency as shown in Fig. 9.13, it usually goes up at a certain point $L_c = \pi/\Delta k$ and down periodically, owing to a sine function as

$$\eta = \frac{P_2(z)}{P_1} = \frac{2\mu_0\omega^2}{cn_1^2 n_2}\frac{d^2 P_1}{S}\frac{\sin^2(\Delta kz/2)}{(\Delta k/2)^2} \quad (9.194)$$

BPM is a process to make $\Delta k = 0$; however, QPM is another method to make $\Delta k = K$ through the d-tensor as a periodic function of the optic axis. Assume that there is no pump depletion (E_1 as constant) and we produce the medium in which d is periodically varying with z as

$$d(z) = d_0 \sin Kz \tag{9.195}$$

If we have multiple periodicities in a function, we can have multiple spatial frequencies, K, simultaneously as

$$K = \frac{2\pi}{\lambda} \tag{9.196}$$

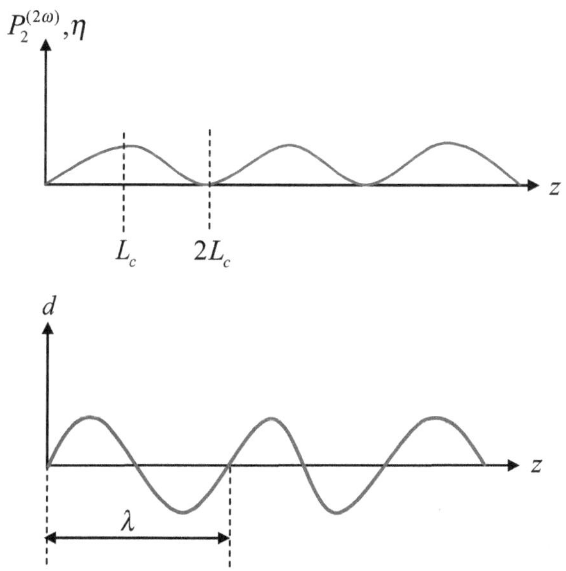

Figure 9.13 The second harmonic power efficiency as it goes up and down periodically (top) and distance is lambda periodic function of optic axis z (bottom).

For QPM, we assume that

$$K = \Delta k = \frac{2\pi}{\lambda} \tag{9.197}$$

$$L_c = \frac{\pi}{\Delta k} \Rightarrow \Delta k = \frac{\pi}{L_c} \Rightarrow \lambda = 2L_c \tag{9.198}$$

$$L_c = \frac{\pi}{\Delta k} = \frac{\pi}{\frac{2\omega}{c}(n_2 - n_1)} = \frac{\pi c}{2\omega(n_2 - n_1)} \tag{9.199}$$

Suppose that we launch the input power of 1 W at the fundamental wavelength of 1060 nm, 1 mm ϕ beam into the $LiNbO_3$ crystal under n_1= 2.1561, n_2 =2.2355, and $d = 30 \times 10^{-12}$ m/V, and the coherence length is 3.3 μm. The maximum SHG exists at 3.3 μm, and it becomes zero again at 6.6 μm without phase matching condition.

The power efficiency of the SHG at the crystal length is calculated as follows:

$$\frac{dE_2}{dz} = i\frac{\omega d}{cn_2}E_1^2 e^{-i\Delta kz} \tag{9.200}$$

$$\frac{dE_2}{dz} = i\frac{\omega}{cn_2}E_1^2\left(d_0 \sin Kz\right)e^{-ik\Delta z} \tag{9.201}$$

$$\frac{dE_2}{dz} = i\frac{\omega d_0}{cn_2}E_1^2\left(\frac{e^{iKz} - e^{-iKz}}{2i}\right)e^{-ik\Delta z} \tag{9.202}$$

$$\frac{dE_2}{dz} = \frac{\omega d_0}{2cn_2}E_1^2\left[e^{i(K-\Delta k)z} - e^{-i(K+\Delta k)z}\right] \tag{9.203}$$

$$\int dE_2 = \frac{\omega d_0}{2cn_2}E_1^2\int e^{i(K-\Delta k)z} - e^{-i(K+\Delta k)z}dz \tag{9.204}$$

$$E_2(z) = \frac{\omega d_0}{2cn_2}E_1^2\left[\frac{e^{i(K-\Delta k)z}}{i(K-\Delta k)}\right]_0^z - \left[\frac{e^{-i(K+\Delta k)z}}{-i(K+\Delta k)}\right]_0^z \tag{9.205}$$

$$E_2(z) = \frac{\omega d_0}{2cn_2}E_1^2\left[\frac{e^{i(K-\Delta k)z}}{i(K-\Delta k)}\right]_0^z - \left[\frac{e^{-i(K+\Delta k)z}}{-i(K+\Delta k)}\right]_0^z \tag{9.206}$$

$$E_2(z) = \frac{\omega(d_0/2)}{cn_2}E_1^2\left[\frac{e^{i(K-\Delta k)z} - 1}{i(K-\Delta k)} + \frac{e^{-i(K+\Delta k)z} - 1}{i(K+\Delta k)}\right] \tag{9.207}$$

$$E_2(z) = \frac{\omega(d_0/2)}{cn_2}E_1^2\left[\frac{e^{i(K-\Delta k)z/2}\left(e^{i(K-\Delta k)z/2} - e^{-i(K-\Delta k)z/2}\right)}{i(K-\Delta k)} + \frac{e^{-i(K-\Delta k)z/2}\left(e^{-i(K-\Delta k)z/2} - e^{i(K-\Delta k)z/2}\right)}{i(K+\Delta k)}\right] \tag{9.208}$$

$$E_2(z) = \frac{\omega(d_0/2)}{cn_2} E_1^2 \left[e^{i(K-\Delta k)z/2} \frac{2\sin(K-\Delta k)z/2}{(K-\Delta k)} + e^{-i(K+\Delta k)z/2} \frac{2\sin(K+\Delta k)z/2}{(K+\Delta k)} \right] \tag{9.209}$$

$$E_2(z) = \frac{\omega(d_0/2)}{cn_2} E_1^2 \left[e^{i(K-\Delta k)z/2} \frac{\sin(K-\Delta k)z/2}{(K-\Delta k)/2} \right] \tag{9.210}$$

$$P_1 = \frac{n_1}{2c\mu_0} |E_1|^2 S \Rightarrow |E_1|^4 = \frac{4c^2\mu_0^2 P_1^2}{n_1^2 S^2} \tag{9.211}$$

$$P_2(z) = \frac{n_2}{2c\mu_0} |E_2(z)|^2 S \tag{9.212}$$

$$P_2(z) = \frac{n_2}{2c\mu_0} \left| \frac{\omega(d_0/2)}{cn_2} E_1^2 \frac{\sin(K-\Delta k)z/2}{(K-\Delta k)/2} \right|^2 S \tag{9.213}$$

$$P_2(z) = \frac{n_2}{2c\mu_0} \left| \frac{\omega(d_0/2)}{cn_2} E_1^2 z \right|^2 S \tag{9.214}$$

$$P_2(z) = \frac{n_2}{2c\mu_0} \frac{\omega^2(d_0/2)^2}{c^2 n_2^2} |E_1|^4 z^2 S \tag{9.215}$$

$$P_2(z) = \frac{n_2}{2c\mu_0} \frac{\omega^2(d_0/2)^2}{c^2 n_2^2} \frac{4c^2\mu_0^2 P_1^2}{n_1^2 S^2} z^2 S \tag{9.216}$$

$$P_2(z) = \frac{\omega^2(d_0/2)^2}{cn_2} \frac{2\mu_0 P_1^2}{n_1^2 S} z^2 \tag{9.217}$$

$$\eta = \frac{P_2(z)}{P_1} = \frac{\omega^2(d_0/2)^2}{cn_2} \frac{2\mu_0 P_1}{n_1^2 S} z^2 \tag{9.218}$$

It is implied that when we periodically vary a nonlinear coefficient (d) with a period that is twice the coherence length z, as shown in Fig. 9.14, it seem to be adding the second harmonic power rather than periodically oscillating due to the following reasons:

- Since the nonlinear polarization is proportional to d, changing the sign of d will change the sign of the nonlinear polarization.

- When the waves propagate at a coherence length L_c and the nonlinear polarization and the electromagnetic field get pi out of phase, we try to force it back in phase by changing the sign of the nonlinear polarization and bring it back in phase.
- When the waves propagate through a coherence length $2L_c$, it generates another phase difference of pi and we bring it back in phase.
- Every time the nonlinear polarization and the electromagnetic wave get out of phase by pi, we force it back in phase by reversing the sign of d, which essentially changes the nonlinear polarization by a factor of -1.
- When we change the sign of d, we change the sign of the nonlinear polarization, meaning that we compensate the phase difference of pi to bring them back in phase.
- Considering the vector diagram of QPM, a periodic modulation of the nonlinearity (d) gives us a small $\vec{k}$ in the moment space (Fig. 9.15). QPM is, therefore, essentially a process in which we overcome the difference between k_1 and k_2 happening in BPM. In this process, the second harmonic is generated by a spatial frequency K. By orientating the medium appropriately, we can change the direction of the nonlinear coefficient term (d).

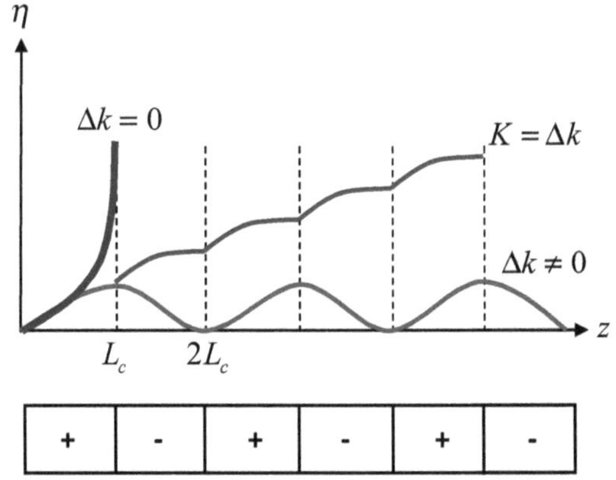

Figure 9.14 The conversion efficiency of the SHG is continuously added as a function of the periodically poled nonlinear crystal length *z*.

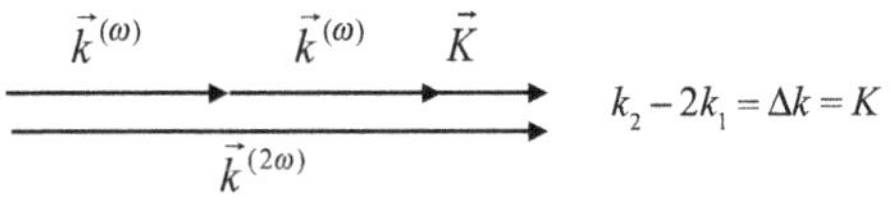

Figure 9.15 Momentum of the photon showing there is an extra small $\vec{k}$ due to the periodic variation of the nonlinear coefficient (*d*).

Exercise 9.3

Suppose that the SHG processes of the lithium niobate crystal satisfy the QPM condition and the finite effective nonlinear polarization is not zero so that a pair of wavelengths at which the amplifier will possibly work is as follows:

Wavelength of	Ordinary refractive index	Extraordinary refractive index
Pump = 0.6 μm	2.296	2.211
Signal = 1 μm	2.236	2.160
Idler = 1.5 μm	2.213	2.140

$\omega_p(\text{ord}) \rightarrow \omega_s(\text{ord}), \omega_i(\text{ord})$ requires $d_{111} = d_{11} = 0$— cannot take place

$$\omega_p(\text{ord}) \rightarrow \omega_s(\text{extraord}), \omega_i(\text{ord}) \text{ requires } d_{131} = d_{15}$$

$$\omega_p(\text{ord}) \rightarrow \omega_s(\text{ord}), \omega_i(\text{extraord}) \text{ requires } d_{113} = d_{15}$$

$\omega_p(\text{ord}) \rightarrow \omega_s(\text{extraord}), \omega_i(\text{extraord})$ requires $d_{133} = d_{13} = 0$— cannot take place

Then we consider for the two remaining processes to find out the required periodicity of lambda of the crystal as follows:

- Selecting $\omega_p(\text{o}) \rightarrow \omega_s(\text{e}), \omega_i(\text{o})$ and $k_p - k_s - k_i = K$, we require the phase matching condition as

$$\frac{2\pi}{\lambda_p} n_0(\lambda_p) - \frac{2\pi}{\lambda_s} n_e(\lambda_s) - \frac{2\pi}{\lambda_i} n_0(\lambda_i) = \frac{2\pi}{\lambda_1}$$

We make the grating periodicity of 5.23 μm and launch $\omega_p(\text{o})$ into the crystal. $\omega_p(\text{o})$ = 0.6 μm under *x*-polarization will split into amplified signal, $\omega_s(\text{e})$ = 1.0 μm under *z*-polarization and generated idler $\omega_i(\text{o})$ = 1.5 μm under *x*-polarization.

- Selecting $\omega_p(o) \rightarrow \omega_s(o), \omega_i(e)$ and $k_p - k_s - k_i = K$, we require the phase matching condition as

$$\frac{2\pi}{\lambda_p} n_0(\lambda_p) - \frac{2\pi}{\lambda_s} n_0(\lambda_s) - \frac{2\pi}{\lambda_i} n_e(\lambda_i) = \frac{2\pi}{\lambda_2}$$

We make a grating periodicity of 6.10 μm and launch $\omega_p(o)$ into the crystal. $\omega_p(o)$ = 0.6 μm under x-polarization will split into $\omega_s(o)$ = 1.0 μm under x-polarization and $\omega_i(e)$ = 1.5 μm under z-polarization.

9.4.4 Spontaneous Polarization

Lithium niobate, as a polarized media, has a spontaneous polarization (Ps) at room temperature, as shown in Fig. 9.16. If we heat this crystal above the Curie temperature, it always loses its spontaneous polarization. Below the Curie temperature, we need to orient all the domains of the crystals. By applying electric fields, these crystals get polarized once again. Since doing lithium niobate with every 3.3 μm for the BPM is a difficult task, QPM is an alternative option with every bigger than 3.3 μm.

The spontaneous polarization, as shown in Fig. 9.17, is connected to the coefficient of d-tensor and the crystal axis. To reverse the sign of nonlinear polarization, we need to push the lithium across the plane below.

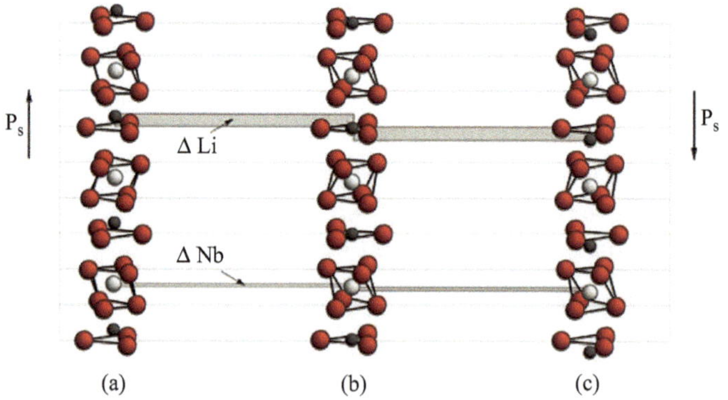

Figure 9.16 Unlike symmetry (b), asymmetry (a and c) in the positioning of lithium atoms showing small circles and medium circles of niobium atoms as well as three large circles of oxygen plane.

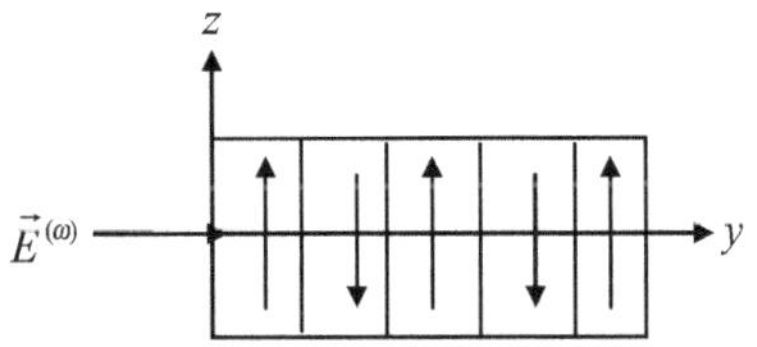

Figure 9.17 Fundamental frequency of input wave into the crystal having the regions with the polarization domain pointing up and down periodically under quasi-phase-match grating. (1) Lithium niobate slap needs every 3.3 μm so that a sign of *d* should be changed, and this lambda periodicity is then suggested to be 6.6 μm. (2) Sticking slap of GaAs every 105 μm is needed and this lambda periodicity is then suggested to be 210 μm.

The spontaneous polarization of the medium is pointing up (down), representing the positive (negative) d-vector component. If we select this period to be lambda, we have a periodic nonlinear variation in the nonlinear coefficient (d) (instead of having sinusoidal variation in coefficient along the propagation direction). We can have a crystal with d-positive and d-negative (by changing the orientation to the crystal axis) in which

$$d(z) = +d \Rightarrow |z| < 0.5l \tag{9.219}$$

$$d(z) = -d \Rightarrow 0.5l < |z| < 0.5\lambda \tag{9.220}$$

Duty cycle is defined as

$$D = l/\lambda \tag{9.221}$$

Since $d(z)$ is a periodic function, we normally make a simple exponential Fourier series expansion of d as a function of z as (Fig. 9.18)

$$d(z+\lambda) = d(z) \tag{9.222}$$

$$d(z) = d_0 \sum_{m=-\infty}^{+\infty} G_m e^{imKz} \tag{9.223}$$

$$d(z+\lambda) = d_0 \sum_{m=-\infty}^{+\infty} G_m e^{imK(z+\lambda)} \tag{9.224}$$

$$d(z+\lambda) = d_0 \sum_{m=-\infty}^{+\infty} G_m e^{imKz} e^{imK\lambda} \tag{9.225}$$

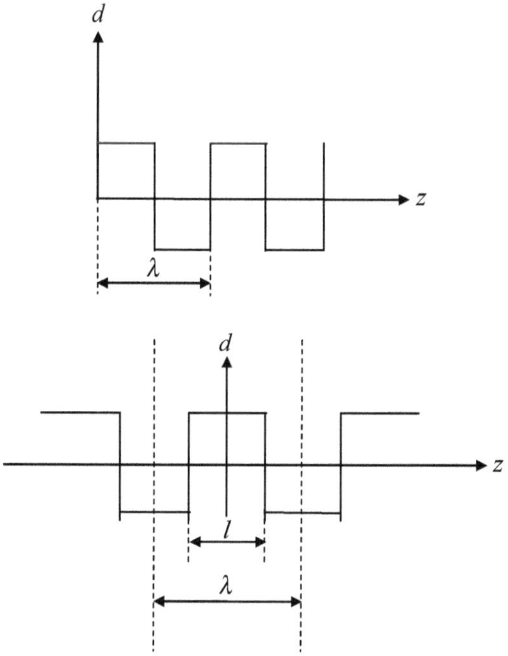

Figure 9.18 The d-tensor varies periodically as a function of the crystal length z (top) and Fourier series of the d-tensor as a function of z in order to cover all spatial frequency components (bottom).

$$d(z+\lambda)=d_0\sum_{m=-\infty}^{+\infty}G_m e^{imKz}e^{im\cdot 2\pi} \tag{9.226}$$

$$d(z+\lambda)=d_0\sum_{m=-\infty}^{+\infty}G_m e^{imKz}=d(z) \tag{9.227}$$

Fourier coefficient as a function of m for the QPM is (Fig. 9.18)

$$G_m=\frac{2}{m\pi}\sin\left(\frac{m\pi l}{\lambda}\right) \tag{9.228}$$

The Fourier term in the expansion is

$$\begin{aligned} d(z)=d_0\Big[&G_0+G_1e^{iKz}+G_{-1}e^{-iKz}+G_2e^{2iKz}+G_{-2}e^{-2iKz}\\ &+G_3e^{3iKz}+G_{-3}e^{-3iKz}+\cdots\Big] \end{aligned} \tag{9.229}$$

Taking a duty cycle of half (50% duty cycle) as l = 0.5λ, we suppress all even terms in the expansion. We have only odd terms in the expansion as

the first-order QPM as $G_1 = \frac{2}{\pi}$

the second-order QPM as $G_2 = 0$

the third-order QPM as $G_3 = -\frac{2}{3\pi}$

the fourth-order QPM as $G_4 = 0$

Substituting the following into exponential Fourier series, we get

$$\frac{dE_2}{dz} = i\frac{\omega d(z)}{cn_2}E_1^2 e^{-i\Delta kz} \tag{9.230}$$

$$\frac{dE_2}{dz} = i\frac{\omega d_0}{cn_2}E_1^2 \sum_{m=-\infty}^{+\infty} G_m e^{i(mK-\Delta k)z} \tag{9.231}$$

$$\begin{aligned}\frac{dE_2}{dz} = i\frac{\omega d_0}{cn_2}E_1^2\Big[G_0 e^{-i\Delta kz} + G_1 e^{i(K-\Delta k)z} + G_{-1}e^{-i(K+\Delta k)z} \\ + G_2 e^{i(2K-\Delta k)z} + G_{-2}e^{-i(2K+\Delta k)z} + \cdots\Big]\end{aligned} \tag{9.232}$$

By integrating the electric field from 0 to z and assuming that the change in the energy of the fundamental frequency is small (E_1 is a constant), we finally get

$$\begin{aligned}\int dE_2 = i\frac{\omega d_0}{cn_2}E_1^2 \int G_0 e^{-i\Delta kz} + G_1 e^{i(K-\Delta k)z} + G_{-1}e^{-i(K+\Delta k)z} \\ + G_2 e^{i(2K-\Delta k)z} + G_{-2}e^{-i(2K+\Delta k)z} dz\end{aligned} \tag{9.233}$$

$$E_2(z) = i\frac{\omega d_0}{cn_2}E_1^2 G_1 \left[\frac{e^{i(K-\Delta k)z}}{i(K-\Delta k)}\right]_0^z \tag{9.234}$$

$$E_2(z) = i\frac{\omega d_0}{cn_2}E_1^2 G \frac{e^{i(K-\Delta k)z} - 1}{i(K-\Delta k)} \tag{9.235}$$

$$E_2(z) = i\frac{\omega d_0}{cn_2}E_1^2 \left[G_1 \frac{e^{i(K-\Delta k)z/2}\left(e^{i(K-\Delta k)z/2} - e^{i(K-\Delta k)z/2}\right)}{i(K-\Delta k)}\right] \tag{9.236}$$

For first-order QPM with $K = \Delta k$, we neglect the contribution of all the finite sine terms and we get

$$E_2(z) = i\frac{\omega d_0}{cn_2}E_1^2\left[G_1 e^{i(K-\Delta k)z/2}\frac{\sin(K-\Delta k)z/2}{(K-\Delta k)/2}\right] \tag{9.237}$$

The power efficiency of the second harmonic at the crystal length under first-order QPM reaches

$$P_1 = \frac{n_1}{2c\mu_0}\left|E_1\right|^2 S \tag{9.238}$$

$$\left|E_1\right|^4 = \frac{4c^2\mu_0^2 P_1^2}{n_1^2 S^2} \tag{9.239}$$

$$P_2(z) = \frac{n_2}{2c\mu_0}\left|E_2(z)\right|^2 S \tag{9.240}$$

$$P_2(z) = \frac{n_2}{2c\mu_0}\frac{\omega^2}{c^2 n_2^2}d_0^2 G_1^2\left|E_1^4\right|\left(\frac{\sin(K-\Delta k)z/2}{(K-\Delta k)/2}\right)^2 S \tag{9.241}$$

$$P_2(z) \simeq \frac{n_2}{2c\mu_0}\frac{\omega^2 d_0^2 G_1^2}{c^2 n_2^2}\frac{4c^2\mu_0^2 P_1^2}{n_1^2 S^2}z^2 S \tag{9.242}$$

$$\eta = \frac{P_2(z)}{P_1} = \frac{\omega^2 d_0^2 G_1^2}{cn_2}\frac{2\mu_0 P_1}{n_1^2 S}z^2 \tag{9.243}$$

This expression is very similar to the situation of the BPM except that

$$\Delta k \to (K - \Delta k)$$

$$d \to d_0 G_1$$

where G_1 is the first Fourier expansion amplitude of $e^{-i\Delta kz}$, which is proportional to the crystal length z. As the crystal length increases, the second harmonic field keeps on increasing rather than being oscillating. However, it increases much slowly than BPM because the effective nonlinear coefficient now becomes $d_0 G_1$. If we select the first-order QPM as $G_1 = 2/\pi$ and neglect the other terms, the effective nonlinear coefficient becomes $d_{\text{eff}} = 2d_0/\pi$, as shown in Table 9.2.

As we go to the higher Fourier coefficients, the amplitude of the Fourier term keeps on dropping down. At the first-order QPM, we need to modify the sign of d-coefficient after the coherence length to get the QPM. If the coherence length is 3.3 μm, we need to have a periodicity of 6.6 μm. Every 3.3 μm, we need to reverse the direction or change the sign of d. If we go to higher-order QPM, the corresponding Fourier coefficients have smaller magnitudes and the effective nonlinear polarization will be smaller. Although it is much easier to make higher-order QPM, the price we pay is upon the second harmonic efficiency.

Table 9.2 Quasi-phase matching with different orders at 50% duty cycle

Quasi-phase matching	K	Required period	SH efficiency
First order	$K=\Delta k=\frac{2\pi}{\lambda}$	$\lambda=\frac{2\pi}{\Delta k}=2L_c$	$\left(\frac{2}{\pi}\right)^2 d_0 \simeq 0.4d_0$
Third order	$3K=\Delta k=\frac{6\pi}{\lambda}$	$\lambda=\frac{6\pi}{\Delta k}=6L_c$	$\left(\frac{2}{3\pi}\right)^2 d_0 \simeq 0.04d_0$

Increasing the electric field of the second harmonic via QPM is required by

$$K=\Delta k \Rightarrow \frac{2\pi}{\lambda}=\frac{2\omega}{c}\left[n(2\omega)-n(\omega)\right] \tag{9.244}$$

The required spatial period of QPM is dependent on the refractive indices and the fundamental frequency of input wave as (Fig. 9.19)

$$\lambda=\frac{\pi c}{\omega\left[n(2\omega)-n(\omega)\right]}=\frac{\pi c}{\omega\left[n_2-n_1\right]} \tag{9.245}$$

The second harmonic power efficiency becomes zero if

$$(K-\Delta k)L/2=\pm\pi \tag{9.246}$$

$$\Delta k(\omega)=K\pm\frac{2\pi}{L} \tag{9.247}$$

$$\Delta k(\omega_0+\Delta\omega)=K\pm\frac{2\pi}{L} \tag{9.248}$$

The second harmonic power efficiency reaches a maximum if

$$\Delta k(\omega_0)=K=\frac{2\pi}{\lambda} \tag{9.249}$$

All imply that the longer the crystal, the narrower the bandwidth of the SHG.

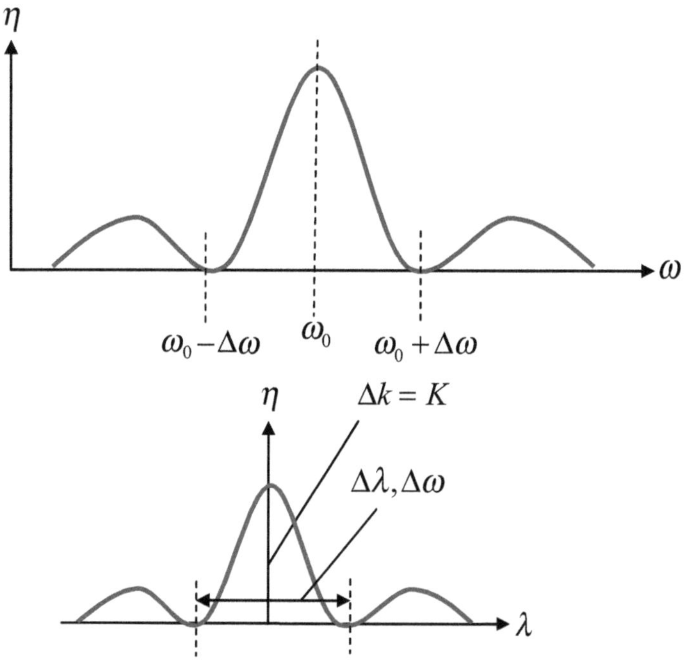

Figure 9.19 Power efficiency of the SHG as a function of the fundamental frequency via QPM (top) and as a function of wavelength (bottom).

When we experimentally put the periodic electrodes (periodic poling lambda) and we apply a strong electric field (i.e., 22 kV/mm) in the downward direction, we can periodically change the orientation of the *z*-axis from pointing up to pointing down (Ps), as shown in Fig. 9.20. Considering a crystal with two different coordinate systems with two parts of crystals, for the first part of the crystal, the laboratory coordinate system and the crystal principal axis exactly match. The *d*-sensor written in the principal axis system is also the *d*-tensor in the laboratory coordinate system. For the second part of the crystal, we need to transform the *d*-tensor from the principal axis system to the laboratory coordinate system. For instance, 25 W power at the wavelength of 589 nm input fundamental power generates the second harmonic power of 3 W.

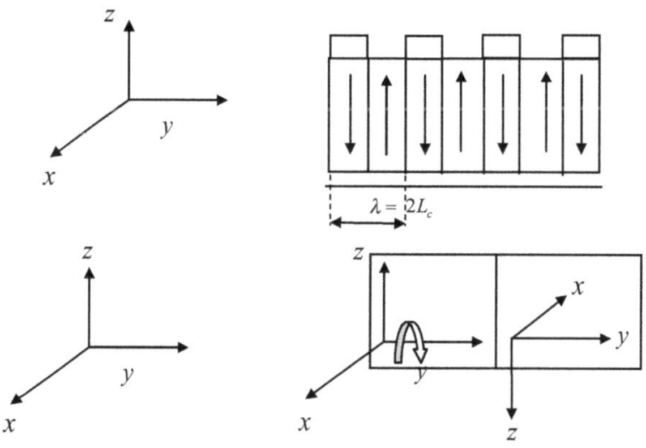

Figure 9.20 (Top) Electric-field-induced periodic poling along the *z*-axis in lithium niobate periodically poled crystals (right) as compared to laboratory coordinate system. (Bottom) Laboratory coordinate system (left) with respect to the principal axes of the crystal.

Problems

9.1 Verify that the following equation is able to produce the second-order nonlinear effect.

$$P^{\mathrm{NL}} = \varepsilon_0 \chi^{(2)} \left[E_0 \cos(kz - \omega t) \right]^2$$

9.2 Show that the following equation can produce *i*th-order nonlinear effect.

$$P^{\mathrm{NL}} = \varepsilon_0 \chi^{(2)} \left[E_1 \cos(k_1 z - \omega t) + E_2 \cos(k_2 z - 2\omega t) \right]^2$$

9.3 Show that the following can produce second harmonic, sum-frequency, and different frequency generations.

$$P^{\mathrm{NL}} = \varepsilon_0 \chi^{(2)} \left[E_1 \cos(k_1 z - \omega_1 t) + E_2 \cos(k_2 z - \omega_2 t) \right]^2$$

9.4 Prove the following equations by starting with the electric field composed of amplitude and phase terms:

$$\frac{du_2}{dz} = \frac{\omega d}{cn_2} u_1^2 \sin\theta$$

$$\frac{d\phi_2}{dz} u_2 = \frac{\omega d}{cn_2} u_1^2 \cos\theta$$

9.5 Find the best duty cycle l/λ in order to make maximum G_2.

9.6 Obtain the values of G_3 and G_4.

9.7 Determine the best duty cycle l/λ that we must use so that we have a finite value of G_2 and we can use the second-order QPM to approach the maximum power efficiency.

9.8 Considering an SHG in lithium niobate crystal propagating along the y-direction with the following details, determine the first-order QPM period required.

Wavelength	**Ordinary refractive index**	**Extraordinary refractive index**
1 μm	2.236	2.160
0.5 μm	2.341	2.249

9.9 Determine the dimension of the second- and third-order nonlinear susceptibility.

9.10 Prove the following equation:

$$\sin^2 \psi_m = \frac{\left[\dfrac{1}{n_0^2(2\omega)} - \dfrac{1}{n_0^2(\omega)}\right]}{\left[\dfrac{1}{n_e^2(\omega)} - \dfrac{1}{n_0^2(\omega)}\right]}$$

9.11 We supply the input power of 1 W at the fundamental wavelength of 1.06 μm, 1 mm ϕ beam into $LiNbO_3$ served as the uniaxial medium. Determine (a) the coherence length if $n_1 = n_0(\omega) = 2.1561$, $n_2 = n_e(2\omega) = 2.2355$, $d = 30 \times 10^{-12}$ m/V, (b) SH power efficiency without and with phase matching.

9.12 What is the condition for the SHG we need by supplying the horizontal polarized light, i.e., Ruby laser at λ_0 = 694.3 nm into the KDP (KH_2PO_4) crystal having $n_0(\omega) = 1.50502$, $n_e(\omega) = 1.46532$, $n_0(2\omega) = 1.53269$, $n_e(2\omega) = 1.49711$?

9.13 When we have the d-tensor as d_{123} and d_{112}, please use the standard convection for the tensor contraction.

9.14 If we launch the "ordinary polarization" at the fundamental frequency along the optic axis into the lithium niobate crystal ($n_o > n_e$) with the d-tensor as

$$d=\begin{pmatrix} 0 & 0 & 0 & 0 & d_{15} & -d_{22} \\ -d_{22} & d_{22} & 0 & d_{15} & 0 & 0 \\ d_{31} & d_{31} & d_{33} & 0 & 0 & 0 \end{pmatrix}$$

where

$$d_{21} = -d_{22};\ d_{32} = d_{31};\ d_{24} = -d_{15};\ d_{16} = -d_{22}$$

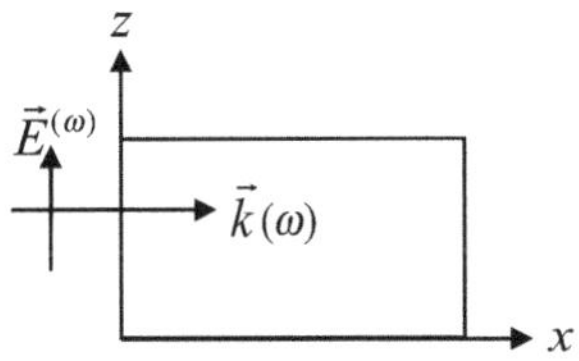

and the ordinary polarization in the z-axis is parallel to the optic axis of the lithium niobate crystal and perpendicular to the k-vector, find out the effective d-tensor.

9.15 Compare the power conversion efficiency at the crystal length.

9.16 Find the wavelength and refractive index of UV light out of the KDP crystal when the SHG satisfies the BPM condition and the finite effective nonlinear polarization is not zero. A pair of wavelengths is composed of the argon ion laser at the wavelength of 351 nm and the oxazine dye laser at the wavelength of 789 nm.

Chapter 10

Other Frequency Mixing Processes

10.1 Introduction

This chapter deals with different frequency and sum frequency generation in addition to the second harmonic generation previously explained in Chapter 9.

10.2 Different Frequency Generation

Upon three wave propagations simultaneously in the anisotropic medium under no pump depletion, three complex electric fields of pump, signal, and idler are possibly written in the order as follows:

$$E_{\mathrm{p}}(z) = E_{\mathrm{p0}} e^{i\phi_{\mathrm{p0}}} \tag{10.1}$$

$$E_{\mathrm{s}}(z=0) = E_{\mathrm{s0}} e^{i\phi_{\mathrm{s0}}} \tag{10.2}$$

$$E_{\mathrm{i}}(z=0) = E_{\mathrm{i0}} e^{i\phi_{\mathrm{i0}}} \tag{10.3}$$

where p is pump, s is signal, and i is idler.

The signal output either gets amplified or attenuated depending on the phase relationship between the idler signal and the fixed pump due to changing energy via a nonlinear coefficient of the crystal according to

Linear and Nonlinear Optics: Materials, Properties, and Applications
Kitsakorn Locharoenrat

ISBN 978-981-4877-15-2 (Hardcover), 978-1-003-05588-4 (eBook)
www.jennystanford.com

- Energy conservation: $\omega_p = \omega_s + \omega_i$
- Phase matching condition: $k_p = k_s + k_i$

so that we can theoretically define a different frequency generation (DFG) and a sum frequency generation (SFG) depending on the pump wavelength tuning method (Fig. 10.1). A pair of wavelengths are tuned by not rotating the crystal, but changing the pump wavelength under energy and momentum conservations as

$$\omega_p = \omega_s + \omega_i \Rightarrow \frac{1}{\lambda_p} = \frac{1}{\lambda_s} + \frac{1}{\lambda_i} \tag{10.4}$$

$$k_p = k_s + k_i \Rightarrow \frac{2\pi}{\lambda_p} n_p = \frac{2\pi}{\lambda_s} n_s + \frac{2\pi}{\lambda_i} n_i \tag{10.5}$$

$$\frac{n(\lambda_p)}{\lambda_p} = \frac{n(\lambda_s)}{\lambda_s} + \frac{n(\lambda_i)}{\lambda_i} \tag{10.6}$$

$$\frac{n_e(\lambda_p, \psi_m)}{\lambda_p} = \frac{n_0(\lambda_s)}{\lambda_s} + \frac{n_0(\lambda_i)}{\lambda_i} \tag{10.7}$$

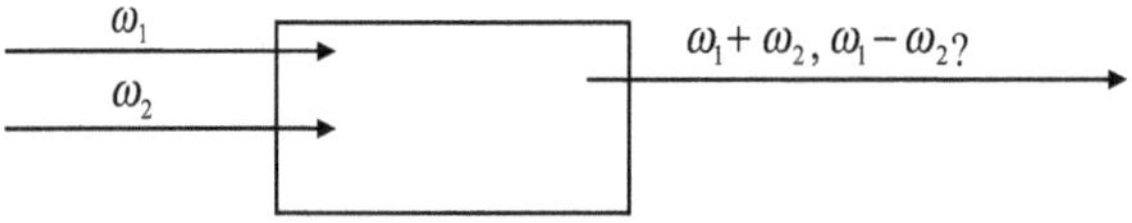

Figure 10.1 Two incident waves at two possible frequencies came out of the crystal ($\omega_1 > \omega_2$).

- Different frequency generation (DFG) is dominant if $\omega_3' = |\omega_1 - \omega_2|$ satisfies $k_3' = |k_1 - k_2|$. If we launch λ_p = 400 nm and λ_s = 700 nm coming in and we assume ω_p is a stronger light beam than ω_s as it is an amplifier for ω_s generating a different frequency λ_i = 933 nm.
- Sum frequency generation (SFG) is dominant if $\omega_3 = \omega_1 + \omega_2$ satisfies $k_3 = k_1 + k_2$. If we launch ω_s = 700 nm and ω_i = 933 nm, we can generate ω_p = 400 nm.

It is not practically possible to get these two output conditions simultaneously for the same set of frequencies at the input conditions. We need to generate one of these processes as (Fig. 10.2)

- Sum frequency generation from ω_s, ω_i to ω_p

- Different frequency generation from ω_p, ω_s to ω_i

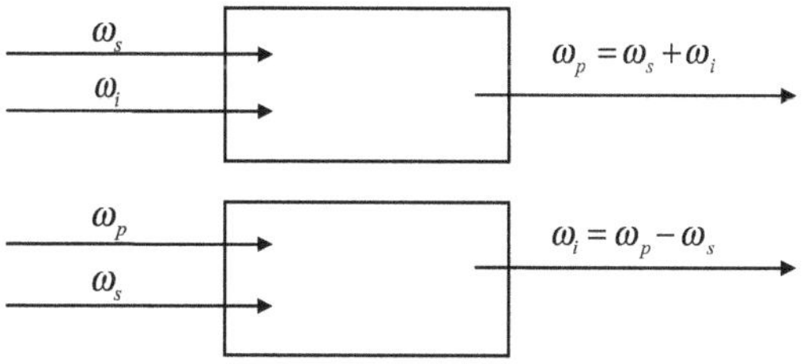

Figure 10.2 (Top) Sum frequency generation is applicable to convert infrared to visible light (if $\omega_s = \omega_i$, we get back to SHG and sub-SH processes). (Bottom) Different frequency generation is applicable to convert visible to infrared light.

According to the nonlinear polarization and the total complex electric fields in the crystal as

$$P^{NL} = 2\varepsilon_0 d E^2 \tag{10.8}$$

$$E = E^{(\omega_p)} + E^{(\omega_s)} + E^{(\omega_i)} \tag{10.9}$$

$$E^{(\omega_p)} = \frac{1}{2}\left[E_p e^{i(k_p z - \omega_p t)} + cc\right] \tag{10.10}$$

$$E^{(\omega_s)} = \frac{1}{2}\left[E_s e^{i(k_s z - \omega_s t)} + cc\right] \tag{10.11}$$

$$E^{(\omega_i)} = \frac{1}{2}\left[E_i e^{i(k_i z - \omega_i t)} + cc\right] \tag{10.12}$$

where E_p is the electric field amplitude of the waves at frequency ω_p, E_s is the electric field amplitude of the waves at frequency ω_s, and E_p is the electric field amplitude of the waves at frequency ω_i.

$$k_p = \frac{\omega_p}{c} n_p \tag{10.13}$$

$$k_s = \frac{\omega_s}{c} n_s \tag{10.14}$$

$$k_i = \frac{\omega_i}{c} n_i \tag{10.15}$$

Each component of the nonlinear polarization generated at these three frequencies is modified as

$$P_{\mathrm{NL}}^{(\omega_{\mathrm{p}})} = \frac{1}{2}\varepsilon_0 d\left[2E_{\mathrm{s}}E_{\mathrm{i}}e^{i\left[(k_{\mathrm{s}}+k_{\mathrm{i}})z-\omega_{\mathrm{p}}t\right]} + cc\right] \tag{10.16}$$

$$P_{\mathrm{NL}}^{(\omega_{\mathrm{s}})} = \frac{1}{2}\varepsilon_0 d\left[2E_{\mathrm{p}}E_{\mathrm{i}}^{*}e^{i\left[(k_{\mathrm{p}}-k_{\mathrm{i}})z-\omega_{\mathrm{s}}t\right]} + cc\right] \tag{10.17}$$

$$P_{\mathrm{NL}}^{(\omega_{\mathrm{i}})} = \frac{1}{2}\varepsilon_0 d\left[2E_{\mathrm{p}}E_{\mathrm{s}}^{*}e^{i\left[(k_{\mathrm{p}}-k_{\mathrm{s}})z-\omega_{\mathrm{i}}t\right]} + cc\right] \tag{10.18}$$

Whereas the wave equations for these nonlinear effects are written as

$$\nabla^2 E^{(\omega_{\mathrm{p}})} - \mu_0\varepsilon(\omega_{\mathrm{p}})\frac{\partial^2 E^{(\omega_{\mathrm{p}})}}{\partial t^2} = \mu_0\frac{\partial^2 P_{\mathrm{NL}}^{(\omega_{\mathrm{p}})}}{\partial t^2} \tag{10.19}$$

$$\nabla^2 E^{(\omega_{\mathrm{s}})} - \mu_0\varepsilon(\omega_{\mathrm{s}})\frac{\partial^2 E^{(\omega_{\mathrm{s}})}}{\partial t^2} = \mu_0\frac{\partial^2 P_{\mathrm{NL}}^{(\omega_{\mathrm{s}})}}{\partial t^2} \tag{10.20}$$

$$\nabla^2 E^{(\omega_{\mathrm{i}})} - \mu_0\varepsilon(\omega_{\mathrm{i}})\frac{\partial^2 E^{(\omega_{\mathrm{i}})}}{\partial t^2} = \mu_0\frac{\partial^2 P_{\mathrm{NL}}^{(\omega_{\mathrm{i}})}}{\partial t^2} \tag{10.21}$$

Neglect the second derivative E with respect to z so that we get a change in the electric field at a particular frequency as the wave propagates as

$$\frac{dE_{\mathrm{p}}}{dz} = i\frac{\omega_{\mathrm{p}}d}{cn_{\mathrm{p}}}E_{\mathrm{s}}E_{\mathrm{i}}e^{-i\Delta kz} \tag{10.22}$$

$$\frac{dE_{\mathrm{s}}}{dz} = i\frac{\omega_{\mathrm{s}}d}{cn_{\mathrm{s}}}E_{\mathrm{p}}E_{\mathrm{i}}^{*}e^{i\Delta kz} \tag{10.23}$$

$$\frac{dE_{\mathrm{i}}}{dz} = i\frac{\omega_{\mathrm{i}}d}{cn_{\mathrm{i}}}E_{\mathrm{p}}E_{\mathrm{s}}^{*}e^{i\Delta kz} \tag{10.24}$$

where

$$\Delta k = k_{\mathrm{p}} - k_{\mathrm{s}} - k_{\mathrm{i}}$$

- Maximum interaction takes place for maximum efficiency of these interactions among three waves for the birefringence phase matching condition as (Fig. 10.3).

$$\Delta k = k_{\mathrm{p}} - k_{\mathrm{s}} - k_{\mathrm{i}} = 0$$

- Maximum interaction takes place for maximum efficiency of these interactions among three waves for the quasi-phase matching condition; d-tensor is a function of the crystal length z as

$$\Delta k = k_p - k_s - k_i = K$$

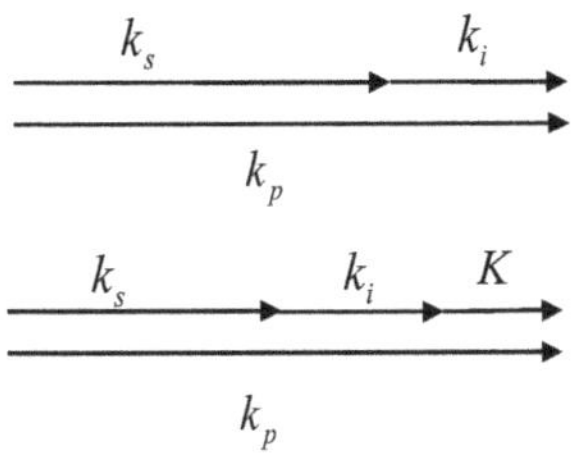

Figure 10.3 Photon picture to satisfy (top) birefringence phase matching condition and (bottom) quasi-phase matching condition. The momentum of the pump photon (the shortest wavelength as extraordinary wave) is the sum of momenta of the signal and idler photons that are generating (the longest wavelength as ordinary waves).

There is a pair of frequencies that will satisfy both conditions in which a different frequency generation will take place. Note that E_p as a real term is constant so that a phase of the pump is equivalent to zero

1. $\Delta k = 0$
2. $\omega_i = \omega_p - \omega_s$

1. The first set reveals the Manley–Rowe relation as follows:

$$\frac{dE_i}{dz} = i\kappa_i E_p E_s^* \tag{10.25}$$

$$\frac{dE_s}{dz} = i\kappa_s E_p E_i^* \tag{10.26}$$

where

$$\kappa_s = \frac{\omega_s d}{cn_s}$$

$$\kappa_i = \frac{\omega_i d}{cn_i} \tag{10.27}$$

Defining a gain coefficient (g), the secondary differential wave equations are finally derived as

$$g^2 = \kappa_s \kappa_i \left| E_p \right|^2 = \kappa_s \kappa_i E_{p0}^2 \tag{10.28}$$

$$\frac{d^2 E_i}{dz^2} = i\kappa_i E_p(-i\kappa_s E_p^* E_i) \tag{10.29}$$

$$\frac{d^2 E_i}{dz^2} \kappa_s \kappa_i \left| E_p \right|^2 E_i = g^2 E_i \tag{10.30}$$

The solution to the idler output is

$$E_i(z) = A \sinh gz + B \cosh gz \tag{10.31}$$

The solution to the signal output is

$$\frac{dE_i}{dz} = i\kappa_i E_p E_s^* \tag{10.32}$$

$$i\kappa_i E_p E_s^*(z) = g(A \cosh gz + B \sinh gz) \tag{10.33}$$

$$E_s^*(z) = \frac{-ig}{\kappa_i E_p}(A \cosh gz + B \sinh gz) \tag{10.34}$$

$$E_s^*(z) = \frac{-i\sqrt{\kappa_s \kappa_i} E_{p0}}{\kappa_i E_{p0} e^{i\phi_{p0}}}\left[A \cosh gz + B \sinh gz\right] \tag{10.35}$$

$$E_s^*(z) = -i\sqrt{\frac{\kappa_s}{\kappa_i}} e^{-i\phi_{p0}}\left[A \cosh gz + B \sinh gz\right] \tag{10.36}$$

$$E_s^*(B=0) = -i\sqrt{\frac{\kappa_s}{\kappa_i}} e^{-i\phi_{p0}} A \quad \Rightarrow A = i\sqrt{\frac{\kappa_i}{\kappa_s}} e^{i\phi_{p0}} E_{s0}^* \tag{10.37}$$

We finally get two solutions as

$$E_i(z) = i\sqrt{\frac{\omega_i n_s}{\omega_s n_i}} e^{i\phi_{p0}} E_{s0}^* \sinh gz + E_{i0} \cosh gz \tag{10.38}$$

saying that at $z = 0$, there is no idler; however, as z increases, the idler power keeps on increasing.

$$E_s^*(z) = -i\sqrt{\frac{\kappa_s}{\kappa_i}} e^{-i\phi_{p0}}\left[\left(i\sqrt{\frac{\kappa_i}{\kappa_s}} e^{i\phi_{p0}} E_{s0}^*\right) \cosh gz + B \sinh gz\right] \tag{10.39}$$

$$E_s^*(z) = E_{s(0)}^* \cosh gz - i\sqrt{\frac{\omega_s n_i}{\omega_i n_s}} e^{-i\phi_{p0}} E_{i(0)} \sinh gz \qquad (10.40)$$

saying that at $z = 0$, there is $E_s(0) = E_{s0}^*$. The signal power increases with z.

Exercise 10.1

Write the electric fields of signal and idler with the initial phase in a matrix form.

Solution

$$\begin{pmatrix} E_s^*(z) \\ E_i(z) \end{pmatrix} = \begin{pmatrix} \cosh gz & -i\sqrt{\frac{\omega_s n_i}{\omega_i n_s}} e^{-i\phi_{p0}} \sinh gz \\ i\sqrt{\frac{\omega_i n_s}{\omega_s n_i}} e^{i\phi_{p0}} \sinh gz & \cosh gz \end{pmatrix} \begin{pmatrix} E_{s0}^* \\ E_{i0} \end{pmatrix}$$

This point reveals that not only does the input pump generate the output idler, but also it amplifies the output signal. The amplification process is independent of the phase of E_{s0}. This is an example of "**phase-insensitive parametric amplifier**" where amplification takes place irrespective of the phase of the input signal. Therefore, every time a signal is generated, an idler is also generated. The increase in the number of signal photons coming out of the crystal is equivalent to the generated idler photon coming out at the idler frequency.

$$\frac{dE_p}{dz} = i\kappa_p E_s E_i \qquad (10.41)$$

$$\frac{dE_s}{dz} = i\kappa_s E_p E_i^* \qquad (10.42)$$

$$\frac{dE_i}{dz} = i\kappa_i E_p E_s^* \qquad (10.43)$$

where

$$\kappa_\alpha = \frac{\omega_\alpha d}{c n_\alpha};\ \alpha \text{ is } p, s, i. \qquad (10.44)$$

$$P_\alpha = \frac{n_\alpha}{2c\mu_0}\left|E_\alpha\right|^2 S \tag{10.45}$$

$$\frac{dP_\mathrm{p}}{dz} = \frac{d}{dz}\left(\frac{n_\mathrm{p}}{2c\mu_0}E_\mathrm{p}E_\mathrm{p}^*S\right) = \frac{n_\mathrm{p}S}{2c\mu_0}\left[E_\mathrm{p}^*\frac{dE_\mathrm{p}}{dz} + E_\mathrm{p}\frac{dE_\mathrm{p}^*}{dz}\right] \tag{10.46}$$

where P_p is the pump power.

$$\frac{dP_\mathrm{p}}{dz} = \frac{n_\mathrm{p}S}{2c\mu_0}\left[i\kappa_\mathrm{p}E_\mathrm{p}^*E_\mathrm{s}E_\mathrm{i} - i\kappa_\mathrm{p}E_\mathrm{p}E_\mathrm{s}^*E_\mathrm{i}^*\right] \tag{10.47}$$

$$\frac{dP_\mathrm{p}}{dz} = \frac{in_\mathrm{p}S}{2c\mu_0}\frac{\omega_\mathrm{p}d}{cn_\mathrm{p}}\left[E_\mathrm{p}^*E_\mathrm{s}E_\mathrm{i} - E_\mathrm{p}E_\mathrm{s}^*E_\mathrm{i}^*\right] \tag{10.48}$$

$$\frac{dP_\mathrm{p}}{dz} = \frac{iSd\omega_\mathrm{p}}{2c^2\mu_0}\left[E_\mathrm{p}^*E_\mathrm{s}E_\mathrm{i} - E_\mathrm{p}E_\mathrm{s}^*E_\mathrm{i}^*\right] \tag{10.49}$$

Similarly, we have

$$\frac{dP_\mathrm{s}}{dz} = \frac{iSd\omega_\mathrm{s}}{2c^2\mu_0}\left[E_\mathrm{p}E_\mathrm{s}^*E_\mathrm{i}^* - E_\mathrm{p}^*E_\mathrm{s}E_\mathrm{i}\right] \tag{10.50}$$

$$\frac{dP_\mathrm{i}}{dz} = \frac{iSd\omega_\mathrm{i}}{2c^2\mu_0}\left[E_\mathrm{p}E_\mathrm{s}^*E_\mathrm{i}^* - E_\mathrm{p}^*E_\mathrm{s}E_\mathrm{i}\right] \tag{10.51}$$

$$-\frac{1}{\omega_\mathrm{p}}\frac{dP_\mathrm{p}}{dz} = \frac{1}{\omega_\mathrm{s}}\frac{dP_\mathrm{s}}{dz} = \frac{1}{\omega_\mathrm{i}}\frac{dP_\mathrm{i}}{dz} \tag{10.52}$$

Equation (10.52) is the Manley–Rowe relation. The rate of reduction in the number of photons at the pump frequency crossing is equal to the rate of increase in the signal photons and is equal to the rate of increase in the idler photon. This means that every time a pump photon is splitting into signal and idler photons. This equation satisfies the energy conservation condition as

$$\frac{d}{dz}(P_\mathrm{p} + P_\mathrm{s} + P_\mathrm{i}) = 0 \tag{10.53}$$

2. The second set reveals the maximum power generating at the signal as follows:

$$\frac{dE_\mathrm{s}}{dz} = i\kappa_\mathrm{s}E_\mathrm{p}E_\mathrm{i}^* \tag{10.54}$$

$$\frac{dE_{\mathrm{i}}}{dz}=i\kappa_{\mathrm{i}}E_{\mathrm{p}}E_{\mathrm{s}}^{*} \tag{10.55}$$

Defining a gain coefficient (g), the secondary differential wave equations are finally derived as

$$\frac{d^2E_{\mathrm{s}}}{dz^2}=i\kappa_{\mathrm{s}}E_{\mathrm{p}}(-i\kappa_{\mathrm{i}}E_{\mathrm{p}}^{*}E_{\mathrm{s}})=\kappa_{\mathrm{s}}\kappa_{\mathrm{i}}\left|E_{\mathrm{p}}\right|^2E_{\mathrm{s}}=g^2E_{\mathrm{s}} \tag{10.56}$$

The solution to the output signal is

$$E_{\mathrm{s}}(z)=A\sinh gz+B\cosh gz \tag{10.57}$$

The solution to the output idler is

$$\frac{dE_{\mathrm{s}}}{dz}=i\kappa_{\mathrm{s}}E_{\mathrm{p}}E_{\mathrm{i}}^{*} \tag{10.58}$$

$$i\kappa_{\mathrm{s}}E_{\mathrm{p}}E_{\mathrm{i}}^{*}(z)=g(A\cosh gz+B\sinh gz) \tag{10.59}$$

$$E_{\mathrm{i}}^{*}(z)=\frac{-ig}{\kappa_{\mathrm{s}}E_{\mathrm{p}}}(A\cosh gz+B\sinh gz) \tag{10.60}$$

$$E_{\mathrm{i}}^{*}(\mathrm{z})=\frac{-i\sqrt{\kappa_{\mathrm{s}}\kappa_{\mathrm{i}}}\,E_{\mathrm{p0}}}{\kappa_{\mathrm{s}}E_{\mathrm{p0}}e^{i\phi_{\mathrm{p0}}}}\left[A\cosh gz+B\sinh gz\right] \tag{10.61}$$

$$E_{\mathrm{i}}^{*}(\mathrm{z})=-i\sqrt{\frac{\kappa_{\mathrm{i}}}{\kappa_{\mathrm{s}}}}e^{-i\phi_{\mathrm{p0}}}\left[A\cosh gz+B\sinh gz\right] \tag{10.62}$$

$$E_{\mathrm{i}}^{*}(B=0)=-i\sqrt{\frac{\kappa_{\mathrm{i}}}{\kappa_{\mathrm{s}}}}e^{-i\phi_{\mathrm{p0}}}A\quad\Rightarrow A=i\sqrt{\frac{\kappa_{\mathrm{s}}}{\kappa_{\mathrm{i}}}}e^{i\phi_{\mathrm{p0}}}E_{\mathrm{i0}}^{*} \tag{10.63}$$

$$E_{\mathrm{s}}(\mathrm{z})=i\sqrt{\frac{\omega_{\mathrm{s}}n_{\mathrm{i}}}{\omega_{\mathrm{i}}n_{\mathrm{s}}}}e^{i\phi_{\mathrm{p0}}}E_{\mathrm{i(0)}}^{*}\sinh gz+E_{\mathrm{s(0)}}\cosh gz \tag{10.64}$$

$$E_{\mathrm{i}}^{*}(\mathrm{z})=-i\sqrt{\frac{\kappa_{\mathrm{i}}}{\kappa_{\mathrm{s}}}}e^{-i\phi_{\mathrm{p0}}}\left[\left(i\sqrt{\frac{\kappa_{\mathrm{s}}}{\kappa_{\mathrm{i}}}}e^{i\phi_{\mathrm{p0}}}E_{\mathrm{i0}}^{*}\right)\cosh gz+B\sinh gz\right] \tag{10.65}$$

$$E_{\mathrm{i}}^{*}(\mathrm{z})=E_{\mathrm{i0}}^{*}\cosh gz-i\sqrt{\frac{\kappa_{\mathrm{i}}}{\kappa_{\mathrm{s}}}}e^{-i\phi_{\mathrm{p0}}}E_{\mathrm{s0}}\sinh gz \tag{10.66}$$

$$P_s(z)=\frac{n_s}{2c\mu_0}\left|E_s\right|^2 S=\frac{n_s}{2c\mu_0}\left|E_{s(0)}\right|^2\cosh^2 gzS \tag{10.67}$$

$$P_s(0)=\frac{n_s}{2c\mu_0}\left|E_s(0)\right|^2 S \tag{10.68}$$

$$\left|E_s(0)\right|^2=\frac{2c\mu_0 P_s(0)}{n_s S} \tag{10.69}$$

$$P_s(z)=\frac{n_s}{2c\mu_0}\frac{2c\mu_0 P_s(0)}{n_s S}\cosh^2 gz\cdot S \tag{10.70}$$

$$P_s(z)=P_s(0)\cosh^2 gz \tag{10.71}$$

In addition to the maximum power generating at the output signal, the input power at the frequency ω_i is written as

$$P_i(z)=\frac{n_i}{2c\mu_0}\left|E_i(z)\right|^2 S \tag{10.72}$$

$$E_i\,(z)=i\sqrt{\frac{\omega_i n_s}{\omega_s n_i}}e^{i\phi_{p0}}E^*_{s(0)}\sinh gz+E_{i(0)}\cosh gz \tag{10.73}$$

$$P_i(z)=\frac{n_i}{2c\mu_0}\left|i\sqrt{\frac{\omega_i n_s}{\omega_s n_i}}e^{i\phi_{p0}}E^*_{s(0)}\right|^2\sinh^2 gz\cdot S \tag{10.74}$$

$$P_i(z)=\frac{n_i}{2c\mu_0}\frac{\omega_i n_s}{\omega_s n_i}\left|E_{s(0)}\right|^2\sinh^2 gz\cdot S \tag{10.75}$$

$$P_i(z)=\frac{n_i}{2c\mu_0}\frac{\omega_i n_s}{\omega_s n_i}\frac{2c\mu_0 P_s(0)}{n_s S}\sinh^2 gz\cdot S \tag{10.76}$$

$$P_i(z)=\frac{\omega_i}{\omega_s}P_s(0)\sinh^2 gz \tag{10.77}$$

$$\Delta P_s=P_s(L)-P_s(0)=P_s(0)(\cosh^2 gL-1) \tag{10.78}$$

$$\Delta P_s=P_s(0)\sinh^2 gL \tag{10.79}$$

$$\Delta P_i=P_i(L)-P_i(0)=\frac{\omega_i}{\omega_s}P_s(0)\sinh^2 gL \tag{10.80}$$

This is a confirmation that the idler frequency is usually smaller than the signal frequency ($\omega_s > \omega_i$); the power in the idler is, hence, smaller than the power in the signal (Fig. 10.4).

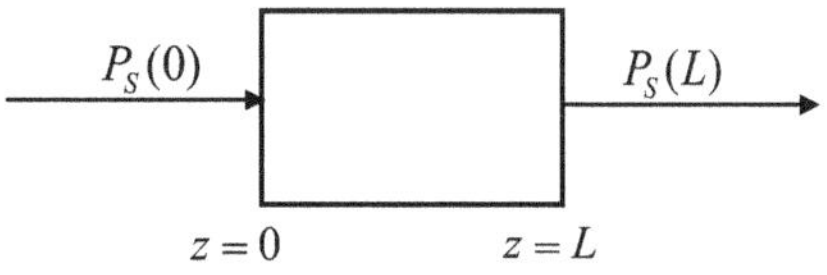

Figure 10.4 Power efficiency in the signal photons.

10.3 Sum Frequency Generation

There is a pair of frequencies that will satisfy both conditions in which a sum frequency generation will take place. It is noted that E_p, as a real term, is constant so that a phase of the pump is equivalent to zero. If we launch the light at ω_s and ω_i coming in and hence satisfies a phase matching condition, we can amplify the light at ω_p. This is another kind of optical amplifier independent of the existence of certain levels of frequencies as long as the crystal has nonlinearity and it is transparent.

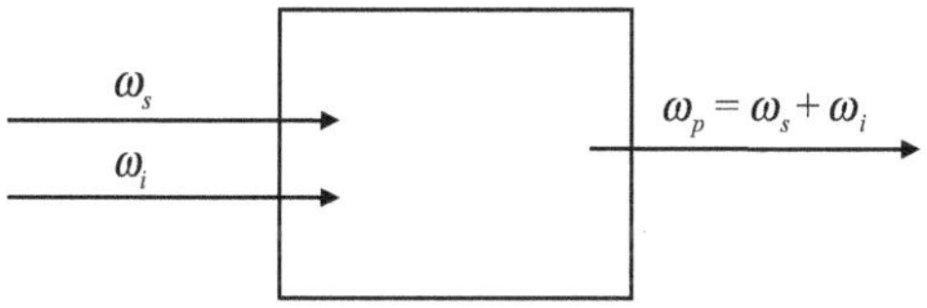

Figure 10.5 Sum frequency generation under no pump depletion. If we make $\omega_s = \omega_i$, we will get back into the degenerate parametric amplifier process where $\omega_p = 2\omega$.

10.3.1 Sum Frequency Generation with Phase Matching

If we are perfectly phase matched ($\Delta k = 0$), two wave solutions according to two wave equations at the input are (Fig. 10.5)

$$\frac{dE_s}{dz} = i\kappa_s E_p E_i^* \tag{10.81}$$

$$\frac{dE_i}{dz} = i\kappa_i E_p E_s^* \tag{10.82}$$

where

$$E_s(0) = E_{s0} e^{i\phi_{s0}} \tag{10.83}$$

$$E_i(0) = E_{i0} e^{i\phi_{i0}} \tag{10.84}$$

$$E_p = E_{p0} e^{i\phi_{p0}} \tag{10.85}$$

$$E_s^*(z) = E_{s0} e^{-i\phi_{s0}} \cosh gz - i\sqrt{\frac{\omega_s n_i}{\omega_i n_s}} e^{-i\phi_{p0}} E_{i0} e^{i\phi_{i0}} \sinh gz \tag{10.86}$$

$$E_i\,(z) = i\sqrt{\frac{\omega_i n_s}{\omega_s n_i}} e^{i\phi_{p0}} E_{s0} e^{-i\phi_{s0}} \sinh gz + E_{i0} e^{i\phi_{i0}} \cosh gz \tag{10.87}$$

Two powers corresponding two wave solutions at the input are

$$P_s(0) = \frac{n_s}{2c\mu_0} |E_s(0)|^2 S = \frac{n_s}{2c\mu_0} E_{s0}^2 S \tag{10.88}$$

$$P_i(0) = \frac{n_i}{2c\mu_0} |E_i(0)|^2 S = \frac{n_i}{2c\mu_0} E_{i0}^2 S \tag{10.89}$$

The number of photons at the signal incident per unit time at the input is

$$N_{s0} = \frac{P_s(0)}{\hbar\omega_s} = \frac{n_s}{2c\mu_0} \frac{1}{\hbar\omega_s} E_{s0}^2 S \tag{10.90}$$

The number of photons at the idler incident per unit time at the input is

$$N_{i0} = \frac{P_i(0)}{\hbar\omega_i} = \frac{n_i}{2c\mu_0} \frac{1}{\hbar\omega_i} E_{i0}^2 S \tag{10.91}$$

$$E_{i0} = \sqrt{\frac{\omega_i n_s}{\omega_s n_i}} E_{s0}; \qquad E_{s0} = \sqrt{\frac{\omega_s n_i}{\omega_i n_s}} E_{i0} \tag{10.92}$$

$$\frac{n_s}{\omega_s} E_{s0}^2 = \frac{n_i}{\omega_i} E_{i0}^2 \tag{10.93}$$

Finally, we reach

$$E_s^*(z) = E_{s0}e^{-i\phi_{s0}}\cosh gz - iE_{s0}e^{-i(\phi_{p0}-\phi_{i0})}\sinh gz \qquad (10.94)$$

$$E_s^*(z) = E_{s0}e^{-i\phi_{s0}}\left[\cosh gz + e^{-i(\phi_{p0}-\phi_{i0}-\phi_{s0}+\frac{\pi}{2})}\sinh gz\right] \qquad (10.95)$$

This is implied that a combination of idler and signal input generates the pump output. This amplification is dependent on the phase of E_{s0}. This is an example of the "**phase-sensitive parametric amplifier**" between the signal and the idler, and the pump with two following situations:

1. As the input signal satisfies $\phi_{p0} - \phi_{s0} - \phi_{i0} = -\frac{\pi}{2}$, the signal gets amplified as

$$E_s^*(z) = E_{s0}e^{-i\phi_{s0}}e^{gz} \qquad (10.96)$$

This point reveals that the pump photons are down converting and generating the signal and idler photons.

2. As the input signal satisfies $\phi_{p0} - \phi_{s0} - \phi_{i0} = +\frac{\pi}{2}$, the signal gets attenuated as

$$E_s^*(z) = E_{s0}e^{-i\phi_{s0}}e^{-gz} \qquad (10.97)$$

This point reveals that both signal and idler photons are combined to generate the pump photons.

10.3.2 Sum Frequency Generation without Phase Matching

If we are not perfectly phase matched ($\Delta k \neq 0$), for getting this amplifier, two wave solutions according to two wave equations at the input are

$$\frac{dE_s}{dz} = i\kappa_s E_p E_i^* e^{i\Delta kz} \qquad (10.98)$$

$$\frac{dE_i}{dz} = i\kappa_i E_p E_s^* e^{i\Delta kz} \qquad (10.99)$$

$$\frac{d^2E_s}{dz^2} = i\kappa_s E_p E_i^*(i\Delta k)e^{i\Delta kz} + i\kappa_s E_p \frac{dE_i^*}{dz}e^{i\Delta kz} \qquad (10.100)$$

$$\frac{d^2E_s}{dz^2} = i\Delta k\frac{dE_s}{dz} + i\kappa_s E_p(-i\kappa_i E_p^* E_s e^{-i\Delta kz})e^{i\Delta kz} \tag{10.101}$$

$$\frac{d^2E_s}{dz^2} = i\Delta k\frac{dE_s}{dz} + \kappa_s\kappa_i \left|E_p\right|^2 E_s = i\Delta k\frac{dE_s}{dz} + g^2E_s \tag{10.102}$$

$$\frac{d^2E_s}{dz^2} - i\Delta k\frac{dE_s}{dz} - g^2E_s = 0 \tag{10.103}$$

Using the following expression, we finally have the wave solution as

$$D^2 - i\Delta kD - g^2 = 0 \tag{10.104}$$

$$D = \frac{i\Delta k \pm \sqrt{-(\Delta k)^2 + 4g^2}}{2} \tag{10.105}$$

$$E_s(z) = Ae^{i\Delta kz/2}e^{\left(\sqrt{g^2-\Delta k^2/4}\right)z} + Be^{i\Delta kz/2}e^{-\left(\sqrt{g^2-\Delta k^2/4}\right)z} \tag{10.106}$$

The pump power using the following condition is written as

$$\frac{\Delta k^2}{4} < g^2 \tag{10.107}$$

$$P_p = \frac{n_p}{2c\mu_0}\left|E_p\right|^2 S \Rightarrow \left|E_p\right|^2 = \frac{2c\mu_0}{n_p}\frac{P_p}{S} \tag{10.108}$$

$$\kappa_s\kappa_i\left|E_p\right|^2 > \frac{\Delta k^2}{4} \Rightarrow \frac{\omega_s d}{cn_s}\frac{\omega_i d}{cn_i}\frac{2c\mu_0}{n_p}\frac{P_p}{S} > \frac{\Delta k^2}{4} \tag{10.109}$$

This point reveals that if we are not perfectly phase matched, the amplifier has certain "bandwidth" over which the pump will get amplified determined by Δk varied with the frequencies as

$$\frac{P_p}{S} > \frac{cn_s n_i n_p}{8\omega_s\omega_i d^2\mu_0}\Delta k^2 \tag{10.110}$$

However, at a particular signal frequency of $\omega_s = \omega_{s0}$, we can reach perfectly phase match.

10.4 Optical Parametric Interactions

10.4.1 Optical Parametric Amplification

Parametric down conversion is a process in which we generate a low frequency from a higher frequency with the sum of the infinite set of combinations of the two frequencies (Fig. 10.6). The input pump at the double frequency (strong beam) and the input signal at the fundamental frequency (very weak beam) simultaneously generate the output signal at the fundamental frequency where the amplification or attenuation relies on the relative phases of the input signal.

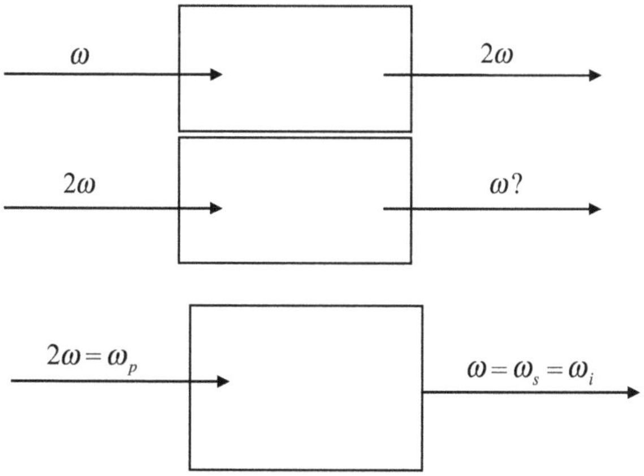

Figure 10.6 The second harmonic generation (top) as compared to the possible parametric down conversion for the generations of fundamental frequency at the same crystal under Δk = 0 (middle) and degenerated optical parametric amplifier with 1 μm pump and 2 μm signal (bottom).

The electric fields of the signal/idler and the pump are written in the order as

$$E_1 = u_1 e^{i\phi_1} \tag{10.111}$$

$$E_2 = u_2 e^{i\phi_2} \tag{10.112}$$

Substituting these aspects into the following wave equation, we get the wave solution as follows:

$$\frac{dE_1}{dz} = i\kappa\ E_2 E_1^* \Rightarrow \Delta k = 0 \tag{10.113}$$

$$\frac{d^2E_1}{dz^2} = i\frac{\omega d}{cn_1}E_2\frac{dE_1^*}{dz} \tag{10.114}$$

$$\frac{d^2E_1}{dz^2} = i\frac{\omega d}{cn_1}E_2\left(-i\frac{\omega d}{cn_1}E_2^*E_1\right) \tag{10.115}$$

$$\frac{d^2E_1}{dz^2} = \frac{\omega^2 d^2}{c^2n_1^2}\left|E_2\right|^2 E_1 \tag{10.116}$$

$$\gamma = \kappa^2\left|E_{20}\right|^2 = \kappa^2 E_{20}^2 \tag{10.117}$$

$$\frac{d^2E_1}{dz^2} = \gamma^2 E_1 \tag{10.118}$$

$$E_1(z) = A\sinh\gamma z + B\cosh\gamma z \tag{10.119}$$

$$E_1(0) = u_{10}e^{i\phi_{10}} = B \tag{10.120}$$

$$E_2(0) = u_2e^{i\phi_{20}} \tag{10.121}$$

$$E_1(z) = A\sinh\gamma z + B\cosh\gamma z \tag{10.122}$$

$$\frac{dE_1}{dz} = A\gamma\cosh\gamma z + \gamma u_{10}\,e^{i\phi_{10}}\sinh\gamma z \tag{10.123}$$

$$\left.\frac{dE_1}{dz}\right|_{z=0} = i\frac{\omega d}{cn_1}E_2(0)E_1^*(0) = i\kappa u_2e^{i\phi_{20}}u_{10}e^{-i\phi_{10}} \tag{10.124}$$

$$A \simeq i\frac{\kappa}{\gamma}u_2u_{10}e^{i(\phi_{20}-\phi_{10})} \tag{10.125}$$

$$E_1(z) = A\sinh\gamma z + B\cosh\gamma z \tag{10.126}$$

$$E_1(z) = \left[i\frac{\kappa}{\gamma}u_2u_{10}e^{i(\phi_{20}-\phi_{10})}\right]\sinh\gamma z + \left[u_{10}e^{i\phi_{10}}\right]\cosh\gamma z \tag{10.127}$$

$$\gamma^2 = \frac{\omega^2 d^2}{c^2 n_1^2}|E_2|^2 = \kappa^2 u_2^2 \tag{10.128}$$

$$E_1(z) = u_{10}e^{i\phi_{10}}\left[\cosh\gamma z + e^{i(\phi_{20}-2\phi_{10}+\frac{\pi}{2})}\sinh\gamma z\right] \tag{10.129}$$

$$E_1(z) = u_{10}e^{i\phi_{10}}\frac{e^{\gamma z}+e^{-\gamma z}}{2} + u_{10}e^{i(\phi_{20}-\phi_{10}+\frac{\pi}{2})}\frac{e^{\gamma z}-e^{-\gamma z}}{2} \tag{10.130}$$

$$E_1(z) = u_{10}e^{i\left(\frac{\phi_{20}}{2}+\frac{\pi}{4}\right)}\cos\left(\frac{\phi_{20}}{2}-\phi_{10}+\frac{\pi}{4}\right)e^{\gamma z} + u_{10}e^{i\left(\frac{\phi_{20}}{2}-\frac{\pi}{4}\right)}$$
$$\cos\left(\frac{\phi_{20}}{2}-\phi_{10}-\frac{\pi}{4}\right)e^{-\gamma z} \tag{10.131}$$

This point is explained as follows:

- If the input signal phase satisfies $\theta = \phi_{20} - 2\phi_{10} + \frac{\pi}{2} = 0$, the fundamental wave grows exponentially in amplitude as z increases as

  $$E_1(z) = u_{10}e^{i\phi_{10}}e^{\gamma z}$$

  so-called optical parametric amplifier. The energy transfers from double to fundamental frequencies, saying that this situation is not allowed if there is no SH incident.
- If the input signal phase satisfies $\theta = \phi_{20} - 2\phi_{10} + \frac{\pi}{2} = \pi$, the fundamental wave decays exponentially in amplitude as the crystal length z increases as

  $$E_1(z) = u_{10}e^{i\phi_{10}}e^{-\gamma z}$$

  so-called the optical parametric attenuator. The energy transfers from fundamental to double frequencies, saying that the fundamental wave goes down in amplitude as the second harmonic generation increases with the crystal length z.

For a sum frequency generation, we assume that ω_s has higher power than ω_i. Herein we can convert or tune light at a higher wavelength to light at a smaller wavelength under the following conditions:

- $\omega_p = \omega_s + \omega_i$
- E_s is almost a constant (real term from a phase of the signal to be zero)
- $\Delta k = 0$

The wave solutions according to two wave equations are calculated as follows:

$$\frac{dE_{\mathrm{p}}}{dz} = i\kappa_{\mathrm{p}} E_{\mathrm{s}} E_{\mathrm{i}} \tag{10.132}$$

$$\frac{dE_{\mathrm{i}}}{dz} = i\kappa_{\mathrm{i}} E_{\mathrm{p}} E_{\mathrm{s}}^{*} \tag{10.133}$$

$$\frac{d^2E_{\mathrm{p}}}{dz^2} = i\kappa_{\mathrm{p}} E_{\mathrm{s}} \left(\frac{dE_{\mathrm{i}}}{dz} \right) = i\kappa_{\mathrm{p}} E_{\mathrm{s}} \left(i\kappa_{\mathrm{i}} E_{\mathrm{s}}^{*} E_{\mathrm{p}} \right) \tag{10.134}$$

$$\frac{d^2E_{\mathrm{p}}}{dz^2} = -\kappa_{\mathrm{p}}\kappa_{\mathrm{i}} \left| E_{\mathrm{s}} \right|^2 E_{\mathrm{p}} = -\delta^2 E_{\mathrm{p}} \tag{10.135}$$

$$E_{\mathrm{p}}(z) = A\cosh\delta z + B\sinh\delta z \tag{10.136}$$

Under the following condition, we finally get the wave solution to the pump as

$$E_{\mathrm{p}}(z=0) = 0 \Rightarrow A = 0 \tag{10.137}$$

$$E_{\mathrm{p}} = B\sinh\delta z \tag{10.138}$$

$$E_{\mathrm{i}}(z=0) = E_{\mathrm{i}}(0) \tag{10.139}$$

$$\frac{dE_{\mathrm{p}}}{dz} = i\kappa_{\mathrm{p}} E_{\mathrm{s}} E_{\mathrm{i}} \tag{10.140}$$

$$B\delta = i\kappa_{\mathrm{p}} E_{\mathrm{s}} E_{\mathrm{i}}(0) \tag{10.141}$$

$$B = \frac{i\kappa_{\mathrm{p}} E_{\mathrm{s}} E_{\mathrm{i}}(0)}{\delta} = \frac{i\kappa_{\mathrm{p}} E_{\mathrm{s}} E_{\mathrm{i}}(0)}{\sqrt{\kappa_{\mathrm{i}}\kappa_{\mathrm{p}} \left| E_{\mathrm{s}} \right|^2}} = i\sqrt{\frac{\kappa_{\mathrm{p}}}{\kappa_{\mathrm{i}}}} E_{\mathrm{i}}(0) \tag{10.142}$$

$$E_{\mathrm{p}}(z) = i\sqrt{\frac{\kappa_{\mathrm{p}}}{\kappa_{\mathrm{i}}}} E_{\mathrm{i}}(0)\sinh\delta z \tag{10.143}$$

Similarly, we finally get the wave solution to the idler as

$$B\delta\cosh\delta z = i\kappa_{\mathrm{p}} E_{\mathrm{s}} E_{\mathrm{i}}(z) \tag{10.144}$$

$$E_{\mathrm{i}}(z) = \frac{1}{i\kappa_{\mathrm{p}} E_{\mathrm{s}}} B\delta\cosh\delta z \tag{10.145}$$

$$E_{\mathrm{i}}(z) = \frac{1}{i\kappa_{\mathrm{p}} E_{\mathrm{s}}} i\sqrt{\frac{\kappa_{\mathrm{p}}}{\kappa_{\mathrm{i}}}} E_{\mathrm{i}}(0)\delta\cosh\delta z \tag{10.146}$$

$$E_{\mathrm{i}}(z)=\frac{1}{i\kappa_{\mathrm{p}}E_{\mathrm{s}}}i\sqrt{\frac{\kappa_{\mathrm{p}}}{\kappa_{\mathrm{i}}}}E_{\mathrm{i}}(0)\sqrt{\kappa_{\mathrm{p}}\kappa_{\mathrm{i}}}E_{\mathrm{s}}\cosh\delta z \tag{10.147}$$

$$E_{\mathrm{i}}(z)=E_{\mathrm{i}}(0)\cosh\delta z \tag{10.148}$$

The gain coefficient is written as

$$P_2=\frac{n_2}{2c\mu_0}\left|E_2\right|^2 S=\frac{n_2}{2c\mu_0}u_2^2 S \tag{10.149}$$

$$u_2=\sqrt{\frac{2c\mu_0 P_2}{n_2 S}} \tag{10.150}$$

$$\gamma=\kappa u_2=\frac{\omega d}{cn_1}\sqrt{\frac{2c\mu_0 P_2}{n_2 S}} \tag{10.151}$$

The output pump power is written as

$$P_{\mathrm{p}}(z)=\frac{n_{\mathrm{p}}}{2c\mu_0}\left|E_{\mathrm{p}}\right|^2 S=\frac{n_{\mathrm{p}}}{2c\mu_0}\left|i\sqrt{\frac{\kappa_{\mathrm{p}}}{\kappa_{\mathrm{i}}}}E_{\mathrm{i}}(0)\right|^2\sinh^2\delta z\cdot S \tag{10.152}$$

$$P_{\mathrm{p}}(z)=\frac{n_{\mathrm{p}}}{2c\mu_0}\frac{\kappa_{\mathrm{p}}}{\kappa_{\mathrm{i}}}\left|E_{\mathrm{i}}(0)\right|^2\sinh^2\delta z\cdot S \tag{10.153}$$

The input idler power is written as

$$P_{\mathrm{i}}(0)=\frac{n_{\mathrm{i}}}{2c\mu_0}\left|E_{\mathrm{i}}(0)\right|^2 S \tag{10.154}$$

$$\left|E_{\mathrm{i}}(0)\right|^2=\frac{2c\mu_0 P_{\mathrm{i}}(0)}{n_{\mathrm{i}}S} \tag{10.155}$$

The maximum power generating at the pump is written as

$$P_{\mathrm{p}}(z)=\frac{n_{\mathrm{p}}}{2c\mu_0}\frac{cn_{\mathrm{i}}}{\omega_{\mathrm{i}}d}\left|\frac{2c\mu_0 P_{\mathrm{i}}(0)}{n_{\mathrm{i}}S}\right|\sinh^2\delta z\cdot S \tag{10.156}$$

$$P_{\mathrm{p}}(z)=\frac{\omega_{\mathrm{p}}}{\omega_{\mathrm{i}}}P_{\mathrm{i}}(0)\sinh^2\delta z \tag{10.157}$$

This point reveals that $\frac{P_{\mathrm{i}}(0)}{\omega_{\mathrm{i}}}$ is the number of photons at the idler frequency entering the crystal per unit time. It is also implied that we have completely converted all the photons at the idler frequency to the pump frequency at the output (Fig. 10.7).

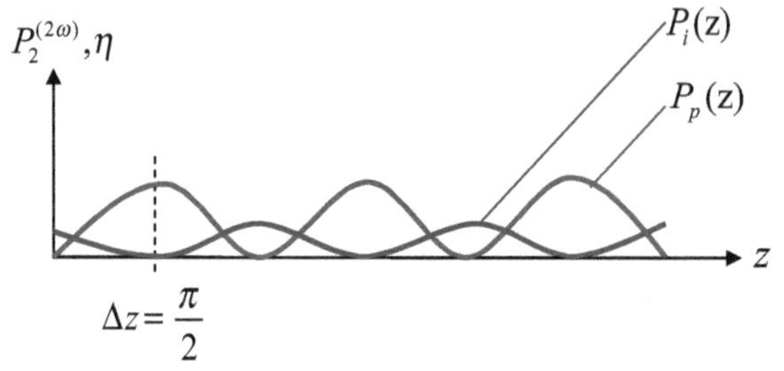

Figure 10.7 Conversion of all the photons at the idler frequency to the pump frequency at the crystal length.

10.4.2 Optical Parametric Oscillation

If we put the parametric amplifier within a laser cavity under a pair of mirrors (Fig. 10.8), we can convert the amplifier into the oscillator, the so-called optical parametric oscillator.

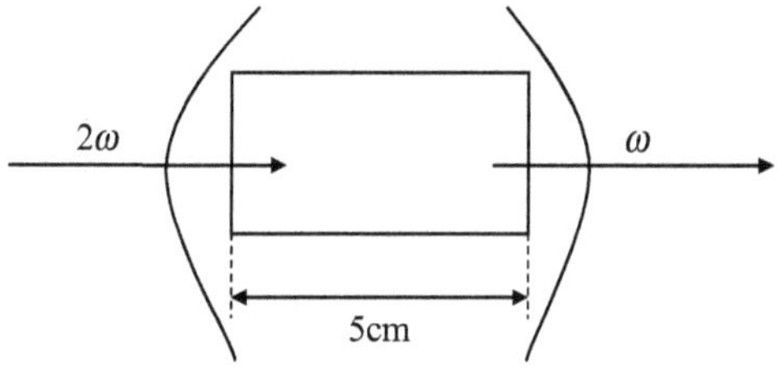

Figure 10.8 The double frequency (i.e., visible laser) passes through a pair of mirrors to generate the fundamental frequency (i.e., infrared laser).

According to the power efficiency (no pump depletion) as

$$\eta = \frac{2\mu_0\omega^2}{cn_1^2 n_2} \frac{d^2 P_1 \ell^2}{A} \frac{\sin^2(\Delta k\ell/2)}{(\Delta k\ell/2)^2} \tag{10.158}$$

we can increase the power efficiency by increasing P_1/A by focusing the input fundamental frequency at the same power of the fundamental. This aspect is carried out by decreasing a cross-sectional area of the input fundamental frequency (Fig. 10.9) via:

$$w_z^2 = w_0^2\left(1+\frac{\lambda^2\ell^2}{\pi^2 w_0^4}\right) \tag{10.159}$$

The beam maintains its approximated cross section at the crystal length of interaction as

$$\ell \simeq 2Z_\mathrm{R} = 2\left[\frac{\pi w_0^2}{\lambda}\right] \tag{10.160}$$

$$A = \pi w_0^2 \tag{10.161}$$

$$\frac{\ell}{A} = \frac{2Z_\mathrm{R}}{\pi w_0^2} = \frac{2}{\lambda} \tag{10.162}$$

where w_0 is the waist size of Gaussian beam at the focal point and Z_R is the Raleigh range.

The constant power efficiency at the crystal length reaches

$$\eta = \frac{2\mu_0\omega^2}{cn_1^2 n_2} d^2 P_1 \frac{2\ell}{\lambda} \frac{\sin^2(\Delta k\ell/2)}{(\Delta k\ell/2)^2} \tag{10.163}$$

This reveals that when we are phase matched with the plane wave, power efficiency increases quadratically with the crystal length of interaction (Fig. 10.10). If we double the length of interaction within the crystal, we will have four times the power efficiency for the optical parametric oscillator. We can also adjust the input frequency for the maximum power efficiency under the phase matching condition.

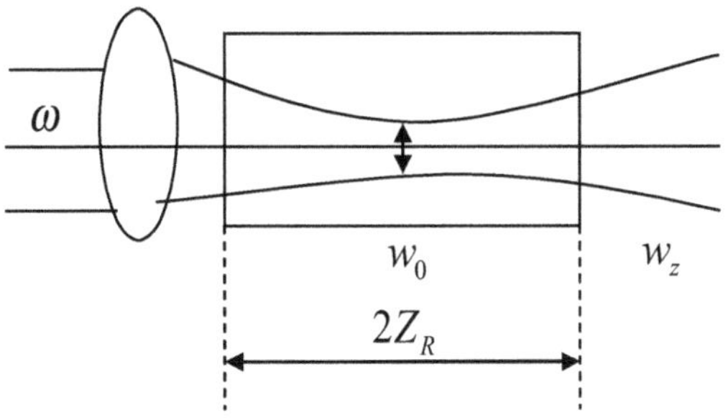

Figure 10.9 Gaussian fundamental beam focused by lens into the crystal expands from the crystal.

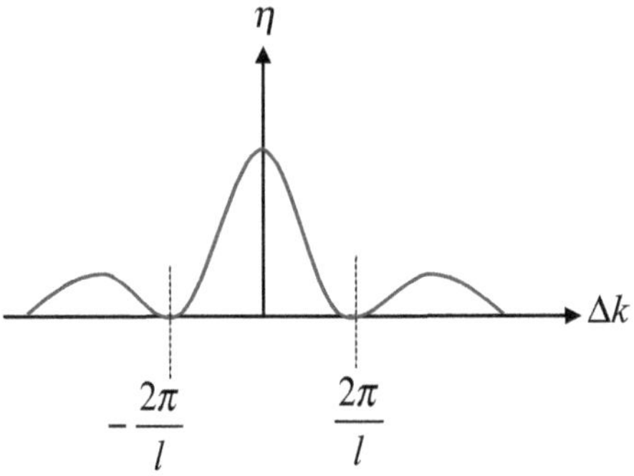

Figure 10.10 Power efficiency as a function of phase difference in a tunable laser.

10.5 Optical Resonant Oscillator

We can convert an amplifier as a source of radiation to an oscillator by the following conditions:

- Put this amplifier within a pair of mirrors.
- Make mirrors having a reflectivity at a certain wavelength and complete transmitting other wavelength.
- Assume the reflectivity of the pump and idler to be zero at two mirrors, while the reflectivity of the signal wavelength is very high.

By interference effect on the thin film dielectric coating, the output wavelength is generated out of the crystal according to the following steps (Fig. 10.11):

- Spontaneously generated signal photons hit mirror M_2 and most of them transmit (99%) and part of them get reflected back (1%) without amplification.
- After hitting M_1 in the reverse direction, partly transparent and partly reflected, now it is used as the pump and then gets amplified ($\omega_s + \omega_p$).
- Signal will now draw the energy from the pump and get amplified as it propagates to M_2. This signal will go back and forth and will increase this signal inside the cavity.
- Idler escapes from the cavity.

- Any amplifier has the gain saturation because we start with a loop gain more than 1. So the amount of loss is less; then gain increases in the amplitude. If we have all the time gain, more than loss, this will keep on increasing to infinity.
- When the signal becomes larger and larger, it will bring down the gain to a value such that the gain is equal to loss. For example, if the loss is the finite reflectivity of 99%, every time we go to one round trip, we loss 2% of that light, so we must gain 2%. The pump will adjust itself to give a gain of 2% for every loss of 2% in one round trip.
- As the signal power increases, the pump power will start to drop; the gain will drop because the gain depends on the pump power.
- We will reach gain saturation where the signal will be such that the round trip gain is equal to the round trip loss, and the continuous signal will come out of a cavity at ω_s, ω_i,.

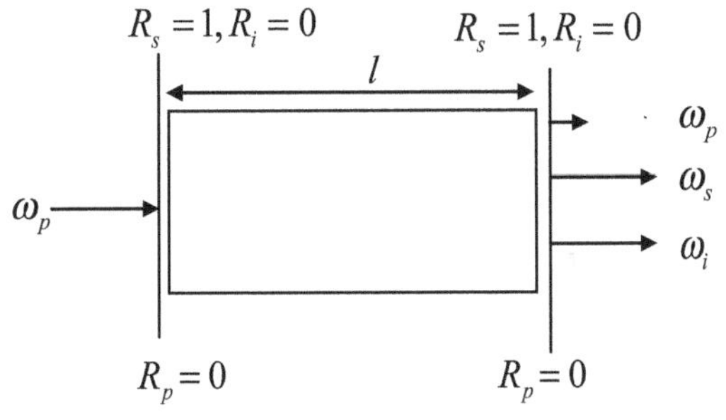

Figure 10.11 By interference effect on the thin film dielectric coating, we can have strong reflectivity at a certain wavelength and weak reflectivity at other wavelength. The mirrors do not perfectly reflect at all wavelengths.

- Assume that we have two mechanisms used for a birefringence phase matching (crystal orientation) and quasi-phase matching (periodic domain reversal) as

$$\omega_p = \omega_s + \omega_i \tag{10.164}$$

$$k_p = k_s + k_i \tag{10.165}$$

$$k_p = k_s + k_i + K \tag{10.166}$$

- ω_p photon comes in and then interacts with the crystal and spontaneous parametric fluorescence or down conversion to a pair of ω_s and ω_i photons. This, just like a spontaneous emission, is brought about by the vacuum fluctuation.

- According to a crystal length, the output wavelength is written as

$$l = \frac{m\lambda_s}{2n_s} = \frac{m}{2}\frac{\lambda_s}{n_s} \quad (10.167)$$

$$\lambda_s = \frac{2n_s l}{m} = \frac{c}{v_s} \Rightarrow v_s = m\frac{c}{2n_s l} \quad (10.168)$$

10.5.1 Single Resonant Oscillator

Single resonant oscillator (SRO) exists if the signal only resonates inside the cavity, but not for idler because mirrors have zero reflectivity at the idler (Fig. 10.12).

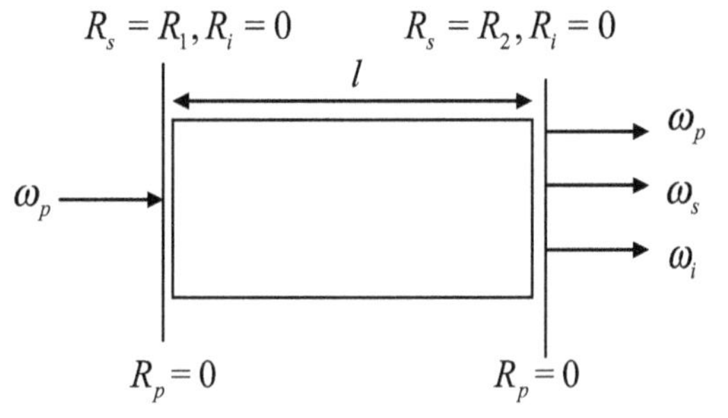

Figure 10.12 Optical parametric oscillator in a single resonator with the mirror reflectivities R_1 and R_2.

Suppose that the mirrors are not introducing any phase change due to the wave propagation via the crystal. The phase change in one round trip inside the cavity is defined as

$$2k_s l = 2m\pi \quad (10.169)$$

$$\frac{n_s\omega_s}{c}l = m\pi \Leftrightarrow \omega_s = m\frac{\pi c}{l n_s} \quad (10.170)$$

The inter-mode frequency at the crystal length for the oscillation within a cavity corresponding to the oscillation frequency is written as

$$\nu_s = \frac{\omega_s}{2\pi} = m\frac{c}{2n_s l} \quad (10.171)$$

$$\Delta\nu_s \simeq \frac{c}{2n_s l} \quad (10.172)$$

For the effective oscillation, Es_4 must be equal to Es_0 under a steady-state situation where the continuous pump comes in and the signal will adjust itself to the values such that there is a complete compensation of loss by the gain (Fig. 10.13).

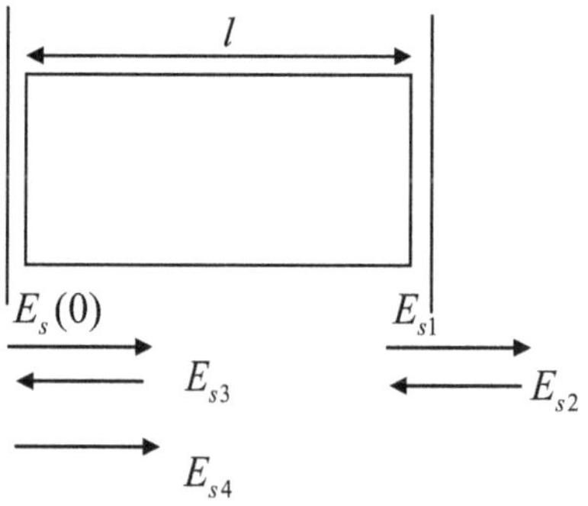

Figure 10.13 The signal field in one cycle in the single resonant oscillator.

The wave solutions corresponding to the wave equations are written as

$$\frac{dE_s}{dz} = i\kappa_s E_p E_i^* \quad (10.173)$$

$$\frac{dE_i}{dz} = i\kappa_i E_p E_s^* \quad (10.174)$$

$$\frac{d^2E_s}{dz^2} = i\kappa_s E_p\left(-i\kappa_i E_p^* E_s\right) = \kappa_s\kappa_i\left|E_p\right|^2 E_s = g^2 E_s \quad (10.175)$$

$$E_s(z) = A\sinh gz + B\cosh gz \quad (10.176)$$

$$E_s(z=0) = E_s(0) \quad \Rightarrow B = E_s(0) \quad (10.177)$$

$$E_i(z=0) = 0 \quad \Rightarrow \frac{\partial E_s}{\partial z} = 0 \quad \Rightarrow A = 0 \quad (10.178)$$

$$E_s(z) = E_s(0)\cosh gz \tag{10.179}$$

This expression for the signal field is implied as follows:

- The signal starts from M_1 and gets amplified on reaching M_2 as

$$E_{s1}(z) = E_s(0)\cosh gl \tag{10.180}$$

- The signal gets reflected back by a factor as

$$E_{s2}(z) = \sqrt{R_2}\,E_s(0)\cosh gl \tag{10.181}$$

- The signal does not get amplified as

$$E_{s3}(z) = \sqrt{R_2}\,E_s(0)\cosh gl \tag{10.182}$$

- After that, the signal gets reflected by M_1 with a factor as

$$E_{s4}(z) = \sqrt{R_1 R_2}\,E_s(0)\cosh gl = E_s(0) \tag{10.183}$$

If the signal starts to oscillate and, hence, we increase the pump power beyond this value, the signal will increase just like laser. Consequently, the gain threshold is written as

$$\cosh(g_{\text{th}}\,\ell) \simeq 1 + \frac{g_{\text{th}}^2\,\ell^2}{2} = \frac{1}{\sqrt{R_1 R_2}} \tag{10.184}$$

$$\left(g_{\text{th}}\right)_{\text{SRO}} = \frac{\sqrt{2}}{\ell}\left(\frac{1}{\sqrt{R_1 R_2}} - 1\right)^{1/2} \tag{10.185}$$

The threshold intensity of the pump corresponding to the pump intensity and gain coefficient (g) is written as

$$I_{\text{P}} = \frac{n_{\text{p}}}{2c\mu_0}\left|E_{\text{p}}\right|^2 \tag{10.186}$$

$$\left|E_{\text{p}}\right|^2 = \frac{2c\mu_0}{n_{\text{p}}} I_{\text{P}} \tag{10.187}$$

$$g^2 = \kappa_s \kappa_i \left|E_{\text{p}}\right|^2 = \frac{\omega_s d}{c n_s}\frac{\omega_i d}{c n_i}\left|E_{\text{p}}\right|^2 = \frac{\omega_s \omega_i d^2}{c^2 n_s n_i}\left|E_{\text{p}}\right|^2 \tag{10.188}$$

$$\frac{\omega_s \omega_i d^2}{c^2 n_s n_i}\frac{2c\mu_0}{n_{\text{p}}} I_{\text{P,th}} = \frac{2}{\ell^2}\left(\frac{1}{\sqrt{R_1 R_2}} - 1\right) \tag{10.189}$$

$$I_{\text{P,th}} = \frac{cn_{\text{p}}n_{\text{s}}n_{\text{i}}}{\mu_0 \omega_{\text{s}} \omega_{\text{i}} d^2 \ell^2}\left(\frac{1}{\sqrt{R_1 R_2}} - 1\right) \tag{10.190}$$

The pump power required for the oscillator to start under the cross-sectional area of S is calculated by

$$P_{\text{P,th}} = I_{\text{P,th}} \times S \tag{10.191}$$

The frequency range (a gain bandwidth) under the following conditions is calculated as (Fig. 10.14)

$$\Delta k = 0$$

$$\omega_{\text{s}} = \omega_{\text{s0}}$$

$$\omega_{\text{i}} = \omega_{\text{i0}}, \tag{10.192}$$

$$\Delta \nu \simeq \frac{c}{l\left[N(\omega_{\text{s0}}) - N(\omega_{\text{i0}})\right]} \tag{10.193}$$

$$\Delta \nu \simeq \frac{c(1-R)}{2\pi n_{\text{s}} l} \tag{10.194}$$

The group index is defined as

$$N(\omega_{\text{s0}}) = n(\omega_{\text{s0}}) + \omega_{\text{s0}} \left.\frac{dn}{d\omega}\right|_{\text{s0}} \tag{10.195}$$

$$N(\omega_{\text{s0}}) - N(\omega_{\text{i0}}) \simeq 0.05 \tag{10.196}$$

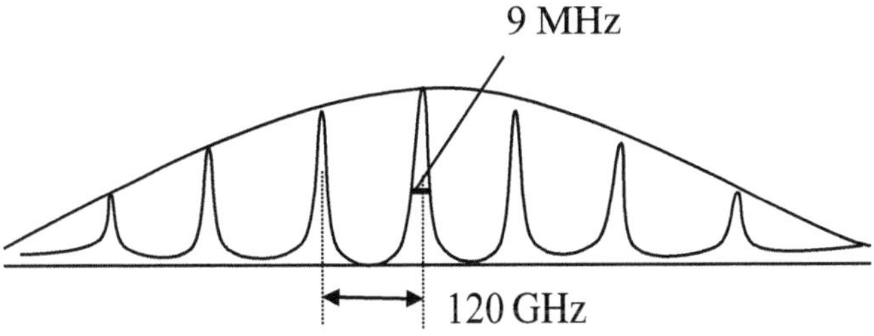

Figure 10.14 Overall big gain bandwidth of 120 GHz and small inter-mode spacing under a crystal length of 5 cm and $N(\omega_{\text{s0}}) - N(\omega_{\text{i0}}) \simeq 0.05$ as well as the inter-mode spacing at approximately 9 MHz ($R = 0.98$).

10.5.2 Double Resonant Oscillator

A double resonant oscillator (DRO) exists when the signal and idler both resonate inside the cavity (Fig. 10.15). However, the power required for operating a DRO is much lower than that required for SRO at some cost in terms of instability of the oscillator itself.

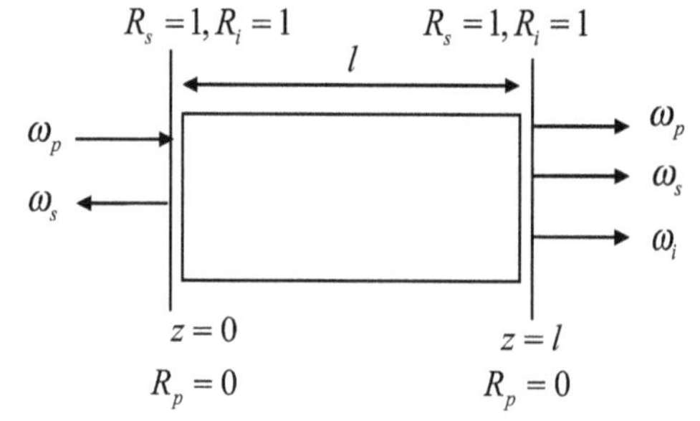

Figure 10.15 The signal and idler fields in a double resonant oscillator.

Assume that the crystal fills the entire cavity; otherwise, the gain will happen in a part of the cavity. The frequency is determined by the entire length of the cavity and the refractive index of the crystal. The wave solutions corresponding to the wave equations are written as

$$\omega_p = \omega_s + \omega_i$$

$$k_p = k_s + k_i$$

$$\Delta k = 0$$

$$E_s(z) = E_s(0)\cosh gz + i\sqrt{\frac{\omega_s n_i}{\omega_i n_s}} E_i^*(0)\sinh gz \tag{10.197}$$

$$E_i^*(z) = -i\sqrt{\frac{\omega_i n_s}{\omega_s n_i}} E_s(0)\sinh gz + E_i^*(0)\cosh gz \tag{10.198}$$

$$g^2 = \kappa_s \kappa_i \left| E_p^2 \right| \tag{10.199}$$

$$\begin{pmatrix} E_s(z) \\ E_i^*(z) \end{pmatrix} = \begin{pmatrix} \cosh gz & i\sqrt{\dfrac{\omega_s n_i}{\omega_i n_s}} \sinh gz \\ -i\sqrt{\dfrac{\omega_i n_s}{\omega_s n_i}} \sinh gz & \cosh gz \end{pmatrix} \begin{pmatrix} E_s(0) \\ E_i^*(0) \end{pmatrix} = \begin{pmatrix} A & B \\ C & D \end{pmatrix} \begin{pmatrix} E_s(0) \\ E_i^*(0) \end{pmatrix} \tag{10.200}$$

The matrix propagation from left to right over a crystal length L is written as

$$\begin{pmatrix} E_s(\text{ref}) \\ E_i^*(\text{ref}) \end{pmatrix} = \begin{pmatrix} \sqrt{R_s} & 0 \\ 0 & \sqrt{R_i} \end{pmatrix} \begin{pmatrix} E_s(\text{inc}) \\ E_i^*(\text{inc}) \end{pmatrix} \tag{10.201}$$

where rt is the round trip, ref is reflection, and inc is incident.

- The matrix for one round trip is written as

$$\begin{pmatrix} E_s(\text{rt}) \\ E_i^*(\text{rt}) \end{pmatrix} = \begin{pmatrix} \sqrt{R_s} & 0 \\ 0 & \sqrt{R_i} \end{pmatrix} \begin{pmatrix} 1 & 0 \\ 0 & 1 \end{pmatrix} \begin{pmatrix} \sqrt{R_s} & 0 \\ 0 & \sqrt{R_i} \end{pmatrix} \begin{pmatrix} A & B \\ C & D \end{pmatrix} \begin{pmatrix} E_s(0) \\ E_i^*(0) \end{pmatrix} \tag{10.202}$$

$$\begin{pmatrix} E_s(\text{rt}) \\ E_i^*(\text{rt}) \end{pmatrix} = \begin{pmatrix} R_s A & R_s B \\ R_i C & R_i D \end{pmatrix} \begin{pmatrix} E_s(0) \\ E_i^*(0) \end{pmatrix} \tag{10.203}$$

- After one round trip, we have

$$\begin{pmatrix} E_s(\text{rt}) \\ E_i^*(\text{rt}) \end{pmatrix} = \begin{pmatrix} E_s(0) \\ E_i^*(0) \end{pmatrix} \tag{10.204}$$

$$\begin{pmatrix} R_s A & R_s B \\ R_i C & R_i D \end{pmatrix} \begin{pmatrix} E_s(0) \\ E_i^*(0) \end{pmatrix} - \begin{pmatrix} E_s(0) \\ E_i^*(0) \end{pmatrix} = 0 \tag{10.205}$$

$$\begin{pmatrix} R_s A - 1 & R_s B \\ R_i C & R_i D - 1 \end{pmatrix} \begin{pmatrix} E_s(0) \\ E_i^*(0) \end{pmatrix} = 0 \tag{10.206}$$

$$\begin{vmatrix} R_s A - 1 & R_s B \\ R_i C & R_i D - 1 \end{vmatrix} = 0 \tag{10.207}$$

- It starts to oscillate and the gain threshold is then written as

$$\cosh g_{\text{th}}\ell = \frac{1+R_s R_i}{R_s + R_i} \tag{10.208}$$

$$\left(g_{\text{th}}\right)_{\text{DRO}} \approx \frac{\sqrt{(1-R_s)(1-R_i)}}{\ell} \tag{10.209}$$

$$\frac{(g_{\text{th}})_{\text{DRO}}}{(g_{\text{th}})_{\text{SRO}}} = \frac{\sqrt{(1-R_s)(1-R_i)}}{\ell} \frac{\ell}{\sqrt{2(1-R_s)}} = \sqrt{\frac{1-R_i}{2}} \tag{10.210}$$

- The ratio of the threshold intensity of the pump (T-factor) is written as

$$T = \frac{(I_{\text{p,th}})_{\text{DRO}}}{(I_{\text{p,th}})_{\text{SRO}}} = \frac{(g_{\text{th}})^2_{\text{DRO}}}{(g_{\text{th}})^2_{\text{SRO}}} = \frac{1-R_i}{2} \tag{10.211}$$

The threshold intensity of the pump requires 100 (T = 1/100). For example, if 10 W pump power is required from the SRO, there will be 100 mW power from the DRO, meaning that 100 times down the power to operate the oscillator and we can almost have a continuous wave oscillation of this parametric oscillator because both signal and idler have high intensity within a cavity and both are responsible for drawing energy from the pump.

By neglecting the frequency dependence on the refractive index, the relationship between the frequency spacing is written as (Fig. 10.16)

$$\omega_s = m\frac{\pi c}{n_s l} \tag{10.212}$$

$$\omega_i = p\frac{\pi c}{n_i l} \tag{10.213}$$

$$\omega_s > \omega_i \,;\, \mathbf{n}_s > n_i \tag{10.214}$$

$$\Delta\omega_s \simeq \frac{\pi c}{n_s l} \tag{10.215}$$

$$\Delta\omega_i \simeq \frac{\pi c}{n_i l} \tag{10.216}$$

$$\Delta\omega_s < \Delta\omega_i \tag{10.217}$$

$$\omega_p = \omega_s + \omega_i \tag{10.218}$$

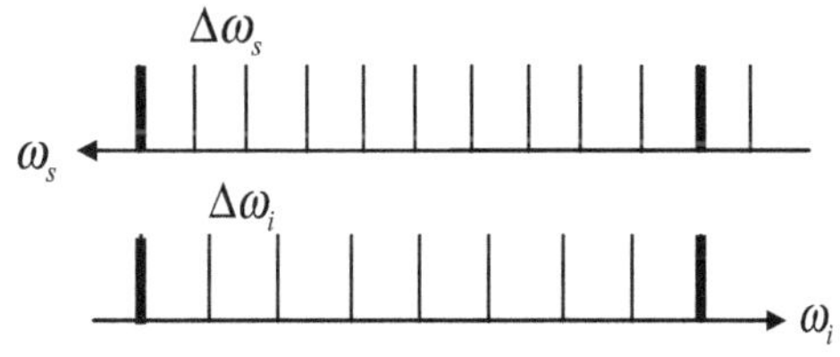

Figure 10.16 ω_s increases in the $-x$-direction and ω_i increases in the $+x$-direction to form a standing wave in the optical cavity.

This point reveals the relationship of the frequency spacing between signal and idler frequencies. The axes of these two are reverse of each other. At any values, the sum of these two values is the same (at a constant ω_p); either the first pair will oscillate or the second pair will oscillate in every range of frequency. Suppose the temperature or length changes; both these combs move because the refractive index depends on the temperature or cavity length. So the oscillator will adjust itself to a pair of frequencies, which satisfy the condition that they are exactly overlapping. This leads to a problem of instability of the oscillator.

Problems

10.1 Clarify that there is no generation of sub-harmonic if we feed the input pump just at the double frequency.

10.2 Prove the following equation.

$$\sin\psi_m = \frac{n_e(2\omega)}{n_0(\omega)}\sqrt{\frac{n_0^2(2\omega)-n_0^2(\omega)}{n_0^2(2\omega)-n_e^2(2\omega)}}$$

10.3 Determine the gain coefficient of lithium niobate having $\lambda_0 = 1$ micron, $d = 30 \times 10^{-12}$ m/V, $n_1 = 2.2$, $P_2 = 1$ W, $S = 0.1$ mm^2.

10.4 Show the difference of the electric field of signal and idler with and without the initial phase in a matrix form.

10.5 Give a few examples of parametric fluorescence, different frequency, and sum frequency generation.

10.6 Check the inter-mode frequency at a crystal length of 5 cm and $n_s = 2$.

10.7 Determine the threshold intensity of the pump from the double resonant oscillator if λ_p, λ_s, λ_i = 0.6, 1.0, 1.5 micron and

$l = 5$ cm with $d = 3 \times 10^{-11}$ m/V ($R_1 = R_2 = 0.98$).

10.8 Find the T-factor when $R_i = 0.98$ from the double resonant oscillator.

Suggested Readings

A G Smith, *Introduction to Arduino*, CreateSpace, North Carolina, 2011.

A P Godse and D A Godse, *Digital & Computer Fundamentals*, Technical Publications, Pune, 2016.

C Gerry and P Knight, *Introductory to Quantum Optics*, Cambridge University Press, UK, 2006.

H Li, L Cai, and Z Chen, Coumarin-derived fluorescent chemosensors, In *Advances in Chemical Sensors*, InTech, New York, 2012.

H M Daluz, *Fundamentals of Fingerprint Analysis*, CRC Press, Florida, 2014.

J R Lakowicz, *Principles of Fluorescence Spectroscopy*, Kluwer Academic/ Plenum Publishers, New York, 1999.

K D Tomsia and E M Goldys, Gold and silver nanowires for fluorescence enhancement, In *Nanowires: Fundamental Research*, InTech, Croatia, 2011.

K-T Chang, *Computation for Bilinear Interpolation: Introduction to Geographic Information Systems*, McGraw-Hill, New York, 2014.

Los Alamos Laboratory, *MCNP – A General Monte Carlo N-Particle Transport Code (Version 5) Volume 1: Overview and Theory*, USA, 2003.

Los Alamos Laboratory, *MCNP – A General Monte Carlo N-Particle Transport Code (Version 5) Volume 2: User's Guide*, USA, 2003.

M Banzi, *Getting Started with Arduino*, Make: Books, California, 2011.

M Fox, *Quantum Optics: An Introduction*, Oxford University Press, UK, 2006.

N T K Thanh, *Clinical Applications of Magnetic Nanoparticles*, CRC Press, Florida, 2018.

O Venkataraman, *Applied Optimization with MATLAB Programming*, Wiley, New York, 2009.

S J Chapman, *MATLAB Programming with Applications for Engineers*, Cengage Learning, New York, 2012.

S Ramanathan, *Thin Film Metal-Oxides*, Springer, New York, 2010.

Solutions to Problems

Chapter 1

1.1 Write Maxwell's equations in the forms of integral and differential equations.

Solution

Faraday's law is written as

$$\oint \vec{E} \cdot d\vec{l} = -\mu \frac{d}{dt} \iint \vec{H} \cdot d\vec{S}$$

$$\nabla \times \vec{E} = -\mu \frac{\partial \vec{H}}{\partial t}$$

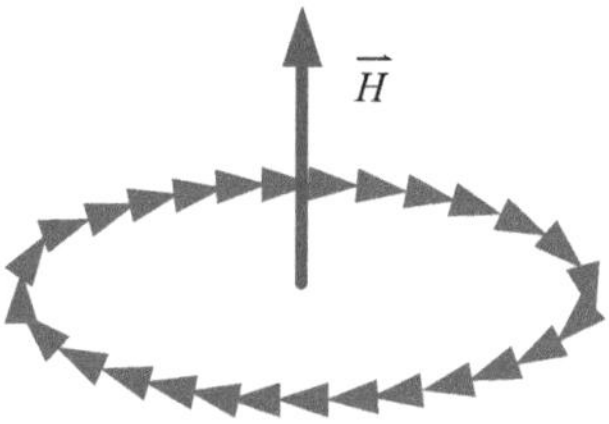

Figure S.1.1

Ampere's law is written as

$$\oint \vec{H} \cdot d\vec{l} = \frac{d}{dt} \iint \vec{J} \cdot d\vec{S} + \varepsilon \frac{d}{dt} \iint \vec{E} \cdot d\vec{S}$$

$$\nabla \times \vec{H} = \vec{J} + \varepsilon \frac{\partial \vec{D}}{\partial t}$$

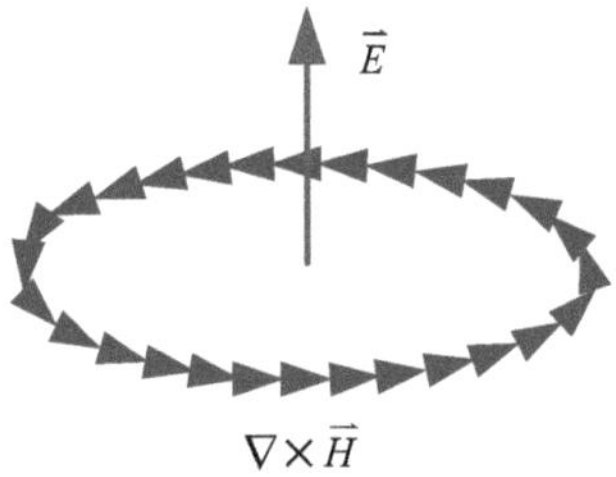

Figure S.1.2

Gauss's law for electric field is written as

$$\oiint \vec{E} \cdot d\vec{S} = \frac{1}{\varepsilon_o} \iiint \rho dV$$

$$\nabla \cdot \vec{E} = \frac{\rho}{\varepsilon_o}$$

Gauss's law for magnetic field is written as

$$\oiint \vec{B} \cdot d\vec{S} = 0$$

$$\nabla \cdot \vec{B} = 0$$

1.2 Show a solution of current density with respect to the electric field based on Drude model.

Solution

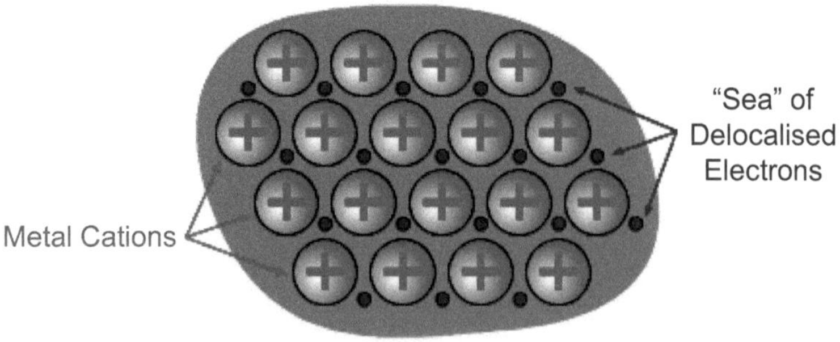

Figure S.1.3

Summation of force on free electrons with respect to the second law of Newton is written as

$$\sum \vec{F} = \vec{F}_{\text{coulomb}} + \vec{F}_{\text{electron collisions}}$$

$$\vec{F}_{\text{coulomb}} = q\vec{E}(t)$$

$$\vec{F}_{\text{electron collisions}} = -m^{*}\Gamma_0 \frac{d\vec{x}(t)}{dt}$$

$$\sum \vec{F} = q\vec{E}(t) - m^{*}\Gamma_0 \frac{d\vec{x}(t)}{dt}$$

$$\sum \vec{F} = m^{*}\vec{a}$$

$$q\vec{E}(t) - m^{*}\Gamma_0 \frac{d\vec{x}(t)}{dt} = m^{*}\frac{d^2\vec{x}(t)}{dt^2}$$

Due to the influence of current density, the wave equation reaches

$$\vec{J}(t) = -Ne\vec{v}(t)$$

$$\frac{d\vec{J}(t)}{dt} + \gamma\vec{J}(t) = \left(\frac{Ne^2}{m^{*}}\right)\vec{E}(t)$$

$$\frac{d\vec{J}_0 e^{-i\omega t}}{dt} + \gamma\vec{J}_0 e^{-i\omega t} = \left(\frac{Ne^2}{m^{*}}\right)\vec{E}_0 e^{-i\omega t}$$

where

$$\vec{E}(t) = \vec{E}_0 e^{-i\omega t}$$

$$\vec{J}(t) = \vec{J}_0 e^{-i\omega t}$$

The wave solution of current density is written as

$$\vec{J}_0 = \left(\frac{Ne^2}{m^{*}(-i\omega + \gamma)}\right)\vec{E}_0$$

$$\vec{J}_0 = \left(\frac{Ne^2}{m^{*}\gamma(1 - i\omega / \gamma)}\right)\vec{E}_0$$

1.3 Draw the surface plasmon resonance of metal nanowires due to the difference of applied electric field.

Solution

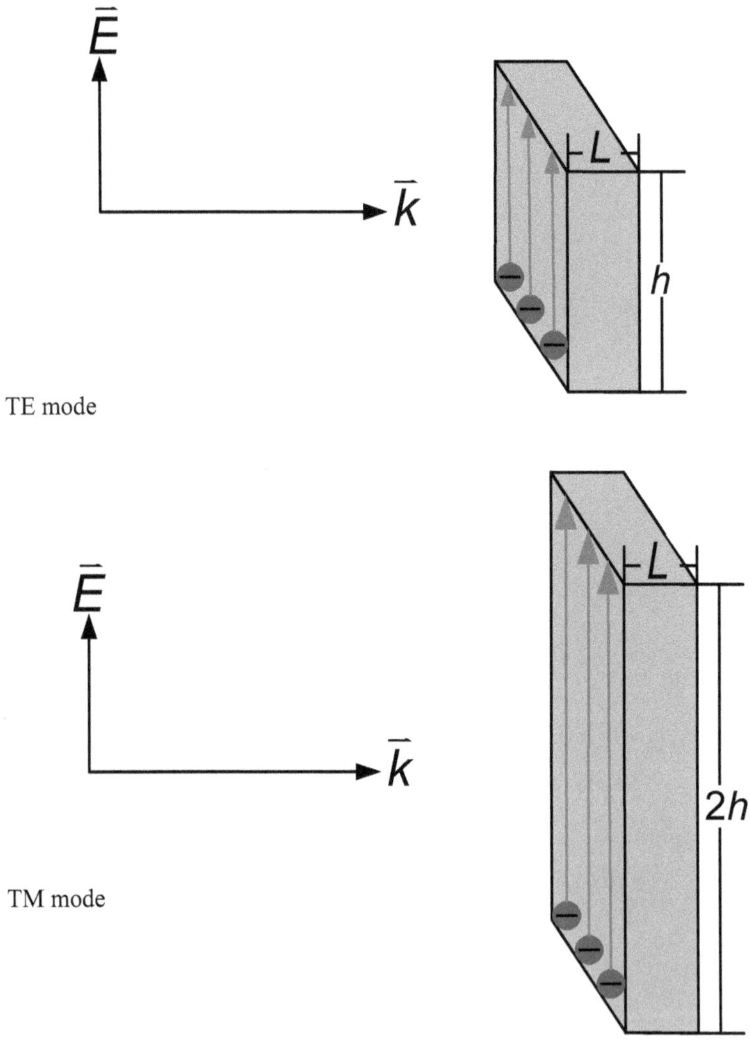

Figure S.1.4

1.4 Draw the electric field profile inversely proportional to the periodicity of metal nanostructures due to the plasmoic effects.

Solution

Figure S.1.5

1.5 Explain the optical line spectra of hydrogen and sodium atoms.

Solution

Emission spectra are divided into three groups as

(1) Continuous spectra are produced by incandescent solids, such as rainbow and black body.
(2) Band spectra are produced by molecules radiating their rotational and/or vibration energies, such as Ca and Ba salts.
(3) Line spectra produced by atoms when electrons change from one to other energy levels, such as H and Na.

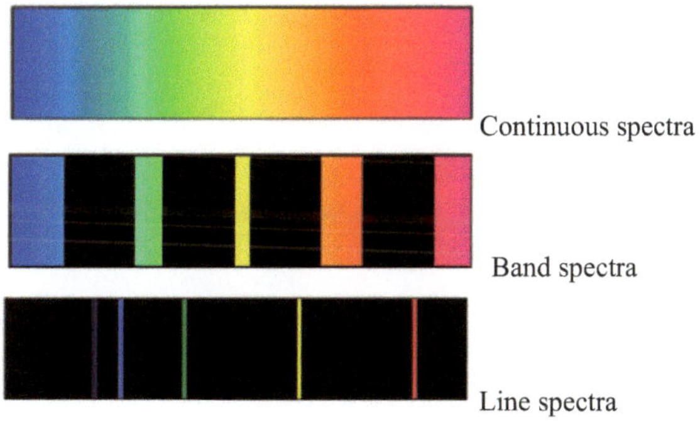

Figure S.1.6

Several line spectra of Hare originated by change of the energy level of one electron in atom. Emission spectra are dependent of the electron transition lines—Lyman, Balmer, Paschen, Brackelt, and Pfund series.

Several line spectra of Na as element in alkali metals are originated by change of the valence energy level of one free electron in atom. Emission spectra are dependent of the electron transition lines.

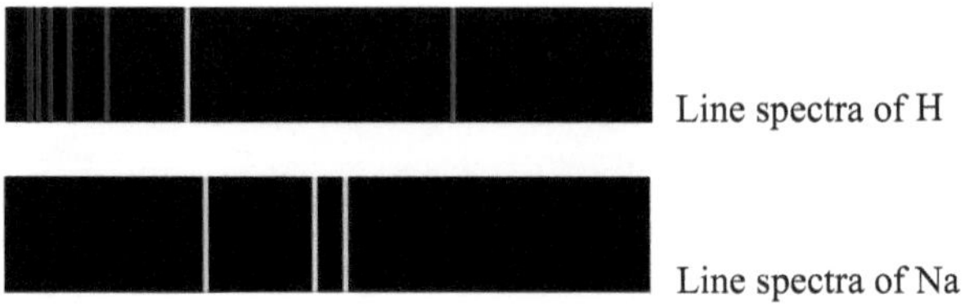

Figure S.1.7

1.6 Show the database to obtain the dielectric constants of some materials.

Solution

Inorganic materials: Ag, Au, Cu, Fe, Pd, Pt, Ti, Zn, etc.
Organic materials: alcohol, ester, benzene, toluene, etc.
Glasses: BK7, BAF10, SF5, SF10, etc.

Source: www.refractiveindex.info
John and Christy, 1972
Hagemannet al., 1974
Windtet al., 1988
Stahrenberget al., 2001
Werner et al., 2009
Wu et al., 2014
Yang et al., 2015
McPeak et al., 2015
Babar and Weaver, 2015

Chapter 2

2.1 Describe the band transition of ZnO in a comparison to Si and Ge.

Solution

Band gap of semiconductors are divided into two groups:

(1) Direct band transition belongs to ZnO, GaAs, and CdS.
Maximum energy level of the valence band aligns with minimum energy level of the conduction band under the same momentum so that the electron–hole recombination easily appears before releasing the photon energy at the valence band.

(2) Indirect band transition belongs to Si and Ge.
Maximum energy level of the valence band does not align with minimum energy level of the conduction band under the different momentum so that the electron–hole recombination hardly appears before the emission of photon energy. Then, electron at the conduction band tries to emit less energy corresponding to the momentum conservation so that the electron will align well with the maximum energy level of the valence band.

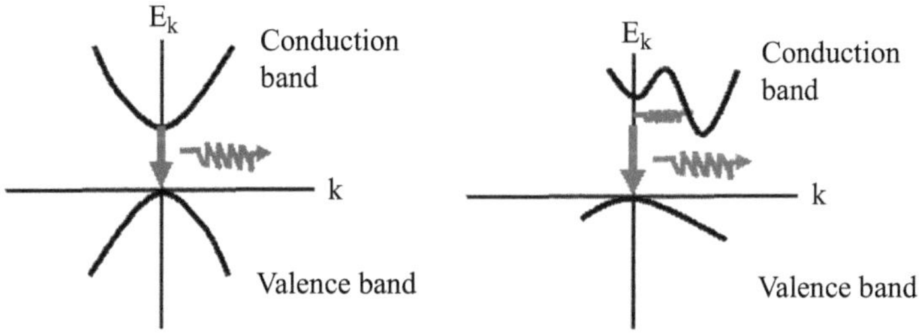

Figure S.2.1 Direct band transition (left) and indirect band transition (right).

2.2 Show a band diagram of metals and semiconductors.

Solution

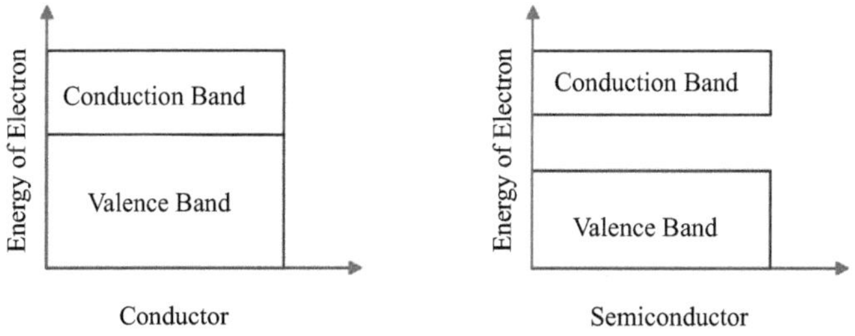

Figure S.2.2

2.3 Clarify the effect of Schottky barrier at interface between metals and semiconductors.

Solution

When metal film is less than a single atomic layer, Fermi levels of metal will pin the Fermi level in semiconductor. Band-bending at the interfaces are dependent of work functions of metals and n-type/p-type of semiconductors according to the following diagram:

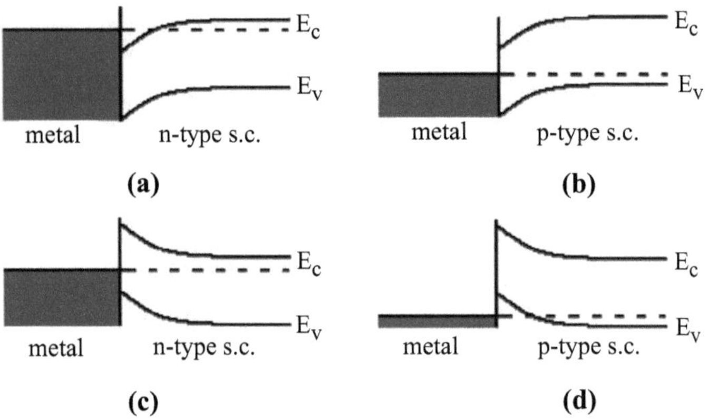

Figure S.2.3

Band diagram of low work function of metal and n-type (a), low work function of metal and p-type (b), high work function of metal and n-type (c), and high work function of metal and p-type (d).

2.4 Explain the absorbance by Beer–Lambert law.

Solution

Beer–Lambert law explains the light attenuation by a light absorbing solution as

$$A = -\log\left[\frac{I}{I_0}\right] = -\log T$$

$$A = \varepsilon C x$$

where

I is intensity of output light
I_0 is intensity of input light
A is absorbance
ε is molar absorption coefficient, 1/Mcm
C is concentration of absorbing solutions, M
L is optical path length of absorbing medium, cm

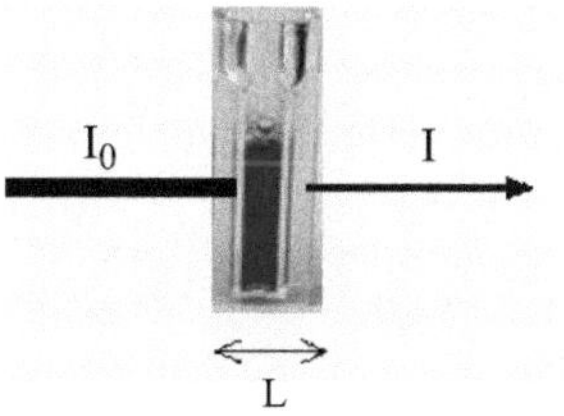

Figure S.2.4

2.5 Show the formation of the reactive oxygen species via electron/hole pairs of ZnO nanoparticles.

Solution

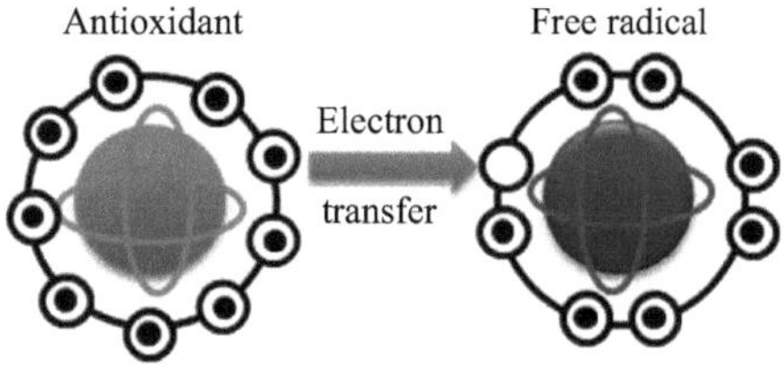

Figure S.2.5

Chapter 3

3.1 Show that how the energy (E) consumed by a light source is determined.

Solution

$$P_{\mathrm{RMS}} = I_{\mathrm{RMS}} \times V_{\mathrm{RMS}}$$

$$E = P_{\mathrm{RMS}} \times t$$

where
P_{RMS} is root mean square power
I_{RMS} is root mean square current
V_{RMS} is a voltage drop across a light source
t is operation time

3.2 Show the material lists to manipulate 3W white-LED in the pulsed mode.

Solution

(a) The driver circuit used to convert a logic level pulse (0–5 V) to a higher voltage pulse (0–12 V) to turn on a high-power LED.

(b) Function generator provided the square wave pulses to the NPN transistor.

(c) 12 V DC power supply applied to the source terminal of the p-channel MOSFET.

(d) Output signal from the drain terminal was fed to a high power LED.

3.3 Explain the fluorescence quantum yield in nontechnical term.

Solution

Fluorescence quantum yield is a ratio of number of photons emitted to the number of photons absorbed. Fluorescence quantum yield ranges between 0 to 1. The higher the fluorescence quantum yield value, the brighter the fluorophore and the stronger the fluorescence intensity is.

3.4 Name other terms describing the fluorescence resonance energy transfer existed via nonradiative dipole–dipole coupling. The energy transfer efficiency is inversely proportional to the third power of a gap distance between donor and acceptor.

Solution

Förster resonance energy transfer (FRET)
Fluorescence resonance energy transfer (FRET)
Resonance energy transfer (RET)
Electronic energy transfer (EET)

Chapter 4

4.1 Show the block diagram of absorption measurement.

Solution

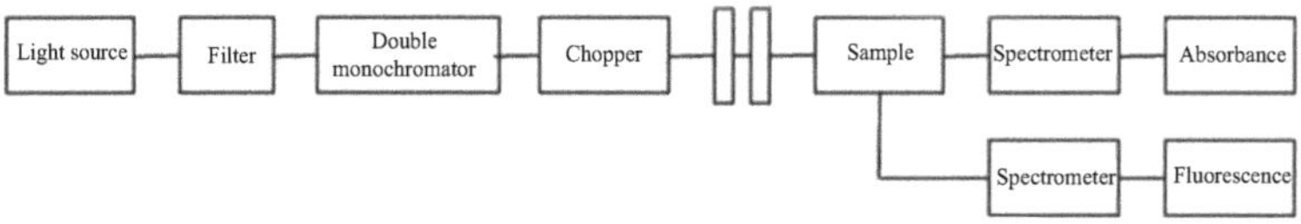

Figure S.4.1

4.2 Draw the simple picture of the spin coating method.

Solution

Figure S.4.2

4.3 Determine the amplitude of electromagnetic field when OLED is operated under frequency of 2.4 GHz and the power intensity of 2.0 W/m^2.

Solution

$$I = \frac{(c/n)}{\mu_0}\left\langle B^2 \right\rangle = \frac{(c/n)}{2\mu_0}\left|B_0\right|^2$$

$$B_0 = \sqrt{\frac{2\mu_0 I}{c}}$$

$$B_0 = \sqrt{\frac{2 \times 1.257 \times 10^{-6} \times 2}{2.998 \times 10^8}} = 4.6 \times 10^{-6}\ T$$

$$E_0 = cB_0 = 2.998 \times 10^8 \times 4.6 \times 10^{-6}$$

$$E_0 = 1.37\ \text{kV/m}$$

4.4 Find a distance from 10 mW OLED as the excitation source to obtain the amplitude of electric field of 0.01 V/m.

Solution

$$I=\frac{P}{A}=\frac{nc\varepsilon_0}{2}|E_0|^2$$

$$\frac{P}{4\pi r^2}=\frac{nc\varepsilon_0}{2}|E_0|^2$$

$$r=\sqrt{\frac{2P}{4\pi nc\varepsilon_0|E_0|^2}}$$

$$r=\sqrt{\frac{2\times 3}{4\pi\times 2.998\times 10^8\times 8.85\times 10^{-12}\times 0.01^2}}=77\text{ m}$$

Chapter 5

5.1 Explain some radioisotopic source to produce the gamma ray.

Solution

Cobalt-60, cesium-137, and europium-152 are the artificial minerals. They were produced in the first nuclear explosion in 1945 at New Mexico. Radioisotopic informations are detailed as the following table:

Mineral	**Half-life, year**	**Energy, MeV**	**Emission probability**
Co-60	5.27	1.173–1.333	1
Cs-137	30.17	0.6617	0.8521
Eu-152	13.33	0.1219–1.408	0.0223–0.2837

5.2 Show the equation of free electrons existed in photoelectric absorption.

Solution

$$E_{e^-}=h\nu-E_b$$

where

E_b is binding energy
$h\nu$ is photon energy
E_{e^-} is energy of photoelectrons or free electrons.

5.3 Write the energy transfer and scattering angle in term of energy and momentum conservation.

Solution

Back scattering is existed if the scattering angle of photon is equivalent to pi angle via the following equation:

$$h\nu' = \frac{h\nu}{1 + \frac{h\nu}{m_0c^2}(1 - \cos\theta)}$$

where
m_0c^2 is rest mass energy of electrons, 0.511 MeV.

5.4 Write the condition of pair production.

Solution

$$h\nu_{\min} = 2m_ec^2 = 1.022\,\text{MeV}$$

5.5 Draw a relationship between photoelectric, Compton, and pair production processes.

Solution

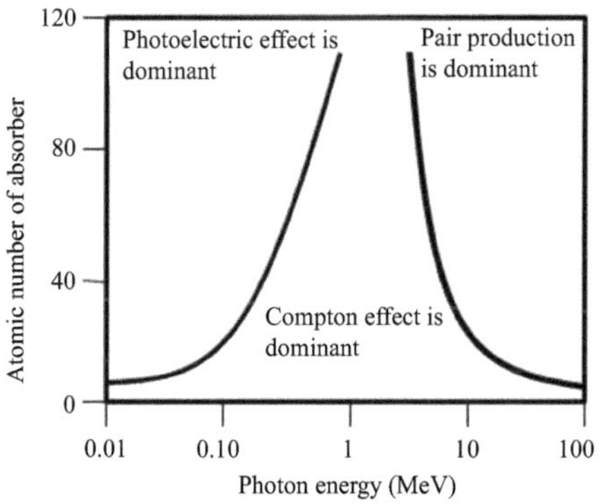

Figure S.5.1

Chapter 6

6.1 List the example of point sources and detectors for the refraction probe.

Solution

Point sources are light emitting diode and laser diode.
Detectors are photodiode and avalanche photodiode.

6.2 Show the step by step of photon propagation in each layer in the human tissue.

Solution

(1) Initial photon position and direction are set as follows:

$$(x = 0, y = 0, z = 0)$$

$$(\mu_x = 0, \mu_y = 0, \mu_z = 1)$$

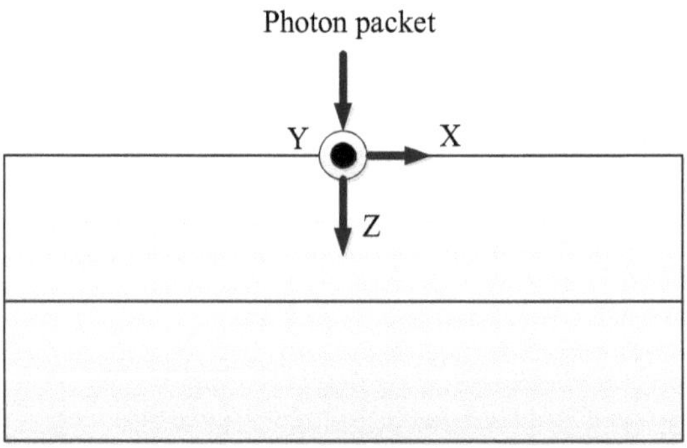

Figure S.6.1

(2) Once the step size, $s = -\ln\sigma / \mu_t$ goes forward, the position and weight of photon are updated as

$$x \rightarrow x + \mu_x \cdot s \qquad y \rightarrow y + \mu_y \cdot s \qquad z \rightarrow z + \mu_z \cdot s$$

$$W \rightarrow W - \Delta W \qquad \Delta W = (\mu_a / \mu_t)W \qquad W = 1 - R_{sp}$$

$$R_{sp} = \frac{(n_{\text{incidence}} - n_{\text{transmission}})^2}{(n_{\text{incidence}} + n_{\text{transmission}})^2}$$

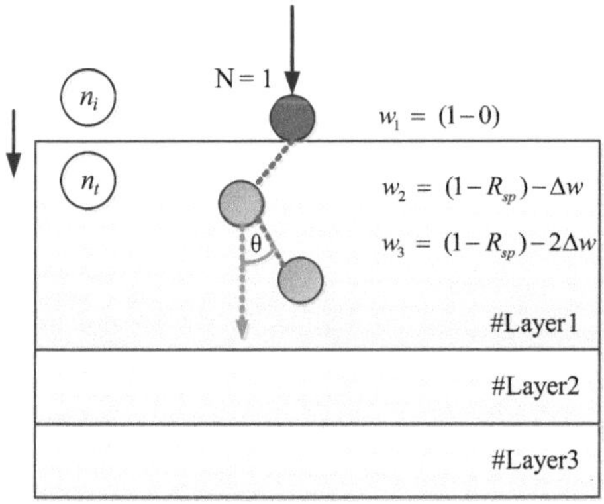

Figure S.6.2

(3) Photon direction is also updated as

$$\mu_x' = \sin\theta\cos\delta$$

$$\mu_y' = \sin\theta\sin\delta$$

$$\mu_z' = \left(\mu_z / |\mu_z|\right)\cos\theta$$

$$\cos\theta = \frac{1}{2\gamma}\left[1+\gamma^2 - \left(\frac{1-\gamma^2}{1-\gamma+2\gamma\sigma}\right)^2\right]$$

$$\delta = 2\pi\sigma$$

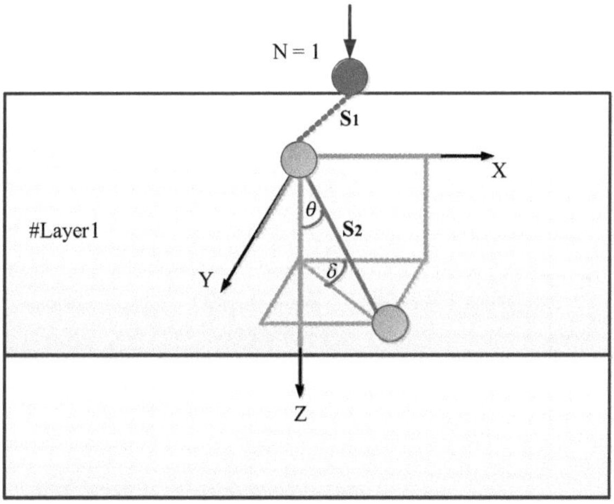

Figure S.6.3

(4) Photon crosses over to the next boundary once step size, *s* hits the first boundary and the internal reflection is defined as

$$R_i = \frac{1}{2}\left(\frac{\sin^2(\lambda_{\text{incidence}} - \lambda_{\text{transmission}})}{\sin^2(\lambda_{\text{incidence}} + \lambda_{\text{transmission}})} + \frac{\tan^2(\lambda_{\text{incidence}} - \lambda_{\text{transmission}})}{\tan^2(\lambda_{\text{incidence}} + \lambda_{\text{transmission}})}\right)$$

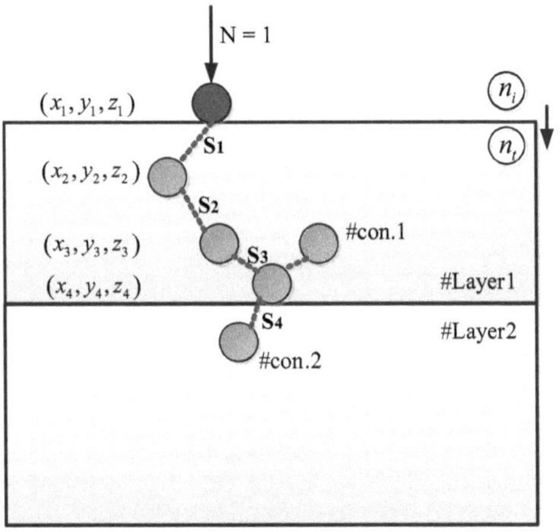

Figure S.6.4

(5) Photon termination is dependent of photon weight, *W* as

$$W = \begin{cases} mW & \text{if } \sigma \le 1/m \\ 0 & \text{if } \sigma > 1/m \end{cases}$$

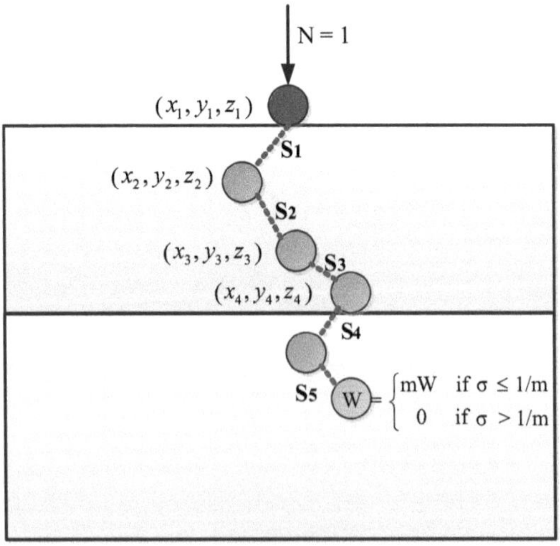

Figure S.6.5

6.3 Find the number of photons coming out of the laser diode having the power of 1 mW and wavelength of 670 nm.

Solution

Photon energy is written as

$$E = h\upsilon = \frac{hc}{\lambda}$$

where

h is Plank's constant, $6.63 \times 10^{-34}\ \mathrm{J \cdot s}$
λ is wavelength
υ is frequency
c is speed of light, $3 \times 10^{8}\ \mathrm{m/s}$

$$E = \frac{6.63 \times 10^{34}\ \mathrm{J \cdot s} \times 3.0 \times 10^{8}\ \mathrm{m/s^{-1}}}{670 \times 10^{-9}\ \mathrm{m}} \qquad = 2.965 \times 10^{-3}\ \mathrm{J \cdot s^{-1}}$$

$$\text{Total energy} \quad = 1.0 \times 10^{-3}\ \mathrm{W} \times \frac{1\ \mathrm{J \cdot s^{-1}}}{1\ \mathrm{W}}$$

$$= 1.0 \times 10^{-3}\ \mathrm{J \cdot s^{-1}}$$

$$\text{Number of photon} = \frac{\text{Total energy}}{\text{Energy of one photon}}$$

$$= \frac{1.0 \times 10^{-3}\ \mathrm{J \cdot s^{-1}}}{2.965 \times 10^{-19}\ \mathrm{J}}$$

$$= 3.4 \times 10^{15}\ \text{photons}$$

6.4 Derive the step size of photon packet from the probability density function.

Solution

Probability density function with respective sampling probability distribution of free path of photon is written as

$$\mathrm{P}(s) = \mu_t \exp(-\mu_t\, s)$$

$$\xi = \int_0^{S_1} \mathrm{P(s)}ds = \int_0^{S_1} \mu_t \exp(-\mu_t\, s)ds = 1 - \exp(\mu_t\, s_1)$$

$$S_1 = \frac{-\ln(1-\xi)}{\mu_t}$$

$$S_1 = \frac{-\ln(\xi)}{\mu_t}$$

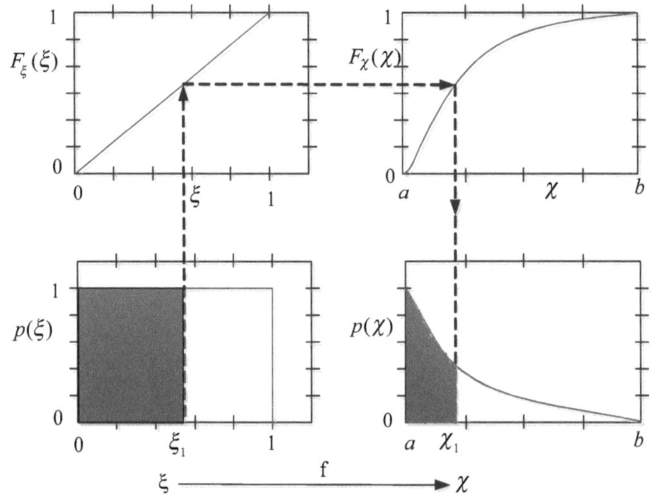

Figure S.6.6

Cumulative probability distribution function and probability density function based on the random variable.

Chapter 7

7.1 Show the difference of Maxwell's equations of electric fields in dielectric and conductive media.

Solution

(a) Maxwell's equations are modified when a plane wave is propagating through a dielectric medium as

$$\vec{\nabla}\cdot\vec{D}=0 \;\Rightarrow\; \vec{\nabla}\cdot\vec{E}=0$$

$$\vec{\nabla}\times\vec{E}=-\dot{B}$$

$$\vec{\nabla}\cdot\vec{B}=0$$

$$\vec{\nabla}\times\vec{H}=\dot{D} \quad\Rightarrow\quad \vec{\nabla}\times\vec{B}=\mu_0\varepsilon_0\dot{E}$$

$$\vec{\nabla}\times\vec{\nabla}\times\vec{E}=\vec{\nabla}(\vec{\nabla}\cdot\vec{E})-\vec{\nabla}^2\vec{E}=-\vec{\nabla}\times\dot{B}$$

$$\vec{\nabla}^2\vec{E}=\vec{\nabla}\times\dot{B}$$

$$\vec{\nabla}^2\vec{E}=\mu_0\varepsilon_0\ddot{E}$$

$$\vec{E}=E_0e^{i(kz-\omega t)}$$

$$k^2\vec{E}=\mu_0\varepsilon_0\omega^2\vec{E}$$

$$v^2=\frac{\omega^2}{k^2}=1/\mu_0\varepsilon_0$$

(b) The electromagnetic wave in a conductive media is originally related to Maxwell's equations as

$$\vec{\nabla}\times\vec{E}=-\dot{B}$$

$$\vec{\nabla}\times\vec{H}=\vec{j}+\dot{D} \quad\Rightarrow\quad \vec{\nabla}\times\vec{B}=\mu\vec{j}+\mu\varepsilon\dot{E} \quad\Rightarrow\quad \vec{\nabla}\times\vec{B}=\mu\sigma\vec{E}+\mu\varepsilon\dot{E}$$

$$\vec{\nabla}\cdot\vec{D}=\rho_{\text{TRUE}}$$

$$\vec{\nabla}\cdot\vec{B}=0$$

$$\vec{\nabla}\times\vec{\nabla}\times\vec{E}=-\vec{\nabla}\times\dot{B}$$

$$\vec{\nabla}\times\vec{\nabla}\times\vec{E}=\vec{\nabla}(\vec{\nabla}\cdot\vec{E})-\vec{\nabla}^2\vec{E}=-\mu\sigma\dot{E}-\mu\varepsilon\ddot{E}$$

$$\vec{\nabla}^2\vec{E}=\mu\sigma\dot{E}+\mu\varepsilon\ddot{E}$$

$$\vec{E}=E_0e^{i(kz-\omega t)}$$

$$-k^2\vec{E}=-i\omega\mu\sigma\vec{E}-\mu\varepsilon\omega^2\vec{E}$$

$$-k^2\vec{E}+i\omega\mu\sigma\vec{E}+\mu\varepsilon\omega^2\vec{E}=0$$

$$k=\alpha+i\beta$$

$$\alpha=\omega\sqrt{\mu\varepsilon\left(\frac{1}{2}+\frac{1}{2}\sqrt{1+\frac{\sigma^2}{\mu^2\varepsilon^2}}\right)}$$

$$\beta=\frac{\varepsilon\mu\sigma}{2\alpha}$$

$$\vec{E}=E_0e^{-\alpha z+i(kz-\omega t)}$$

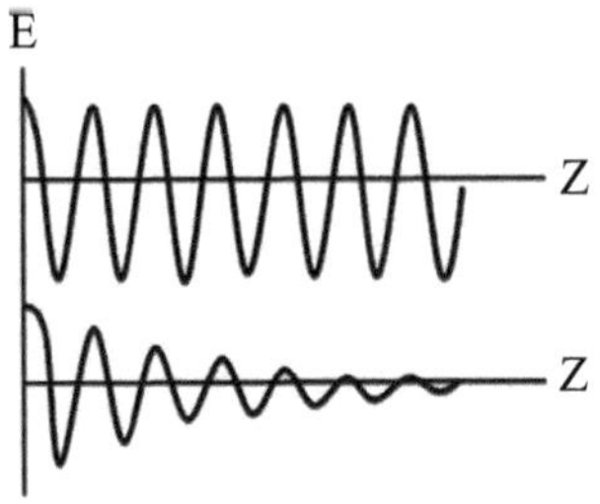

Figure S.7.1

Undamped (top) and damped (bottom) oscillations of the electric field with respect to the optic axis.

7.2 Explain the polarization states applicable to the optic devices.

Solution

(a) Polarization by polarizer

Unpolarized electric field transmitted through three polarizers. The first and last polarizers are orthogonal orientation. The intensity I_1 from the first polarizer after passing through each polarizer is written as

$$I_2 = I_1 \cos^2\theta$$

$$I_3 = I_2 \cos^2\left(\frac{\pi}{2} - \theta\right) = I_1 \cos^2\theta \cos^2\left(\frac{\pi}{2} - \theta\right) = \frac{I_1}{4}\sin^2(2\theta)$$

(b) Polarization by calcite crystal

When the unpolarized light is incident in the polarizing angle, the refracted and reflected lights are perpendicular to each other.

(c) Polarization by polaroid in sunglass

Upon the direction of emitted radiation, the emitted light parallel to the orientation of incident light is completely unpolarized.The emitted light perpendicular to the orientation of incident light is completely polarized. The emitted light scattered in other orientations of incident light are partially polarized.

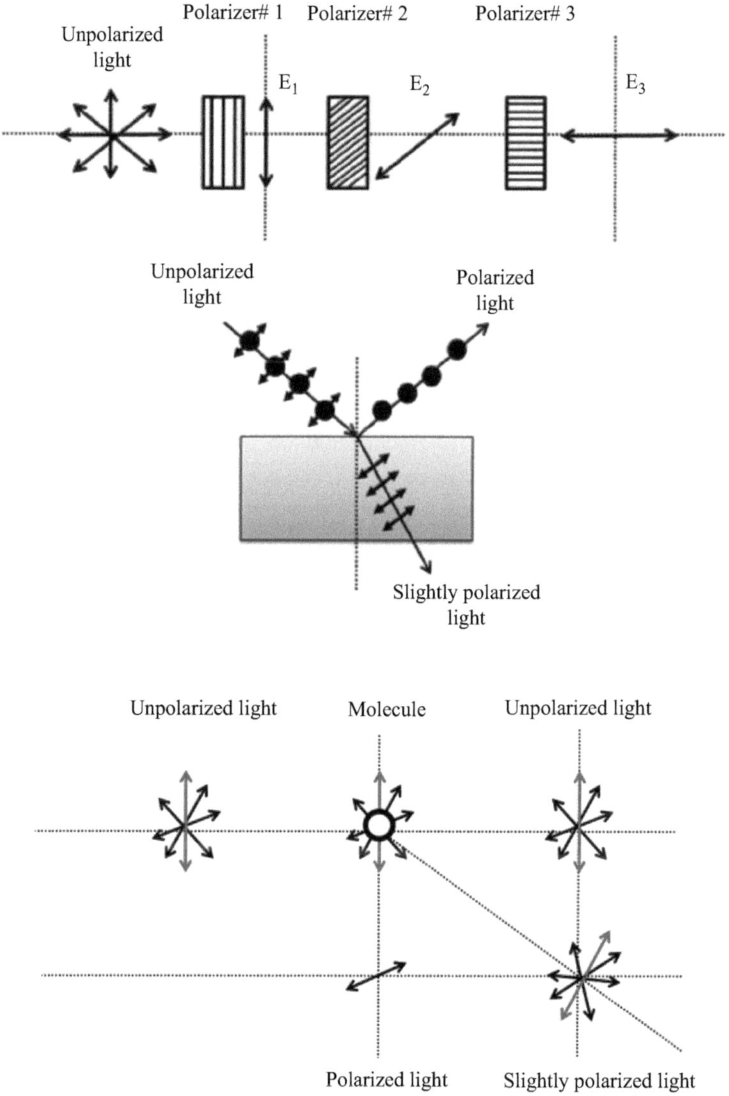

Figure S.7.2 Polarization by polarizer (top), refraction (middle), and scattering (bottom).

7.3 Draw the ellipse polarization state with respect to the circular polarization state in arbitrary direction if a net energy propagates along the optic axis.

Solution

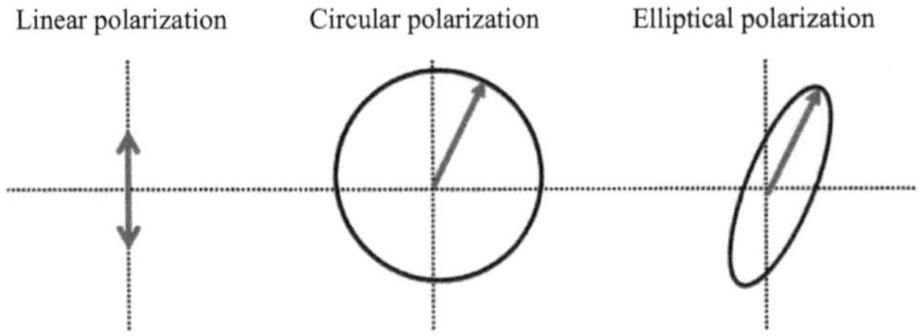

Figure S.7.3

7.4 Determine the intensity of a plane wave from the following relation:

$$I = \langle \vec{S} \rangle = \tfrac{1}{2}\mathrm{Re}\left[\vec{E} \times \vec{H}^*\right]$$

Solution

$$I = \langle \vec{S} \rangle = \tfrac{1}{2}\mathrm{Re}\left[\vec{E} \times \frac{1}{\mu_0 \omega}\vec{k} \times \vec{E}^*\right]$$

$$I = \langle \vec{S} \rangle = \tfrac{1}{2}\mathrm{Re}\left[\vec{E} \times \left(\vec{k} \times \vec{E}^*\right)\right]\frac{1}{\mu_0 \omega}$$

$$I = \langle \vec{S} \rangle = \frac{1}{2\mu_0 \omega}\mathrm{Re}\left[\vec{E}(\vec{k} \cdot \vec{E}^*) - \vec{E} \cdot (\vec{k}\, \vec{E}^*)\right]$$

$$I = \langle \vec{S} \rangle = \frac{1}{2\mu_0 \omega}\mathrm{Re}\left[|E|^2\, \vec{k} - 0\right]$$

7.5 Draw the quadratic and cubic polarizations as a function of the external electric field.

Solution

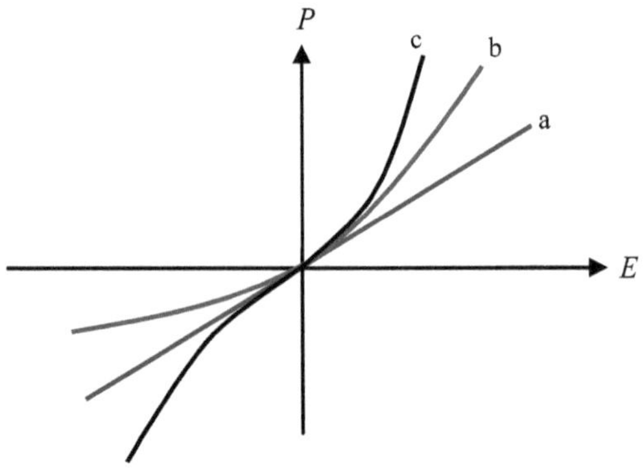

Figure S.7.4

Linear (a), 2rd order nonlinear (b), and 3rd nonlinear (c) polarizations response to the applied electric field.

7.6 Prove a relationship between refractive index and dielectric constant. Assume that the permeability of medium is equivalent to 1.

Solution

$$n = \frac{c}{v}$$

$$n = \sqrt{\frac{\mu\varepsilon}{\mu_0\varepsilon_0}}$$

$$n = \sqrt{\mu_r\varepsilon_r}$$

$$n = \sqrt{\varepsilon_r}$$

7.7 Prove that the speed of light is 3×10^8 m/s.

Solution

Substituting the dielectric permittivity of vacuum $\varepsilon_0 = 8.854 \times 10^{-12}$ F/m and the magnetic permeability of vacuum $\mu_0 = 1.257 \times 10^{-6}$ H/m into the following equation:

$$c^2 = \frac{1}{\varepsilon_0\mu_0}$$

we will get a speed of light at 2.998×10^8 m/s.

7.8 Determine the light intensity traveling in all directions at a distance of 20 km far apart from the wave source having the incident power of 100 kW.

Solution

Substituting the spherical area of light of $A = 4\pi R^2 = 4\pi(20 \times 1000 \text{ m})^2$ and power of $P = 100 \times 1000$ W into the following equation:

$$I = P / A$$

we will get the light intensity at 19.8 μW/m^2.

7.9 Find the time domain expression of the electric field when the magnetic field of a plane wave propagates in nonmagnetic medium is given by

$$\vec{H} = 0.50\, e^{i(10z - 6\pi \times 10^8 t)}\ \text{A/m}$$

Solution

Dielectric constant is calculated by

$$\vec{H} = H_0 e^{i(kz - \omega t)}$$

$H_0 = 0.50$, $\omega = 6\pi \times 10^8$, and $k = 10$

$$k = \omega\sqrt{\mu\varepsilon} = \omega\sqrt{\mu_0\varepsilon_0\varepsilon_r}$$

$$k = 6\times 10^8 \sqrt{\mu_0\varepsilon_0\varepsilon_r} = 6\times 10^8 \frac{\sqrt{\varepsilon_r}}{c}$$

$$10 = 2\pi\sqrt{\varepsilon_r} \;\Rightarrow\; \sqrt{\varepsilon_r} = 2.53$$

Intrinsic impedance is calculated by

$$\eta = \sqrt{\frac{\mu}{\varepsilon}} = \sqrt{\frac{\mu_0\mu_r}{\varepsilon_0\varepsilon_r}} \;\Rightarrow\; \sqrt{\frac{\mu_0}{\varepsilon_0}} \times \frac{1}{\sqrt{\varepsilon_r}}$$

$$\eta = \frac{120\pi}{\sqrt{\varepsilon_r}} = \frac{120\pi}{\sqrt{2.53}} = 237.10\,\Omega$$

Time domain expression of the electric field becomes

$$\vec{E} = \eta\vec{H} = 237.10 \times \left(0.50\, e^{i(10z - 6\pi\times 10^8 t)}\right)$$

$$\vec{E} = 118.50\, e^{i(10z - 6\pi\times 10^8 t)}\ \text{V/m}$$

7.10 Determine the B-vector and E-vector when the plane wave $\vec{E} = E_0\, e^{i(kx - \omega t)}$ with the amplitude of 42 V/m passing through UHV chamber in x-direction.

Solution

$$E_0 = \sqrt{E_{0x}^2 + E_{0y}^2 + E_{0z}^2} = \sqrt{0 + E_{0y}^2 + E_{0z}^2} \Rightarrow \sqrt{0 + 2E_{0y}^2}$$

$$E_0 = 45 = \sqrt{0 + 2E_{0y}^2} = \sqrt{0 + 2E_{0z}^2}$$

$$E_{0y} = E_{0z} = 30\ \text{V/m}$$

$$\vec{E}_z = 30\, e^{i(kx-\omega t)}$$

$$\vec{B}_y = \frac{\vec{E}_z}{c}\, e^{i(kx-\omega t)}$$

$$\vec{B}_y = \frac{30}{3\times 10^8}\, e^{i(kx-\omega t)} \Rightarrow \vec{B}_y = 3\times 10^8\, e^{i(kx-\omega t)}$$

7.11 Determine root mean square value of the magnetic field at the user when this user is located at 1.8 m far apart from light source with power of 250 W.

Solution

$$I = \frac{P}{A} = \frac{P}{4\pi r^2}$$

$$I = \frac{P}{4\pi r^2} = \frac{n}{2c\mu_0}|E_0|^2 \Rightarrow \frac{1}{c\mu_0}\vec{E}_{\mathrm{RMS}}$$

$$\vec{E}_{\mathrm{RMS}} = \sqrt{\frac{Pc\mu_0}{4\pi r^2}} = \sqrt{\frac{250\times 2.998\times 10^8 \times 4\pi \times 10^{-7}}{4\pi \times 1.8^2}} = 48\ \mathrm{V/m}$$

$$\vec{B}_{\mathrm{RMS}} = \frac{\vec{E}_{\mathrm{RMS}}}{c} = \frac{48}{2.998\times 10^8} = 1.6\times 10^{-7}\ \mathrm{T}$$

7.12 Quartz crystal has $n_o(\omega) = 1.541$, $n_e(\omega) = 1.550$, $n_o(2\omega) = 1.566$, $n_e(2\omega) = 1.577$. Draw the index surfaces of the negative as compared to the positive uniaxial crystals applicable to the birefringence phase matching.

Solution

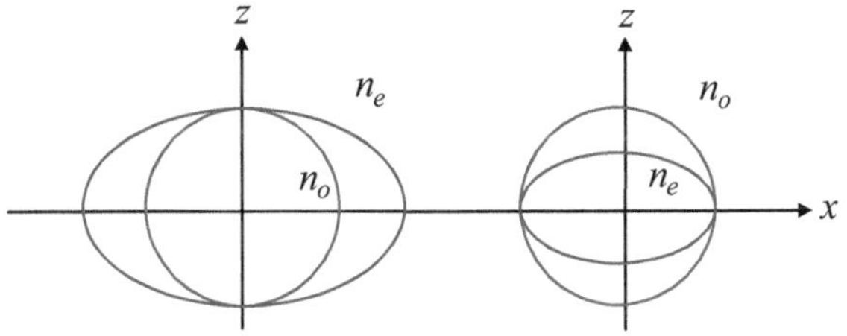

Figure S.7.5 Positive (left) and negative (right) uniaxial crystals.

Chapter 8

8.1 Prove the following equation:

$$\kappa_y \kappa_x E_x + \left(\frac{K_y}{n^2} - \kappa_x^2 - \kappa_z^2 \right) E_y + \kappa_y \kappa_z E_z = 0$$

Solution

Substituting the displacement field into those two equations, we will have

$$\vec{D} = \overline{\overline{\varepsilon}} \vec{E} = \varepsilon_0 \overline{\overline{K}} \vec{E}$$

$$\vec{k} \times \vec{H} = -\omega \varepsilon_0 \overline{\overline{K}} \vec{E}$$

$$\vec{k} \times \left(\vec{k} \times \vec{E} \right) = \omega \mu_0 \left(\vec{k} \times \vec{H} \right) = -\omega^2 \varepsilon_0 \mu_0 \overline{\overline{K}} \vec{E}$$

$$\vec{k} \left(\vec{k} \cdot \vec{E} \right) - k^2 \vec{E} = -\frac{\omega^2}{c^2} \overline{\overline{K}} \vec{E}$$

Component of electric field in y-direction is calculated as

$$\left(k_x E_x + k_y E_y + k_z E_z \right) k_y - k^2 E_y = -\frac{\omega^2}{c^2} K_y E_y$$

$$\frac{\omega^2}{c^2} n^2 \left(\kappa_x E_x + \kappa_y E_y + \kappa_z E_z \right) \kappa_y - \frac{\omega^2}{c^2} n^2 E_y = -\frac{\omega^2}{c^2} K_y E_y$$

$$\left(\kappa_x E_x + \kappa_y E_y + \kappa_z E_z \right) \kappa_y - E_y = -\frac{K_y E_y}{n^2}$$

$$\kappa_x \kappa_y E_x + E_y \left(\kappa_y^2 - 1 + \frac{K_y}{n^2} \right) + \kappa_y \kappa_z E_z = 0$$

$$\kappa_x \kappa_y E_x + E_y \left(\frac{K_y}{n^2} - \kappa_x^2 - \kappa_z^2 \right) + \kappa_y \kappa_z E_z = 0$$

8.2 Show that how the electric field propagates in y–z plane if a plane wave propagates at a certain angle ψ with respect to the optic axis.

Solution

$$E_y\left(\frac{K_y}{n^2}-\cos^2\psi\right)+\sin\psi\cos\psi E_z=0$$

$$\sin\psi\cos\psi E_y+\left(\frac{K_z}{n^2}-\sin^2\psi\right)E_z=0$$

They are rewritten in a matrix form as

$$\begin{pmatrix}\frac{K_y}{n^2}-\cos^2\psi & \sin\psi\cos\psi\\ \sin\psi\cos\psi & \frac{K_z}{n^2}-\sin^2\psi\end{pmatrix}\begin{pmatrix}E_y\\E_z\end{pmatrix}=\begin{pmatrix}0\\0\end{pmatrix}$$

$$\left(\frac{K_y}{n^2}-\cos^2\psi\right)\left(\frac{K_z}{n^2}-\sin^2\psi\right)-\sin^2\psi\cos^2\psi=0$$

$$\frac{K_yK_z}{n^4}-\frac{1}{n^2}\left(K_y\sin^2\psi+K_z\cos^2\psi\right)=0$$

$$\frac{1}{n^2}=\frac{\cos^2\psi}{K_y}+\frac{\sin^2\psi}{K_z}$$

$$\frac{1}{n_e^2(\psi)}=\frac{\cos^2\psi}{n_o^2}+\frac{\sin^2\psi}{n_e^2}$$

The possible wave solution is

$n_o=\sqrt{K_y}$, $n_e=\sqrt{K_z}$ $\qquad E_x=0,\ E_y\neq 0,\ E_z\neq 0$

8.3 Plane wave propagates along a direction given by a unit vector of $\frac{\sqrt{3}}{2}\hat{x}+\frac{1}{2}\hat{z}$ into the uniaxial medium with n_0 = 2.3 and n_e = 2.2. Determine the angle made by $\vec{k}$ of the ordinary wave as compared with the optic axis.

Solution

$$\frac{1}{n_e^2(\psi)}=\frac{\cos^2\psi}{n_o^2}+\frac{\sin^2\psi}{n_e^2}$$

$$\psi = \tan^{-1}\left(\frac{\sqrt{3}/2}{1/2}\right) = 60^0$$

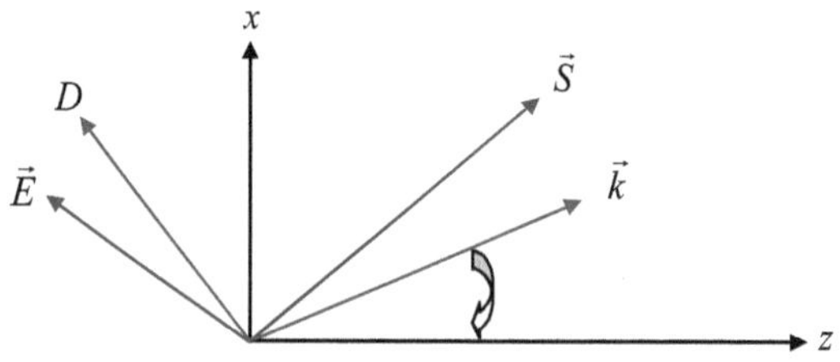

Figure S.8.1 Plane wave propagates along a direction given by the unit vector.

8.4 If we select the arbitrary direction of the electric vector and a net energy propagates along *z*-direction, please show the details of electric fields in *x*- and *y*-axis.

Solution

The electric field components E_x and E_y move along the *x*–*y* axes at the beginning point with

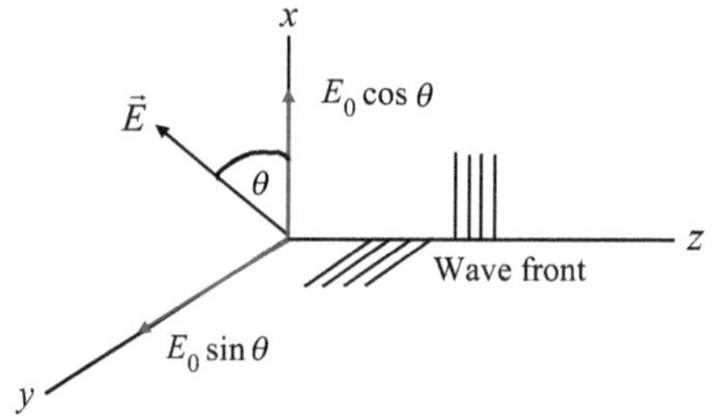

Figure S.8.2 $\vec{E}(z=0) = \left(\hat{x}E_0\cos\theta + \hat{y}E_0\sin\theta\right)e^{-i\omega t}$

However, at a certain distance of *z*, we are trying to add two polarization light waves with a finite phase difference as

$$\vec{E}(z) = \hat{x}E_0\cos\theta e^{i\left(\frac{\omega\sqrt{K_x}}{c}z-\omega t\right)} + \hat{y}E_0\sin\theta e^{i\left(\frac{\omega\sqrt{K_y}}{c}z-\omega t\right)}$$

so-called the ellipse polarization state meaning the two components of the electric waves are not equal (K_x is not equal to K_y).

Chapter 9

9.1 Verify that the following equation is able to produce the second-order nonlinear effect:

$$P^{\mathrm{NL}} = \varepsilon_0 \chi^{(2)} \left[E_0 \cos(kz - \omega t) \right]^2$$

Solution

Due to trigonometric function, we get

$$\cos^2 \theta = \frac{1+\cos 2\theta}{2}$$

$$P^{\mathrm{NL}} = \varepsilon_0 \chi^{(2)} \left[E_0 \cos(kz - \omega t) \right]^2$$

$$P^{\mathrm{NL}} = {}^1\!/_2\, \varepsilon_0 \chi^{(2)} E_0^2 + {}^1\!/_2\, \varepsilon_0 \chi^{(2)} E_0^2 \cos 2(kz - \omega t)$$

The second term on the right equation reveals the double frequency.

9.2 Show that the following equation can produce *i*th order nonlinear effect:

$$P^{\mathrm{NL}} = \varepsilon_0 \chi^{(2)} \left[E_1 \cos(k_1 z - \omega t) + E_2 \cos(k_2 z - 2\omega t) \right]^2$$

Solution

$$P^{\mathrm{NL}} = \varepsilon_0 \chi^{(2)} \Big[E_1^2 \cos^2(k_1 z - \omega t) + E_2^2 \cos^2(k_2 z - 2\omega t)$$
$$+ 2E_1 E_2 \cos(k_1 z - \omega t) \cos(k_2 z - 2\omega t) \Big]$$

Using the following trigonometric functions, we will get

$$\cos^2 \theta = \frac{1+\cos 2\theta}{2}, \ \sin^2 \theta = \frac{1-\cos 2\theta}{2}$$

$$2\cos\theta_1 \cos\theta_2 = \cos(\theta_1 + \theta_2) + \cos(\theta_1 - \theta_2)$$

$$P^{\mathrm{NL}} = \frac{1}{2} \varepsilon_0 \chi^{(2)} \{1 + \cos 2(k_1 z - \omega t)\} E_1^2 + \frac{1}{2} \varepsilon_0 \chi^{(2)} \{1 + \cos 2(k_2 z - 2\omega t)\} E_2^2$$

$$+ \varepsilon_0 \chi^{(2)} E_1 E_2 \cos\left[(k_1 + k_2) z - 3\omega t\right] + \varepsilon_0 \chi^{(2)} E_1 E_2 \cos\left[(k_2 - k_1) z - \omega t)\right]$$

The 1st, 2nd, 3rd, and 4th terms represent second harmonic, four-wave, sum frequency, and different frequency generations, respectively.

9.3 Show that the following equation is able to produce second harmonic, sum frequency, and different frequency generations:

$$P^{\mathrm{NL}} = \varepsilon_0 \chi^{(2)} \left[E_1 \cos(k_1 z - \omega_1 t) + E_2 \cos(k_2 z - \omega_2 t) \right]^2$$

Solution

$$P^{\mathrm{NL}} = \varepsilon_0 \chi^{(2)} \Big[E_1^2 \cos^2(k_1 z - \omega_1 t) + E_2^2 \cos^2(k_2 z - \omega_2 t)$$

$$+2E_1 E_2 \cos(k_1 z - \omega_1 t)\cos(k_2 z - \omega_2 t) \Big]$$

Using the following trigonometric functions, we will get

$$\cos^2\theta = \frac{1+\cos 2\theta}{2}, \ \sin^2\theta = \frac{1-\cos 2\theta}{2}$$

$$2\cos\theta_1 \cos\theta_2 = \cos(\theta_1 + \theta_2) + \cos(\theta_1 - \theta_2)$$

$$P^{\mathrm{NL}} = \frac{1}{2}\varepsilon_0 \chi^{(2)} \{1+\cos 2(k_1 z - \omega_1 t)\} E_1^2 + \frac{1}{2}\varepsilon_0 \chi^{(2)} \{1+\cos 2(k_2 z - \omega_2 t)\} E_2^2$$

$$+\varepsilon_0 \chi^{(2)} E_1 E_2 \cos\left[(k_1 + k_2)z - (\omega_1 + \omega_2)t\right]$$

$$+\varepsilon_0 \chi^{(2)} E_1 E_2 \cos\left[(k_2 - k_1)z - (\omega_1 - \omega_2)t)\right]$$

The 1st, 2nd, 3rd, and 4th terms represent second harmonic, sum frequency, and different frequency generations, respectively.

9.4 Prove the following equations by starting with the electric field composed of amplitude and phase terms:

$$\frac{du_2}{dz} = \frac{\omega d}{c n_2} u_1^2 \sin\theta$$

$$\frac{d\phi_2}{dz} u_2 = \frac{\omega d}{c n_2} u_1^2 \cos\theta$$

Solution

$$E_2(z) = u_2(z) e^{i\phi_2(z)}$$

$$\frac{dE_2}{dz} = \frac{du_2}{dz} e^{i\phi_2} + i\frac{d\phi_2}{dz} u_2 e^{i\phi_2} = i\frac{\omega d}{c n_2} u_1^2 e^{2i\phi_1}$$

$$\frac{du_2}{dz} + i\frac{d\phi_2}{dz} u_2 = i\frac{\omega d}{c n_2} u_1^2 e^{-i(\phi_2 - 2\phi_1)}$$

$$\frac{du_2}{dz} + i\frac{d\phi_2}{dz}u_2 = i\frac{\omega d}{cn_2}u_1^2 e^{-i\theta(z)} \quad \Rightarrow \theta(z) = \phi_2 - 2\phi_1$$

$$\frac{du_2}{dz} + i\frac{d\phi_2}{dz}u_2 = i\frac{\omega d}{cn_2}\left(u_1^2 \cos\theta - iu_1^2 \sin\theta\right)$$

$$\frac{du_2}{dz} + i\frac{d\phi_2}{dz}u_2 = i\frac{\omega d}{cn_2}u_1^2 \cos\theta + \frac{\omega d}{cn_2}u_1^2 \sin\theta$$

$$\frac{du_2}{dz} = \frac{\omega d}{cn_2}u_1^2 \sin\theta$$

$$\frac{d\phi_2}{dz}u_2 = \frac{\omega d}{cn_2}u_1^2 \cos\theta$$

9.5 Find a best duty cycle l/λ in order to make maximum G_2.

Solution

Fourier coefficient as a function of *m* for the quasi-phase matching is written as

$$G_m = \frac{2}{m\pi}\sin\left(\frac{m\pi l}{\lambda}\right)$$

$$G_2 = \frac{2}{2\pi}\sin\left(\frac{2\pi l}{\lambda}\right)$$

$$G_2 = 0.318\sin\left(2\pi \cdot \frac{l}{\lambda}\right)$$

$$G_2 = 0.318\sin\left(2\pi \cdot \frac{1}{4}\right)$$

$$G_2 = 0.318 \Leftarrow \frac{l}{\lambda} = \frac{1}{4}$$

9.6 Obtain the values of G_3 and G_4.

Solution

Fourier coefficient as a function of *m* for the quasi-phase matching is written as

$$G_m = \frac{2}{m\pi}\sin\left(\frac{m\pi l}{\lambda}\right)$$

$$G_3 = \frac{2}{3\pi}\sin\left(\frac{3\pi}{}\cdot\frac{l}{\lambda}\right)$$

$$G_3 = \frac{2}{3\pi}\sin\left(3\pi\cdot\frac{1}{4}\right)$$

$$G_3 = 0.338$$

$$G_4 = \frac{2}{4\pi}\sin\left(4\pi\cdot\frac{l}{\lambda}\right)$$

$$G_3 = \frac{2}{4\pi}\sin\left(4\pi\cdot\frac{1}{4}\right)$$

$$G_4 = 0$$

9.7 Determine the best duty cycle l/λ that we must use so that we have a finite value of G_2 and we can use the second order quasi-phase matching to approach the maximum power efficiency.

Solution

Suppose that $D = l/\lambda$ and $\lambda = \pi$

$$G_m = \frac{2}{m\pi}\sin\left(\frac{m\pi l}{\lambda}\right)$$

$$G_2 = \frac{2}{2\pi}\sin\left(2\pi\cdot\frac{l}{\lambda}\right) \Rightarrow \frac{2}{2\pi}\sin(2\pi\cdot D)$$

$$G_2 = 0.7072\sin(2\pi\cdot D)$$

The best duty cycle l/λ to have a finite value of G_2 is 30% as details in the following table:

Duty cycle, *D*	G_2
0.01	0.063
0.02	0.125
0.03	0.187
0.04	0.249
0.05	0.309

Duty cycle, D	G_2
0.10	0.588
0.20	0.951
0.30	**0.957**
0.31	0.930
0.32	0.905
0.33	0.876
0.34	0.844
0.35	0.809
0.40	0.588
0.50	0.000

9.8 Considering a second harmonic generation in lithium niobiate crystal propagating along y-direction with below details, determine the first order quasi-phase matching period required.

Wavelength	Ordinary refractive index	Extraordinary refractive index
1 μm	2.236	2.160
0.5 μm	2.341	2.249

Solution

Increasing the electric field of second harmonic via quasi-phase matching is required by

$$n_1 = n_0(\omega) = 2.236, \qquad n_2 = n_e(2\omega) = 2.249$$

The coherence length is

$$L_c = \frac{\pi}{\Delta k} = \frac{\pi}{\frac{2\omega}{c}(n_2 - n_1)} = \frac{\pi c}{2\omega(n_2 - n_1)}$$

$$L_c = \frac{\pi\lambda}{4\pi(n_2 - n_1)}$$

$$L_c = \frac{\pi \times 1 \times 10^{-6}}{4\pi(2.249 - 2.236)} = 19.23\ \mu\text{m}$$

9.9 Determine a dimension of the second and third order nonlinear susceptibility.

Solution

For linear polarization, we have

$$P = \varepsilon_0 \chi^{(1)} E$$

$$[P] = \frac{C}{m^2}, \quad [\mathrm{E}] = \frac{N}{C}$$

Since the first order nonlinear susceptibility is dimensionless, a dimension of ε_0 is

$$[\varepsilon_0] = \frac{C}{m^2} \cdot \frac{C}{N} = \frac{C^2}{m^2 N}$$

For second order nonlinear polarization, we have

$$P = \varepsilon_0 \chi^{(2)} E^2$$

$$[\chi^{(2)}] = \frac{[P]}{[\varepsilon_0][E^2]}$$

$$[\chi^{(2)}] = \frac{C / m^2}{\dfrac{C^2}{m^2 N} \cdot \dfrac{N^2}{C^2}}$$

$$[\chi^{(2)}] = \frac{C}{N} = \frac{m}{V}$$

For third order nonlinear polarization, we have

$$P = \varepsilon_0 \chi^{(3)} E^3$$

$$[\chi^{(3)}] = \frac{[P]}{[\varepsilon_0][E^3]}$$

$$[\chi^{(3)}] = \frac{C / m^2}{\dfrac{C^2}{m^2 N} \cdot \dfrac{N^3}{C^3}}$$

$$[\chi^{(3)}] = \frac{C / m^2}{\dfrac{1}{m^2} \cdot \dfrac{N^2}{C}} = \frac{C^2}{N^2} = \frac{\mathrm{m}^2}{\mathrm{V}^2}$$

9.10 Prove the following equation:

$$\sin^2\psi_m = \frac{\left[\dfrac{1}{n_0^2(2\omega)} - \dfrac{1}{n_0^2(\omega)}\right]}{\left[\dfrac{1}{n_e^2(\omega)} - \dfrac{1}{n_o^2(\omega)}\right]}$$

Solution

$$\frac{1}{n_e^2(\omega,\psi_m)} = \frac{\cos^2\psi_m}{n_o^2(\omega)} + \frac{\sin^2\psi_m}{n_e^2(\omega)} = \frac{1}{n_o^2(2\omega)}$$

$$\frac{1}{n_0^2(2\omega)} = \frac{1-\sin^2\psi_m}{n_o^2(\omega)} + \frac{\sin^2\psi_m}{n_e^2(\omega)}$$

$$\frac{1}{n_0^2(2\omega)} = \frac{1}{n_o^2(\omega)} + \frac{\sin^2\psi_m}{n_e^2(\omega)} - \frac{\sin^2\psi_m}{n_o^2(\omega)}$$

$$\frac{1}{n_0^2(2\omega)} - \frac{1}{n_o^2(\omega)} = \sin^2\psi_m\left[\frac{1}{n_e^2(\omega)} - \frac{1}{n_o^2(\omega)}\right]$$

$$\sin^2\psi_m = \frac{\left[\dfrac{1}{n_0^2(2\omega)} - \dfrac{1}{n_0^2(\omega)}\right]}{\left[\dfrac{1}{n_e^2(\omega)} - \dfrac{1}{n_o^2(\omega)}\right]}$$

9.11 We supply the input power of 1 W at a fundamental wavelength of 1.06 μm, 1 mm ϕ beam into $LiNbO_3$ served as the uniaxial medium. Determine (a) the coherence length if $n_1 = n_0(\omega)$ = 2.1561, $n_2 = n_e(2\omega)$ = 2.2355, d = 30 × 10^{-12} m/V, (b) SH power efficiency without and with phase matching.

Solution

(a) The coherence length is

$$L_c = \frac{\pi}{\Delta k} = \frac{\pi}{\dfrac{2\omega}{c}(n_2 - n_1)} = \frac{\pi c}{2\omega(n_2 - n_1)} = 3.3\ \mu\text{m}$$

Maximum SHG exists at 3.3 μm and it becomes to be zero again at 6.6 μm.

(b) According to SHG power efficiency as

$$\eta = \frac{2\mu_0\omega^2}{cn_1^2 n_2} \cdot \frac{d^2 P_1}{S} \cdot \frac{\sin^2(\Delta kz/2)}{(\Delta k/2)^2}$$

$$\eta = 2.3\times 10^{-6} \cdot \frac{P_1}{S} \cdot \frac{\sin^2(\Delta kz/2)}{(\Delta k/2)^2}$$

$$\eta_{\max} = 2.3\times 10^{-6} \cdot \frac{P_1}{S} \cdot \frac{4}{(\Delta k)^2} \Rightarrow \sin = 1$$

and if the incident power, P_1 = 1 W and $S = \pi\left(0.5\times 10^{-3}\right)^2$ m^2, we finally get $\eta_{\max} = 13\times 10^{-12}$. So, if we supply 1 W of power at the fundamental frequency, we are generating 13 pW of power at SH frequency (a very low efficiency).

Suppose a crystal length z = 1 cm to achieve Δk = 0, we get

$$\eta = 2.3\times 10^{-6} \cdot \frac{P_1}{S} \cdot z^2 \simeq 2.9\times 10^{-4}$$

with input power of 1 W, we still generate the low output SH power of 0.29 mW.

9.12 What is condition for the second harmonic generation we need by supplying the horizontal polarized light (i.e., Ruby laser) at λ_0 = 694.3 nm into KDP (KH_2PO_4) crystal having $n_0(\omega)$= 1.50502, $n_e(\omega)$=1.46532, $n_o(2\omega)$=1.53269, $n_e(2\omega)$=1.49711?

Solution

The maximum conversion efficiency of SHG is existed when we select the certain angle of wave propagation of ψ_m = 50.5° with the optic axis (OA). Under this orientation, the critical phase matching condition is conserved as $n_0(\omega) = n_e(2\omega, \psi_m)$.

9.13 When we have d-tensor as d_{123} and d_{112}, please use the standard convection for the tensor contraction.

Solution

Since the last two indices are contracted, we will get

$$d_{123} = d_{14}$$

$$d_{112} = d_{16}$$

9.14 If we launch the "ordinary polarization" at fundamental frequency along the optic axis into lithium niobate crystal (n_0>n_e) with d-tensor as

$$d = \begin{pmatrix} 0 & 0 & 0 & 0 & d_{15} & -d_{22} \\ -d_{22} & d_{22} & 0 & d_{15} & 0 & 0 \\ d_{31} & d_{31} & d_{33} & 0 & 0 & 0 \end{pmatrix}$$

where

$$d_{21} = -d_{22};\ d_{32} = d_{31};\ d_{24} = d_{15};\ d_{16} = -d_{22}$$

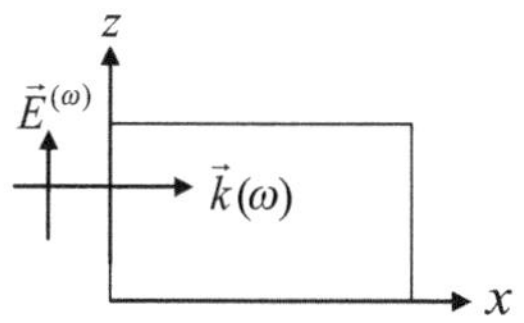

and the ordinary polarization in z-axis is parallel to the optic axis of the lithium niobate crystal and perpendicular to k-vector. Find out the effective d-tensor.

Solution

There is one component that is the electric field of the fundamental waves entering the crystal as

$$E_x^{(\omega)} = E_y^{(\omega)} = 0$$

$$E_z^{(\omega)} = E_1$$

Nonlinear polarization generated by this electric field is

$$\begin{pmatrix} \tilde{P}_x^{(2\omega)} \\ \tilde{P}_y^{(2\omega)} \\ \tilde{P}_z^{(2\omega)} \end{pmatrix} = \varepsilon_0 \begin{pmatrix} 0 & 0 & 0 & 0 & d_{15} & -d_{22} \\ -d_{22} & d_{22} & 0 & d_{15} & 0 & 0 \\ d_{31} & d_{31} & d_{33} & 0 & 0 & 0 \end{pmatrix} \begin{pmatrix} 0 \\ 0 \\ E_1^2 \\ 0 \\ 0 \\ 0 \end{pmatrix}$$

$$\tilde{P}_x^{(2\omega)} = \tilde{P}_y^{(2\omega)} = 0$$

$$\tilde{P}_z^{(2\omega)} = \varepsilon_0 d_{33} E_1^2$$

This number is a very strong nonlinear polarization, however we cannot reach the phase matching $n_0(\omega) \neq n_e(2\omega)$ generated by the different velocities of these two waves under this orientation.

9.15 Compare the power conversion efficiency at a crystal length.

Solution

In case of $\Delta k \neq 0$, SHG power conversion efficiency is dependent of sine function as

$$\eta = \frac{P_2(z)}{P_1} = \frac{2\mu_0 \omega^2}{cn_1^2 n_2} \cdot \frac{d^2 P_1}{S} \cdot \frac{\sin^2(\Delta kz/2)}{(\Delta k/2)^2}$$

In case of $\Delta k = 0$, a maximum SHG power conversion efficiency grows along the crystal length to be z^2 as

$$\eta = \frac{2\mu_0 \omega^2}{cn_1^2 n_2} \cdot \frac{d^2 P_1}{S} \cdot z^2$$

In case of $\Delta k = K$, a maximum SHG power conversion efficiency is relied on a periodic length and crystal length as

$$\eta = \frac{P_2(z)}{P_1} = \frac{\omega^2 (d_0/2)^2}{cn_2} \cdot \frac{2\mu_0 P_1}{n_1^2 S} z^2$$

9.16 Find the wavelength and refractive index of UV light out of KDP crystal when the second harmonic generation satisfy the birefringence phase matching conditionand the finite effective nonlinear polarization not to be zero. A pair of wavelengths are composed of the argon ion laser at wavelength of 351 nm and the oxazine dye laser at wavelength of 789 nm.

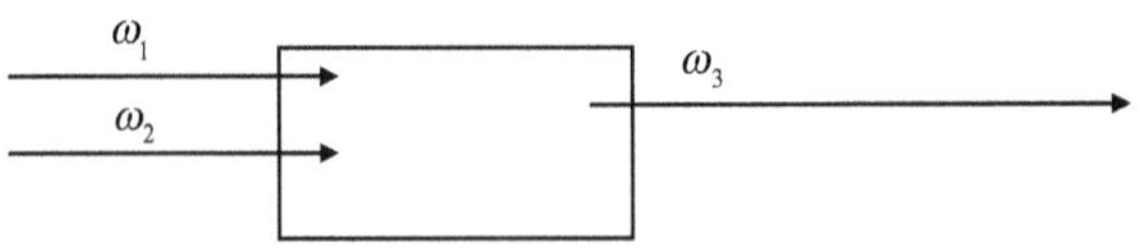

Two incident waves at two frequencies came out of the KDP crystal ($\omega_1 > \omega_2$).

Solution

The second harmonic generation process needs to satisfy the energy conservation as follows:

$$\omega_3 - \omega_2 - \omega_1 = 0$$

$$\frac{2\pi}{\lambda_3} - \frac{2\pi}{\lambda_2} - \frac{2\pi}{\lambda_1} = 0$$

$$\frac{2\pi}{\lambda_3} - \frac{2\pi}{351} - \frac{2\pi}{789} = 0$$

$$\lambda_3 = 243\ \text{nm}$$

The second harmonic generation process needs to satisfy the phase matching condition as follows:

$$k_e^{2\omega} - k_0^{\omega} - k_0^{2\omega} = 0$$

$$\frac{2\pi}{\lambda_3} \cdot n_e(2\omega) - \frac{2\pi}{\lambda_1} \cdot n_0(\omega) - \frac{2\pi}{\lambda_2} \cdot n_0(2\omega) = 0$$

$$\frac{2\pi}{351} \times 1.53236 + \frac{2\pi}{789} \times 1.50226 = \frac{2\pi}{243} \cdot n_e(2\omega)$$

$$n_e(2\omega) = 1.52264$$

Chapter 10

10.1 Clarify that there is no generation of sub-harmonic if we feed the input pump just at the double frequency.

Solution

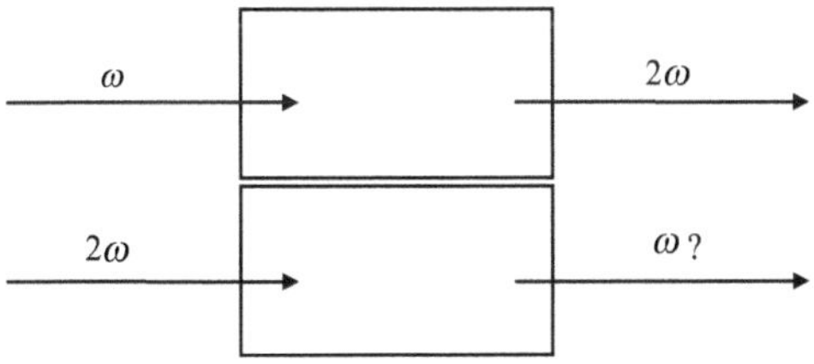

Figure S.10.1

The second harmonic generation (top) as compared to the possible parametric down-conversion for the generations of ω at the same crystal under $\Delta k = 0$ (below).

If there is SH light with certain electric field amplitude of E_2 and no incidence of any fundamental wave at the input of a crystal: $E_2(z)$ = constant and $E_1(0) = 0$), we will have

$$\frac{dE_1}{dz} = i\frac{\omega d}{cn_1}E_2E_1^*$$

$$\frac{d^2E_1}{dz^2} = i\frac{\omega d}{cn_1}E_2\frac{dE_1^*}{dz}$$

$$\frac{d^2E_1}{dz^2} = i\frac{\omega d}{cn_1}E_2\left(-i\frac{\omega d}{cn_1}E_2^*E_1\right)$$

$$\frac{d^2E_1}{dz^2} = \frac{\omega^2 d^2}{c^2 n_1^2}\left|E_2\right|^2 E_1$$

Define the gain coefficient γ then, we get

$$\gamma^2 = \frac{\omega^2 d^2}{c^2 n_1^2}\left|E_2\right|^2$$

$$\frac{d^2E_1}{dz^2} = \gamma^2 E_1$$

The solution is

$$E_1(z) = A\sinh\gamma z + B\cosh\gamma z$$

Therefore, at the amplitudes A=B=0, we get $E_1(z)=0$ showing that there is no generation of sub-harmonic. ω can get converted in 2ω, however, we cannot covert 2ω at ω because there is nonlinearity.

10.2 Prove the following equation:

$$\sin\psi_m = \frac{n_e(2\omega)}{n_0(\omega)}\sqrt{\frac{n_0^2(2\omega) - n_0^2(\omega)}{n_o^2(2\omega) - n_e^2(2\omega)}}$$

Solution

$$\frac{1}{n_e^2(2\omega,\psi_m)} = \frac{\cos^2\psi_m}{n_o^2(2\omega)} + \frac{\sin^2\psi_m}{n_e^2(2\omega)} = \frac{1}{n_o^2(\omega)}$$

$$\frac{1}{n_o^2(\omega)} = \frac{1-\sin^2\psi_m}{n_0^2(2\omega)} + \frac{\sin^2\psi_m}{n_e^2(2\omega)}$$

$$\frac{1}{n_o^2(\omega)} = \frac{1}{n_0^2(2\omega)} + \frac{\sin^2\psi_m}{n_e^2(2\omega)} - \frac{\sin^2\psi_m}{n_0^2(2\omega)}$$

$$\frac{1}{n_o^2(\omega)} - \frac{1}{n_0^2(2\omega)} = \frac{\sin^2\psi_m}{n_e^2(2\omega)} - \frac{\sin^2\psi_m}{n_0^2(2\omega)}$$

$$\frac{1}{n_o^2(\omega)} - \frac{1}{n_0^2(2\omega)} = \sin^2\psi_m\left[\frac{1}{n_e^2(2\omega)} - \frac{1}{n_0^2(2\omega)}\right]$$

$$\sin^2\psi_m = \frac{\left[\dfrac{1}{n_o^2(\omega)} - \dfrac{1}{n_0^2(2\omega)}\right]}{\left[\dfrac{1}{n_e^2(2\omega)} - \dfrac{1}{n_0^2(2\omega)}\right]}$$

$$\sin\psi_m = \frac{n_e(2\omega)}{n_0(\omega)}\sqrt{\frac{n_0^2(2\omega)-n_0^2(\omega)}{n_o^2(2\omega)-n_e^2(2\omega)}}$$

10.3 Determine a gain coefficient of lithium niobate having $\lambda_0 = 1$ micron, $d = 30 \times 10^{-12}$ m/V, $n_1 = 2.2$, $P_2 = 1$ W, $S = 0.1$ mm^2.

Solution

According to

$$\gamma = \kappa u_2 = \frac{\omega d}{c n_1}\sqrt{\frac{2c\mu_0 P_2}{n_2 S}}$$

we can achieve a gain coefficient at 10 m^{-1}. If a crystal length is 5 cm, we have the output power of about $e^{\kappa u_2 z} = e^{10\times 0.05} = e^{0.5} = 1.6$ or 60 % increase of power that is small value.

10.4 Show the difference of the electric field of signal and idler with and without the initial phase in a matrix form.

Solution

$$\begin{pmatrix} E_s^*(z) \\ E_i(z) \end{pmatrix} = \begin{pmatrix} \cosh gz & -i\sqrt{\dfrac{\omega_s n_i}{\omega_i n_s}} e^{-i\phi_{p0}} \sinh gz \\ i\sqrt{\dfrac{\omega_i n_s}{\omega_s n_i}} e^{i\phi_{p0}} \sinh gz & \cosh gz \end{pmatrix} \begin{pmatrix} E_s^*(0) \\ E_i(0) \end{pmatrix}$$

$$\begin{pmatrix} E_s^*(z) \\ E_i(z) \end{pmatrix} = \begin{pmatrix} \cosh gz & -i\sqrt{\dfrac{\omega_s n_i}{\omega_i n_s}} \sinh gz \\ i\sqrt{\dfrac{\omega_i n_s}{\omega_s n_i}} \sinh gz & \cosh gz \end{pmatrix} \begin{pmatrix} E_s^*(0) \\ E_i(0) \end{pmatrix}$$

10.5 Give a few examples of parametric fluorescence, different frequency, and sum frequency generation.

Solution

If we supply ω_p = 400 nm so-called a parametric down conversion (parametric fluorescence), we will get $\omega_s = \omega_i$ = 800 nm. If we supply ω_p = 400 nm and ω_s = 700 nm coming in and we assume ω_p is stronger light beam than ω_s, it is amplifier for ω_s and generating a different frequency ω_i = 933 nm so-called DFG. If we supply ω_s = 700 nm and ω_i = 933 nm, we can generate ω_p = 400 nm so-called SFG as shown in the below figures.

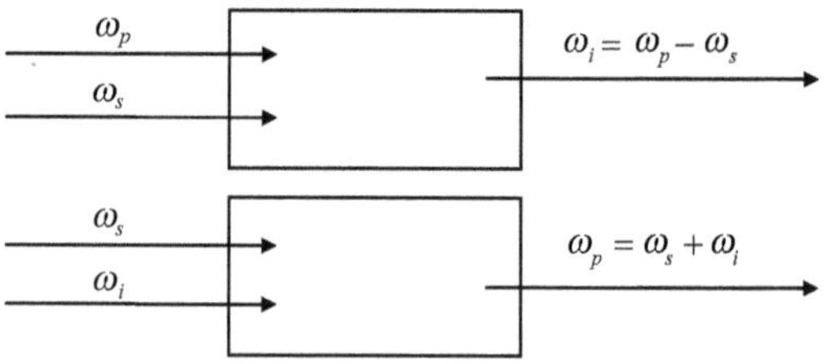

Figure S.10.2

DFG to convert Vis to IR (top) and SFG to convert IR to Vis (bottom: if $\omega_s = \omega_i$ we get back to SHG and sub-SH process).

10.6 Check the inter-mode frequency at a crystal length of 5 cm and $n_s = 2$.

Solution

According to

$$\Delta \nu_s = \frac{c}{2ln_s}$$

$$\Delta \nu_s = \frac{3\times 10^8}{2\times 5\times 10^{-2}\times 2} = 1.5\times 10^9\,\text{Hz} = 1.5\,\text{GHz}$$

we will get the inter-mode frequency of 1.5 GHz required for the oscillation within a cavity.

10.7 Determine the threshold intensity of the pump from the double resonant oscillator if $\lambda_p, \lambda_s, \lambda_i = 0.6,\ 1.0,\ 1.5$ micron and $l = 5$ cm with $d = 3 \times 10^{-11}$ m/V ($R_1 = R_2 = 0.98$).

Solution

According to

$$I_{\text{P,th}} = \frac{cn_p n_s n_i}{\mu_0 \omega_s \omega_i d^2 \ell^2}\left(\frac{1}{\sqrt{R_1 R_2}} - 1\right)$$

we will get the threshold intensity of the pump as

$$I_{\text{P,th}} = \frac{\left(3\times 10^8\right)\left(8\times 1.5\times 10^{-12}\right)}{\left(4\pi\times 10^{-7}\right)\left(4\pi^2\right)\left(9\times 10^{16}\right)\left(9\times 10^{-22}\right)\left(25\times 10^{-4}\right)}\left(\frac{0.02}{0.98}\right)$$

$$I_{\text{P,th}} = 6\times 10^6\,\text{W/m}^2$$

If cross-section area $S = 1$ mm^2, we will get the pump power of $P_{\text{P,th}} = I_{\text{P,th}} \times S = 6\times 10^6 \times 10^{-6} = 6\,\text{W}$ required for the oscillator to start.

When $\Delta k = 0$ at $\omega_s - \omega_{s0}$ and $\omega_i = \omega_{i0}$, thc frcquency range (a gain bandwidth) is calculated as

$$\Delta \nu \simeq \frac{c}{l\left[N(\omega_{s0}) - N(\omega_{i0})\right]}$$

Here the group index is

$$N(\omega_{s0}) = n(\omega_{s0}) + \omega_{s0}\left.\frac{dn}{d\omega}\right|_{s0}$$

$$N(\omega_{s0}) - N(\omega_{i0}) \simeq 0.05$$

$$\Delta\nu \simeq \frac{3\times10^{8}}{5\times10^{-2}[0.05]} = 12\times10^{10}\,\text{Hz} = 120\,\text{GHz}$$

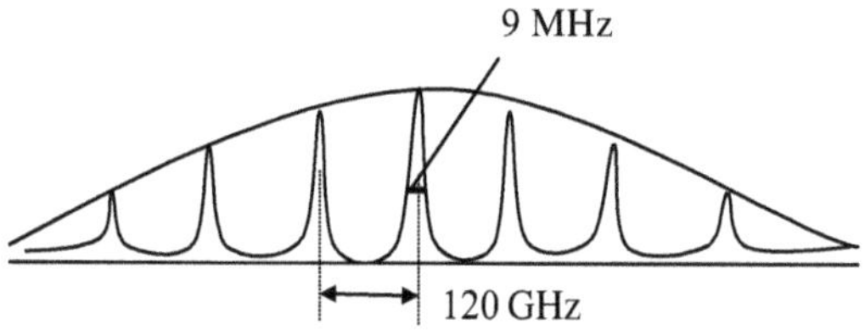

Figure S.10.3

Overall big gain bandwidth and small inter-mode spacing.
If $R = 0.98$, using the following equation, we will get

$$\Delta\nu \simeq \frac{c(1-R)}{2\pi n_s l}$$

the inter-mode spacing at 8.8 MHz.

10.8 Find T-factor when $R_i = 0.98$ from the double resonant oscillator.

Solution

According to

$$T = \frac{(I_{\text{P,th}})_{\text{DRO}}}{(I_{\text{P,th}})_{\text{SRO}}} = \frac{(g_{\text{th}})^2_{\text{DRO}}}{(g_{\text{th}})^2_{\text{SRO}}} = \frac{1-R_i}{2}$$

The threshold intensity of the pump requires of 100 ($T = 1/100$). For example, if 10 W pump power required from the single resonant oscillator, there will be 100 mW power from the double resonant oscillator, meaning that 100 times down the power to operate the oscillator and we can almost have a continuous wave oscillation of this parametric oscillator because both signal and idler are having high intensity with in a cavity and both of them are responsible for drawing energy from the pump.

Appendix A

```
%%%Calculation of mass and linear attenuation coefficient%%%
%%%%%%%%%%%%%%%%%%%%%%%%%%%%%%%%%%%%%%%%%%%%%%%%%%%%%%%%%%%%

function [mac1,mac2,u1L1,u1L2] = linearAttenCoef1E2L
      (material1,material2,E)

switch material1        % Condition command for the 1st
   layer
case 'H'            % In case of Hydrogen
            p1 = 8.38E-05 ;  % density of Hydrogen
             (g/cm^3)
            p = p1 ;
            i = 1 ;% the initial value for (for-loop)
             in Energy column
            f = 35 ; % the final value for (for-loop)
             in Energy column
% importing file energy and u/p of Hydrogen from
   Microsoft Excel
            H = importfile('92 elements.xlsx','1-21',
             'A3:B38');
            A = H ;

        % Function file for calculation
        % mass and linear attenuation coefficient
(mac1,u1L12)
        % for the 1st layer
            [mac1,u1L1] = material1E1L(E,p,i,f,A);
case 'He'
            p1 = 1.66E-04 ;
            p = p1 ;
            i = 1 ;
            f = 35 ;
            He = importfile1('92 elements.xlsx','1-
                 21','D3:E38');
            A = He ;
```

```
                [mac1,u1L1] = material1E1L(E,p,i,f,A);
                                  .
                                  .
                            94 elements
                                  .
                                  .
        End

switch material2        % Condition command for the 2nd
    layer
case 'H'                % In case of Hydrogen
                p2 = 8.38E-05 ;  % density of Hydrogen
                     (g/cm^3)
                p = p2 ;
                i = 1 ;% the initial value for (for-loop)
                     in Energy column
                f = 35 ;% the final value for (for-loop)
                     in Energy column
            % importing file energy and u/p of Hydrogen
               from Microsoft Excel
                H = importfile('92 elements.xlsx','1-
                     21','A3:B38');
                A = H ;
            % Function file for calculation
            % mass and linear attenuation coefficient
              (mac1,u1L12)
% for the 2nd layer
                [mac2,u1L2] = material1E2L(E,p,i,f,A);
case 'He'
                p2 = 1.66E-04 ;
                p = p2 ;
                i = 1 ;
                f = 35 ;
                He = importfile1('92 elements.xlsx','1-
                     21','D3:E38');
                A = He ;
                [mac2,u1L2] = material1E2L(E,p,i,f,A);
                                  .
                                  .
                            94 elements
                                  .
                                  .
```

```
end
end

%%%%%%%%%%%%%%%%%%%%%%%%%%%%%%%%%%%%%%%%%%%%%%%%%%%%%%

function [mac1,u1L1] = material1E1L(E,p,i,f,A)
   y = E ;
   % for-loop command for running in energy column
for m = i : f
   % If-condition for logarithmic interpolation
      Energy = A(n,1);
if y  == Energy                       % process 1
         mac1 = A(n,2)  ;
         lac1 = mac1*p ;
         u1L1 = lac1 ;
elseif E > A(m,1) && E < A(m+1,1)
    mac1 = ((A(m,2))*(log10(A(m+1,1))-log10(E)) +...
               (A(m+1,2))*(log10(E)-log10(A(m,1))) ) /...
( log10(A(m+1,1)) - log10(A(m,1)))   ;
         lac1 = mac1*p ; % mass attenuation become
                linear attenuation
         u1L1 = lac1 ;   % linear attenuation
                coefficient (cm^-1)
end
end

end

%%%%%%%%%%%%%%%%%%%%%%%%%%%%%%%%%%%%%%%%%%%%%%%%%%%%%%

function [mac2,u1L2] = material1E2L(E,p,i,f,A)
  y = E ;
   % for-loop command for running in energy column
for m = i : f
   % If-condition for logarithmic interpolation
      Energy = A(n,1);
if y  == Energy                       % process 1
         mac2 = A(n,2)  ;
         lac2 = mac2*p ;
         u1L2 = lac2 ;
elseif E > A(m,1) && E < A(m+1,1)
         mac2 = ((A(m,2))*(log10(A(m+1,1))-log10(E)) +...
               (A(m+1,2))*(log10(E)-log10(A(m,1))) ) /...
```

```
( log10(A(m+1,1)) - log10(A(m,1)))   ;
          Lac2 = mac2*p ; % mass attenuation become
                 linear attenuation
          u1L2 = lac2 ;   % linear attenuation
                 coefficient (cm^-1)
end
end

end

%%%%%%%%%%%%%%%%%%%%%%%%%%%%%%%%%%%%%%%%%%%%%%%%%%%%%%
```

```
%%%%%%%%%%% Calculation of Buildup factor %%%%%%%%%%
%%%%%%%%%%%%%%%%%%%%%%%%%%%%%%%%%%%%%%

function [B1L1,B1L2] = Buildupfactor1E2L(material1,mate
   rial2,E,lamda1,lamda2)
switch material1     % Condition command for the 1st
   layer
case 'Be'        % In case of Beryllium
          % importing file buildup factor of Hydrogen
             from Microsoft Excel
            Be = importfile1('23 element for Buildup
                 factor.xlsx','Be','A2:Z18');
B = Be
          % Function file for calculation
          % Build up factor (B1L1) for the 1st layer
            B1L1 = materialforB1E1L(E,lamda1,B) ;
case 'B'
            B = importfile2('23 element for Build up
                factor.xlsx','B','A2:Z18');
            B1L1 = materialforB1E1L(E,lamda1,B) ;
                              .
                              .
                     94 elements
                              .
                              .
end

switch material2     % Condition command for the 2nd
   layer
case 'Be'        % In case of Beryllium
          % importing file buildup factor of Hydrogen
             from Microsoft Excel
Be = importfile1('23 element for Buildup factor.
   xlsx','Be','A2:Z18');
 B = Be ;
          % Function file for calculation
          % Build up factor (B1L2) for the 1st layer
     B1L2 = materialforB1E1L(E,lamda2,B) ;
case 'B'
           B = importfile2('23 element for Build up
               factor.xlsx','B','A2:Z18');
```

```
            B1L2 = materialforB1E1L(E,lamda2,B) ;
                              .
                              .
                        94 elements
                              .
                              .
end
end

%%%%%%%%%%%%%%%%%%%%%%%%%%%%%%%%%%%%%%%%%%%

function B1L1 = materialforB1E1L(E,PD1,B)

    % (for-loop) command  for running the data
    % in the first column(0.5-40 mean free path [17
       rows])
    % by the initial value at 1 and the final value 17
for m = 1 : 17
      % (for-loop) command  for running the data
      % in the first row( 0.015-15 MeV [26 columns])
      % by the initial value at 1 and the final value 26
forn = 1 : 26
         % if-condition for calculation
         % logarithmic-bilinear interpolation
mfp = A(m,1);
energy = A(1,n);
ifPD1==mfp&&E==energy   % the 1st condition
         z = A(m,n)  ;
         B1L1 = z ; % Exposure buildup factor in the 1st
           layer
elseifPD1>A(m,1)&& PD1<A(m+1,1)&&E==energy  % the 2nd
    condition
deltaz =(A(m+1,n)-A(m,n))/(A(m+1,1)-A(m,1))*( PD1 -
    A(m,1));
         z = A(m,n)+deltaz ;
         B1L1 = z ; % Exposure buildup factor in the 1st
           layer
elseif  E>A(1,n)&&E<A(1,n+1)&& PD1==mfp% the 3rd
    condition
deltaz =(A(m,n+1)-A(m,n))/(A(1,n+1)-A(1,n))*( E -
    A(1,n));
```

```
          z = A(m,n)+deltaz ;
          B1L1 = z ; % Exposure buildup factor in the 1st
            layer
       % the 4th condition
elseif PD1>B(m,1)&& PD1<B(m+1,1)&&E>A(1,n)&&E<A(1,n+1)
   R1 = ((B(m,n))*(log10(B(m+1,1))-log10(PD1)) +...
             (B(m+1,n))*(log10(PD1)-log10(B(m,1))) )
               /...
( log10(B(m+1,1)) - log10(B(m,1)))  ;
          R2 = ((B(m,n+1))*(log10(B(m+1,1))-log10(PD1))
             +...
             (B(m+1,n+1))*(log10(PD1)-log10(B(m,1)))
             ) /...
( log10(B(m+1,1)) - log10(B(m,1)))  ;
Bfactor = ((R1)*(log10(B(1,n+1))-log10(E)) +...
      (R2)*(log10(E)-log10(B(1,n))) ) /...
( log10(B(1,n+1)) - log10(B(1,n)))  ;
        B1L1 = Bfactor ; % Exposure buildup factor in
             the 1st layer
end
end

end

%%%%%%%%%%%%%%%%%%%%%%%%%%%%%%%%%%%%%%%%%%%

function B1L2 = materialforB1E2L(E,PD2,B)

    % (for-loop) command  for running the data
    % in the first column(0.5-40 mean free path [17 rows])
    % by the initial value at 1 and the final value 17
for m = 1 : 17
      % (for-loop) command  for running the data
      % in the first row( 0.015-15 MeV [26 columns])
      % by the initial value at 1 and the final value 26
for n = 1 : 26
         % if-condition for calculation
         % logarithmic-bilinear interpolation
mfp = A(m,1);
energy = A(1,n);
ifPD2==mfp&&E==energy   % the 1st condition
```

```
          z = A(m,n)  ;
          B1L2 = z ; % exposure buildup factor in the 2nd
                 layer
elseifPD2>A(m,1)&& PD2<A(m+1,1)&&E==energy  % the 2nd
    condition
deltaz =(A(m+1,n)-A(m,n))/(A(m+1,1)-A(m,1))*( PD2 -
    A(m,1));
          z = A(m,n)+deltaz ;
          B1L2 = z ; % exposure buildup factor in the 2nd
                 layer
elseif  E>A(1,n)&&E<A(1,n+1)&& PD2==mfp    % the 3rd
    condition
deltaz =(A(m,n+1)-A(m,n))/(A(1,n+1)-A(1,n))*( E -
    A(1,n));
          z = A(m,n)+deltaz ;
          B1L2 = z ; % exposure buildup factor in the 2nd
                 layer
elseifPD2>B(m,1) &&PD2<B(m+1,1) && E>A(1,n) &&
    E<A(1,n+1)
           R1 = ((B(m,n))*(log10(B(m+1,1))-log10(PD2))
                 +...
                 (B(m+1,n))*(log10(PD2)-log10(B(m,1))) )
                   /...
( log10(B(m+1,1)) - log10(B(m,1)))  ;
           R2 = ((B(m,n+1))*(log10(B(m+1,1))-log10(PD2))
                   +...
                 (B(m+1,n+1))*(log10(PD2)-log10(B(m,1)))
                   ) /...
( log10(B(m+1,1)) - log10(B(m,1)))  ;
Bfactor = ((R1)*(log10(B(1,n+1))-log10(E)) +...
                 (R2)*(log10(E)-log10(B(1,n))) ) /...
( log10(B(1,n+1)) - log10(B(1,n)))  ;
         B1L2 = Bfactor ;% exposure buildup factor in
                the 2ndlayer
end

end
end

%%%%%%%%%%%%%%%%%%%%%%%%%%%%%%%%%%%%%%%%%%%
```

```
%%%%% Solution of optimization by contour graphic %%%%%
%%%%%%%%%%%%%%%%%%%%%%%%%%%%%%%%%%%%%%%%%%%%%%%%%%%%%%%

create array as x1 at initial value 0.1, final value
   15,increment at 1
x1=0:1:20;
% create array as x1 at initial value 0.1, final value
   15,increment at 1
x2=0:1:20;
%  make coordinate point for the contour graph by
   intersection between x1,x2
[X1,X2]=meshgrid(x1,x2);

E = 0.662 ;          % Photon energy of Cs-137 (MeV)
material1 = 'Pb';    % the first materail as iron
material2 = 'Fe';    % the second material as lead
Io = 100;            % the initial intensity of
                       radiation (mSv/hr)
l1 = 10;             % the thickness (cm) for the 1st
                       layer
l2 = 10;              % the thickness (cm) for the 2nd
                       layer

% function file for calculation of mass and linear
   attenuation coefficient
[mac1,mac2,u1L1,u1L2] = linearAttenCoef1E2L(material1,m
   aterial2,E);
PD1 = u1L1*l1;   % a number of mean free path in the
   1st layer
PD2 = u1L2*l2;   % a number of mean free path in the
   2nd layer
% function file for calculation of buildup factor
[B1L1,B1L2] = EABuildupfactor1E2L(material1,material2,E
   ,PD1,PD2);
% function file for radiation intensity (I)
% as the objective function
f1 = objectivefunc(X1,X2,B1L1,B1L2,u1L1,u1L2,Io);

% function file for the thickness sum as
% the 1st constrained function
ineq1 = ineqconstraint1(X1,X2);
```

```
% function file for weight of cylindrical container as
% the 2nd constrained function
ineq2 = ineqconstraint2(X1,X2);

% function file for the price of cylindrical container as
% the 3rd constrained function
ineq3 = ineqconstraint3(X1,X2);

% command for plotting contour graph of the objective
    function
[C,han] = contour(x1,x2,f1,[0.1,0.1],'k-');
set(han,'LineWidth',1.5)
hold on

% command for plotting contour graph of the 1st
    constrained function
[C1,han1] = contour(x1,x2,ineq1,[20,20],'k:');
set(han1,'LineWidth',1.5)

% command for plotting contour graph of the 2nd
    constrained function
[C2,han2] = contour(x1,x2,ineq2,[1000,1000],'k-.');
set(han2,'LineWidth',1.5)
% command for plotting contour graph of the 3rd
    constrained function
[C3,han3] = contour(x1,x2,ineq3,[600,600],'k--');
set(han2,'LineWidth',1.5)

% command for complement of graph
% identify name of axis , size and weight of word on
    x-axis
xlabel('x1: lead(cm)','FontName','times',...
       'FontSize',18,'FontWeight','b');
% identify name of axis , size and weight of word on
    y-axis
ylabel('x2: Iron(cm)','FontName','times',...
       'FontSize',18,'FontWeight','b');
% identify name of axis , size and weight of word on
    title of graph
% set the step of number on x-axis
set(gca,'xtick',[0 1 2 3 4 5 6 7 8 9 10 11 12 13 14 15
    16 17 18 19,... 20],'FontSize',18)
```

```
% set the step of number on y-axis
set(gca,'ytick',[0 1 2 3 4 5 6 7 8 9 10 11 12 13 14 15
   16 17 18 19,... 20],'FontSize',18)

%%%%%%%%%%%%%%%%%%%%%%%%%%%%%%%%%%%%%%%%%%%%%%%%%%%

function f1 = objectivefunc(X1,X2,B1L1,B1L2,u1L1,u1L2,Io)

f1 = Io*(B1L1./exp(u1L1.*X1)).*( B1L2./exp(u1L2.*X2)) ;

end

%%%%%%%%%%%%%%%%%%%%%%%%%%%%%%%%%%%%%%%%%%%%%%%%%%%

function ineq1 = ineqconstraint1(X1,X2)
cair = 7 ;
 ineq1 = cair X1 + X2 ;

end

%%%%%%%%%%%%%%%%%%%%%%%%%%%%%%%%%%%%%%%%%%%%%%%%%%%
%%%%%%%%%%%%%%%%%%%%%

functionineq2 = ineqconstraint2(X1,X2)
cair = 7  ;
pIr  = 7.874 ; % g/cm^3
pPb  = 11.35 ; % g/cm^3
pAir = 0.001225 ; % g/cm^3

ineq2= ( 2*pi*(1/1000) ) .* ( (pAir*(cair^3)) +,...
        ( pIr.*( (((X1+cair).^3) - (cair^3) ) ))  +,...
        ( pPb.*( (((X1+X2+cair).^3) - (X1+cair).^3 ) ))
          );

end

%%%%%%%%%%%%%%%%%%%%%%%%%%%%%%%%%%%%%%%%%%%%%%%%%%%

functionineq3 = ineqconstraint3(X1,X2)
cair = 7 ;
pIr  = 7.874 ;
pPb  = 11.35 ;
Ircost = 0.028 ;  % Iron price in unit USD/lb
```

```
Lecost = 0.97 ;  % Lead price in unit USD/lb

Ineq3 = ( 4.4*pi*(1/1000) ) .* ( ( Lecost.*pIr.*(
         (((X1+cair).^3) -,...
        (cair^3) ) )) + ( Ircost.*pPb.*(
         (((X1+X2+cair).^3) -,...
        (X1+cair).^3 ) )) );

end

%%%%%%%%%%%%%%%%%%%%%%%%%%%%%%%%%%%%%%%%%%%%%%%%%%%%%%%
```

Appendix B

```
%Define constants

N=1; % number of photons
m=100; %1 chance in m of photon packet surviving
hist= 1;
histogramZ=zeros();
histogramX=zeros();
A=0;
ni=1;
nt=1.39;

Xvec = zeros();
Yvec = zeros();
Zvec = zeros();

current_layer=0;
layer=0;
pre_layer=0;
for j=1:N

%Initialize
x=0;
y=0;
z=0;
mu_x=0;
mu_y=0;
mu_z=1;
W=1;

%inital loss of W due to specular reflectance
W=W-(ni-nt)^2/(ni+nt)^2; %Specular reflectance
s=0;
dead=0; %photon is initally not dead

%Switch Case Conditions for Layer Parameters be
```

```
Nsteps=1;

while dead==0;

%Find distance to boundary db
%Check if photon will hit boundary
%Compute distance to boundary

if mu_z==0;
db=inf;
elseif mu_z<0;

db=(0-z)/mu_z ;

elseif mu_z>0;
db=inf; %(If layer is finite change condition)
end


if (0 <= z)&&(z < 0.12)

mu_a=2.7; %absorption coefficient
mu_s=187; %scattering coefficient
g=0.81; %expected value of cos theta %
ni=1; %refractive index of air
nt=1.39; %refractive index of tissue
mu_t=mu_a+mu_s;
Rsp=(ni-nt)^2/(ni+nt)^2;

if s==0;
%Launch photon
s=-log(rand)/mu_t;
end

if z+s>=0.12
    z=0.12;
    s=0;
end

layer=1;

elseif (0.12<= z)&&(z<.5)
```

```
mu_a=3.6; %absorption coefficient [1/cm]
mu_s=171; %scattering coefficient [1/cm]
g=0.85; %expected value of cos theta %
ni=1.39; %refractive index of air
nt=1.44; %refractive index of tissue
mu_t=mu_a+mu_s;
Rsp=(ni-nt)^2/(ni+nt)^2;

if s==0;
%Launch photon
s=-log(rand)/mu_t;
end

if z+s>=0.5
    z=0.5;
    s=0;
end

layer=2;

%============================================

%-----------------------------------------------------
elseif (.5<= z)

mu_a=11.2; %absorption coefficien
mu_s=530; %scattering coefficien
g=.9; %expected value of cos theta %
ni=1.44; %refractive index of air
nt=1.37; %refractive index of tissue
mu_t=mu_a+mu_s;
Rsp=(ni-nt)^2/(ni+nt)^2;

if s==0;
%Launch photon
s=-log(rand)/mu_t;
end

layer=3;

%============================================
```

```
end
%
% if z+(s*mu_z) >= 0.51 % FOR GOLD LAYER
%
%     A=z+(s*mu_z)-0.5;
%
% end

current_layer=layer;

if current_layer~=pre_layer

    W=W-Rsp; %Specular reflectance

end

pre_layer=current_layer;

%Decide if step size is greater than db
if db<=s
%Move photon to boundary
s=s-db;
x=x+mu_x*db;
y=y+mu_y*db;
z=z+mu_z*db;
ai=acos(abs(mu_z));
ac=asin(ni/nt);
if ai>ac;
r_ai=1;
else
at=asin((ni*sin(ai))/nt);
r_ai=1/2*(((sin(ai-at)^2)/(sin(ai+at)^2))+((tan(ai-
   at)^2)/(tan(ai+at)^2)));
end
%Compute probabilty of a photon packet being internally
   reflected
%or transmitted
%if rand<=r_ai
if  rand<=r_ai
mu_z=-1*mu_z;
else
%Transmits;
```

```
dead=1;
end
else
%For photon that does not hit boundary
x=x+(s*mu_x);
y=y+(s*mu_y);
z=z+(s*mu_z);
%z=z+(s*mu_z)-A;
s=0;
% A=0;
%Find new direction cosines
theta=acos((1/(2*g))*(1+(g^2)-((1-g^2)/
(1-g+2*g*rand))^2));
psi=2*pi*rand;
if abs(mu_z)>0.99999
mu_x2=sin(theta)*cos(psi);
mu_y2=sin(theta)*sin(psi);
mu_z2=sign(mu_z)*cos(theta);
else
mu_x2=(sin(theta)*(mu_x*mu_z*cos(psi)-mu_y*sin(psi))/
    sqrt(1-mu_z^2))+mu_x*cos(theta);
mu_y2=(sin(theta)*(mu_y*mu_z*cos(psi)+mu_x*sin(psi))/
    sqrt(1-mu_z^2))+mu_y*cos(theta);
mu_z2=-sin(theta)*cos(psi)*(sqrt(1-mu_
    z^2))+mu_z*cos(theta);
end
mu_x=mu_x2;
mu_y=mu_y2;
mu_z=mu_z2;

dW=(mu_a/mu_t)*W;

% Xvec(1,Nsteps+1) = Xvec(1,Nsteps) + x;
% Yvec(1,Nsteps+1) = Yvec(1,Nsteps) + y;
% Zvec(1,Nsteps+1) = Zvec(1,Nsteps) + z;

% Xvec(N,Nsteps+1) = Xvec(N,Nsteps+1)+ x;
% Yvec(N,Nsteps+1) = Yvec(N,Nsteps+1)+ y;
% Zvec(N,Nsteps+1) = Zvec(N,Nsteps+1)+ z;

Xvec(1,Nsteps+1) = x;
```

```
Zvec(1,Nsteps+1) = z;

Nsteps=Nsteps+1;


%Reduce weight
W=W-dW;
%Roulette
if W<=0.0001
if rand <= 1/m
W=m*W;
else
W=0;
dead=1;
end
end
end

histogramZ(hist) = z ;

histogramX(hist) = x;

hist= hist +1 ;

if (0<= z)&&(z<.02)
histogramXX(hist)=x;
end


end


%% MAX/MIN positions for plot viewing
xMax= max(Xvec); xMin = min(Xvec);
yMax= max(Yvec); yMin = min(Yvec);
zMax= max(Zvec); zMin = min(Zvec);

if abs(xMin) > xMax
xlim= abs(xMin);
else
xlim= xMax;
end
```

```
if abs(zMin) > zMax
zlim= abs(zMin);
else
zlim= zMax;
end

if abs(yMin) > xMax
ylim= abs(yMin);
else
ylim= yMax;
end

xlim= xlim+1;
zlim= zlim+1;
ylim= ylim+1;

Xo = 0;Zo= 0;
Xf = Xvec(end);Zf = Zvec(end);
% Plotting (Line 3-D Plot)
figure(1)
xlabel('wideness(cm)');

ylabel('Skin Depth (cm)');
plot(Xo,Zo,'ro'); % marks origin
holdon
plot(Xo,Zo,'r*'); % marks origin
holdon

plot(Xvec, -Zvec,'b.') % plots path
holdon
plot([-3 3],[0 0.0001],'r')
holdon
plot([-3 3],[-0.12 -0.1200001],'r')

holdon
plot([-3 3],[-0.5 -0.500001],'r')

axis([-xlim xlim -zlim zlim]);

holdon

%legend('\color{red} layer1','\color{blue} layer2','\
    color{green} layer3 ');
```

```
holdon
gridon



end

figure(4);
histogram(histogramZ);

figure(5);
histogram(histogramX);

figure(6);
histogram(histogramXX);
```

Appendix C

```
#include <hidboot.h>
#include <usbhub.h>
#include <Wire.h>
#include <LiquidCrystal_I2C.h>
#ifdef dobogusinclude
#include <spi4teensy3.h>
#include <SPI.h>
#include <string.h>
#endif
#define DISPLAY_WIDTH 20
#define MIDA 8
#define RELAY1  7
const int signalPin= 2;
int signal;
int i=0;
int analogValue = 0;
int potPin1 = 1;
int potPin2 = 2;
char animal[MIDA];
float voltage = 0;
byte batteryA[8] = {
    B11111,
    B11111,
    B11111,
    B11100,
    B11101,
    B11101,
    B11101,
```

```
    B11101
};
byte batteryB[8] = {
    B11101,
    B11101,
    B11101,
    B11101,
    B11100,
    B11111,
    B11111,
    B11111
};
byte batteryC[8] = {
    B11111,
    B11111,
    B11111,
    B00000,
    B00000,
    B00000,
    B00000,
    B00000
};
byte batteryD[8] = {
    B00000,
    B00000,
    B00000,
    B00000,
    B00000,
    B11111,
    B11111,
    B11111
};
```

```
byte batteryE[8] = {
    B11111,
    B11111,
    B11111,
    B00111,
    B00111,
    B00111,
    B00111,
    B00111
};
byte batteryF[8] = {
    B00111,
    B00111,
    B00111,
    B00111,
    B00111,
    B11111,
    B11111,
    B11111
};
byte batteryG[8] = {
    B00000,
    B00000,
    B00000,
    B00000,
    B00000,
    B11100,
    B11100,
    B11100
};
byte batteryH[8] = {
    B11100,
```

```
    B11100,
    B11100,
    B00000,
    B00000,
    B00000,
    B00000,
    B00000
};
LiquidCrystal_I2C lcd(0x3F, 20, 4);
class KbdRptParser : public KeyboardReportParser
{
    void PrintKey(uint8_t mod, uint8_t key);
  protected:
    void OnControlKeysChanged(uint8_t before, uint8_t
      after);
    void OnKeyDown  (uint8_t mod, uint8_t key);
    void OnKeyUp    (uint8_t mod, uint8_t key);
    void OnKeyPressed(uint8_t key);
};
void KbdRptParser::PrintKey(uint8_t m, uint8_t key)
{
  MODIFIERKEYS mod;
  *((uint8_t*)&mod) = m;
};
void KbdRptParser::OnKeyDown(uint8_t mod, uint8_t
   key)
{
  uint8_t c = OemToAscii(mod, key);
  if (c)
    OnKeyPressed(c);
}
void KbdRptParser::OnControlKeysChanged(uint8_t
```

```
   before, uint8_t after) {
  MODIFIERKEYS beforeMod;
  *((uint8_t*)&beforeMod) = before;
  MODIFIERKEYS afterMod;
  *((uint8_t*)&afterMod) = after;
}
void KbdRptParser::OnKeyUp(uint8_t mod, uint8_t key)
{
}
void KbdRptParser::OnKeyPressed(uint8_t key)
{
  static uint32_t next_time = 0;
  static uint8_t current_cursor = 0;
    if( millis() > next_time ) {
      lcd.clear();
      current_cursor = 0;
      delay( 5 );
      lcd.setCursor( 0,0 );
    }
    next_time = millis() + 200;
    if( current_cursor++ == ( DISPLAY_WIDTH + 1 )) {
      lcd.setCursor( 0,1 );
    }
    lcd.setCursor(0,0);
    lcd.print("PET  NAME GENDER AGE PRICE:");
    animal[i]=(char)key;
    i++;
    data();
};
USB     Usb;
HIDBoot<USB_HID_PROTOCOL_KEYBOARD>
```

```
   HidKeyboard(&Usb);
uint32_t next_time;
KbdRptParser Prs;
void data(){
    if(i==8){
 if(animal[0]==’1’ && animal[1]==’5’ && animal[2]==’8’
&& animal[3]==’8’ && animal[4]==’9’ && animal[5]==’3’
&& animal[6]==’2’){

         lcd.setCursor(0,1);
         lcd.print(“CAT  KIT  FEMALE  4”);
         lcd.setCursor(0,2);
         lcd.print(“PRICE: 3000 BAHT”);
       }
  else   if   (animal[0]==’1’   &&   animal[1]==’8’   &&
animal[2]==’5’  &&  animal[3]==’2’  &&  animal[4]==’6’
&& animal[5]==’9’ && animal[6]==’8’){

          lcd.setCursor(0,1);
          lcd.print(“DOG  REX   MALE   2”);
          lcd.setCursor(0,2);
          lcd.print(“PRICE: 2000 BAHT”);
       }
```

```
  else  if(animal[0]==’1’  &&  animal[1]==’4’  &&
animal[2]==’8’ && animal[3]==’7’ && animal[4]==’5’
&& animal[5]==’2’ && animal[6]==’6’){

        lcd.setCursor(0,1);
        lcd.print(“BIRD BEN  FELAME  2”);
        lcd.setCursor(0,2);
        lcd.print(“PRICE: 900 BAHT”);
      }
  else  if(animal[0]==’1’  &&  animal[1]==’5’  &&
animal[2]==’8’ && animal[3]==   ’9’              &&
animal[4]==’6’ && animal[5]==’5’ && animal[6]==’7’){

        lcd.setCursor(0,1);
        lcd.print(“FISH MIKE FEMALE  2”);
        lcd.setCursor(0,2);
        lcd.print(“PRICE:600 BAHT”);

      }
       for(i=0;i<8;i++){
        lcd.setCursor(i,3);
        lcd.print(animal[i]);
      }
       i=0;
delay(4000);
```

```
        lcd.clear();
        lcd.setCursor(0,0);
lcd.print("PET  NAME GENDER AGE PRICE:");
        lcd.setCursor(0,2);
        lcd.print("Barcode is ready");
       }
     }
void setup()
{
  Serial.begin(9600);
  pinMode(signalPin, INPUT);
  lcd.begin();
  lcd.backlight();
  lcd.createChar(2,batteryA);
  lcd.createChar(3,batteryB);
  lcd.createChar(4,batteryC);
  lcd.createChar(5,batteryD);
  lcd.createChar(6,batteryE);
  lcd.createChar(7,batteryF);
  lcd.createChar(8,batteryG);
  lcd.createChar(9,batteryH);
  Wire.beginTransmission(0x68);
  Wire.write(0x07);
  Wire.write(0x10);
  Wire.endTransmission();
  lcd.setCursor(0, 0);
  lcd.print("  Barcode Scanner");
  lcd.setCursor(0, 1);
  lcd.print("     Copy Right 2016");
  lcd.setCursor(0, 2);
  lcd.print(" Faculty of Science");
  lcd.setCursor(0, 3);
```

```
  lcd.print("        KMITL");
  delay(5000);
  lcd.clear();
#if !defined(__MIPSEL__)
  while (!Serial);
#endif
  if (Usb.Init() == -1)
  Serial.println("OSC did not start.");
  delay( 200 );
  next_time = millis() + 5000;
  HidKeyboard.SetReportParser(0, &Prs);
  lcd.clear();
  lcd.noAutoscroll();
  lcd.print("Barcode is ready");
  delay( 200 );
  if(i!=8){
  i=0;
  }
}
void loop()
{
  signal= analogRead(signalPin);   //arduino reads
the value from the infrared sensor
  Serial.println(signal);          //prints out
analog value
  analogValue = analogRead(0);
  voltage = 0.0048*analogValue;
  pinMode(RELAY1, OUTPUT);
  if( voltage <= 5.0 && voltage > 2.8){
  if (signal > 600){
  digitalWrite(RELAY1,1);         // Turns ON Relays 1
```

```
      delay(3500);
    }
    else
    {
      digitalWrite(RELAY1,0);          // Turns Relay Off
    }
    Usb.Task();
    delay(50);
     }
  else if ( voltage <=2.7 ){
      lcd.clear();
      lcd.setCursor(3, 0);
      lcd.print("Please charge!");
  lcd.setCursor(6, 1);
      lcd.write(2);
      lcd.setCursor(7, 1);
      lcd.write(4);
      lcd.setCursor(8, 1);
      lcd.write(4);
      lcd.setCursor(9, 1);
      lcd.write(4);
      lcd.setCursor(10, 1);
      lcd.write(4);
      lcd.setCursor(11, 1);
      lcd.write(6);
      lcd.setCursor(12, 1);
      lcd.write(8);
      lcd.setCursor(6, 2);
      lcd.write(3);
      lcd.setCursor(7, 2);
      lcd.write(5);
      lcd.setCursor(8, 2);
```

```
  lcd.write(5);
  lcd.setCursor(9, 2);
  lcd.write(5);
  lcd.setCursor(10, 2);
  lcd.write(5);
  lcd.setCursor(11, 2);
  lcd.write(7);
  lcd.setCursor(12, 2);
  lcd.write(9);
  lcd.setCursor(3, 3);
  lcd.print(" Low Battery!");
  delay(5000);
  lcd.clear();
}
```

Index

ablation 49, 58, 59
absorbance 8, 14–17, 23, 24, 27, 28, 34–36, 39, 59–61, 64, 66, 68, 69, 71, 79, 80, 83, 96, 97, 260
absorbance peak 34, 47, 57, 59, 64, 65, 80, 83
absorption 4, 5, 10, 16, 67, 70, 75, 122
absorption intensity 14, 15, 47, 59, 79
acceptor 39, 57, 63, 75, 85, 88, 262
adsorption capacity 78, 79, 81, 82
amplification 172, 225, 231, 233, 240
amplifier 172, 192, 207, 220, 229, 231, 232, 238, 240, 241, 294
amplitude 98, 158–160, 185, 186, 191, 212, 215, 235, 241, 276, 282, 292
analysis of variance (ANOVA) 17, 36, 37
ANOVA *see* analysis of variance
Avogadro's number 101

barcode 126–129, 141, 142, 324, 325
barcode scanner 126, 143, 324
  laser 141, 142, 144
barcode scanning module 128, 141, 144
barrier 88, 93
beam 106, 127, 129, 194, 204, 216, 233, 239, 287
  collimation 102
  fundamental 239
  laser 127
birefringence phase matching (BPM) 158, 169, 191, 192, 202, 206, 208, 212, 222, 223, 241, 277, 290
black body 257
BPM *see* birefringence phase matching
buildup factor 102, 114, 301, 305

carrier 4, 29
carrier balance 87, 88, 93
carrier injection 88
cathode 89–91
cavity 238, 240–242, 246, 248, 249, 295, 296
CCD camera 43
cell 47, 49, 51, 58–62, 64
cell-killing approach 51, 61
cell-killing effects 64
charger coupled device (CCD) scanner 127
command 102, 302, 303, 306
command patterns 104
Compton effect 265
Compton scattering 100, 101
conduction band 4, 50, 52, 55, 59, 62, 75, 101, 258, 259
conductive media 158, 270, 271
contour graph 111, 112, 305, 306
conversion efficiency 206
coumarin-153 16, 57, 68–70
crystal 192–194, 202, 207–209, 213–216, 219–221, 225, 229, 233, 237, 239–242, 246, 288, 289, 292
  anisotropic 151, 155, 161, 196
  birefringence 192, 194
  cubic 202

lithium niobite 202
nonlinear 202
quartz 158, 277
current density 46, 87, 92–94, 97, 98, 254, 255

detector 102, 141, 147, 266
device 93, 119
biomedical 10, 119
compact 127
diagnostic 120
multilayers 90
optic 158, 272
photonic 4
sensing 20
DFG *see* different frequency generation
dielectric constant 9, 46, 156, 158, 258, 275, 276
dielectric permittivity 275
different frequency generation (DFG) 219–221, 223, 225, 227, 294
donor 10, 20, 39, 57, 63, 75, 85, 262
double frequency 169–171, 174–177, 181, 182, 184, 192–194, 198, 202, 233, 235, 238, 281
double resonant oscillator (DRO) 246, 248, 249, 295, 296
DRO *see* double resonant oscillator
Drude model 46, 254
DSSC *see* dye-sensitized solar cell
duty cycle 73, 81, 83–85, 209, 210, 213, 216, 283–285
dye 27–29, 44, 45, 57, 69, 74, 78, 79, 81, 83
adsorbate 81
fluorescence 22
metal-fluorescence 44
organic 9, 10, 20, 54, 57, 73, 75
ruthenium-based 57
dye molecule 4, 5, 9, 10, 17, 22, 28–30, 37, 39, 41, 57, 75, 78, 81, 83
fluorescence lifetime of 39
fluorescent quantum yield of 17, 37
metal-fluorescence 43
dye removal 73, 77–84
dye-sensitized solar cell (DSSC) 47, 56
dye wastewater 73, 78

electric field 9, 46, 151–155, 158–164, 167, 170–172, 177, 181, 185, 186, 199–201, 213–215, 254, 255, 274–276, 278, 280, 289
electromagnetic field 6, 8, 11, 13–15, 98, 154, 155, 160, 206, 263
electromagnetic wave 151, 170, 171, 183, 184, 206, 271
embedding method 113
energy conservation 156, 171, 189, 220, 226, 291
energy gap 50, 59, 62
energy transfer 101, 118, 235, 265
fluorescence resonance 4, 16, 20, 75, 85, 262
nonradiative 37, 75
plasmon resonance 10, 30
exposition 22, 44, 46, 77, 81, 83, 85

Faraday's law 253
FDTD method *see* finite-difference time-domain method
Fermi energy 54, 55, 67
Fermi level 94, 259
ferrous ion chelating 58–60, 62, 63
finite-difference time-domain method (FDTD method) 12, 13

fluorescence 9, 10, 15, 16, 18–20, 23, 27, 28, 31, 34–36, 38–40, 42, 43, 68, 69, 75, 78–80, 83
  intrinsic 10, 30
  metal-enhanced 16
  parametric 242, 249, 294
  yellow 22
fluorescence enhancement 10, 43–46, 251
fluorescence enhancer 20, 44
fluorescence intensity 4, 5, 10, 18, 21, 22, 28–30, 32–34, 36, 38–42, 69–70, 79, 80, 262
  integrated 17, 34, 76, 80
fluorescence intensity ratio 29, 30, 39, 40
fluorescence peak 28, 34, 79, 83
fluorescence polarization (FP) 17, 18, 38–40
fluorescence quantum yield 16, 34, 73, 75, 76, 80, 85, 262
fluorescence quenching 39, 43, 45, 46, 75, 80
fluorescence sensor 4, 20
fluorescence spectra 16, 28, 34, 38, 43, 53
fluorescent lamp 64, 65
fluorescent quantum yield (FQY) 17, 35–37
fluorophore 16, 17, 20, 36, 39, 262
Förster resonance energy transfer 262
Fourier coefficient 210, 212, 213, 283
Fourier expansion amplitude 212
Fourier term 210, 212
FP *see* fluorescence polarization
FQY *see* fluorescent quantum yield

gain bandwidth 245, 295, 296
gain coefficient 223, 227, 237, 244, 292, 293
Gaussian beam 239
Gauss's law 254

hole 52, 62, 75, 84, 92–94
hole transport layer (HTL) 88, 89, 91
HOMO 57, 92
HTL *see* hole transport layer
human skin 119, 121, 131, 139–141
human tissue 119, 131, 133, 148, 266
human tissue model 123, 130, 132

idler 207, 219, 224, 225, 229, 231, 236, 240, 242, 246, 248, 249, 293, 296
idler frequency 225, 229, 237, 238, 249
idler photons 223, 225, 226, 231
indium tin oxide (ITO) 88, 89, 92
ITO *see* indium tin oxide

laser 145, 172, 244
  argon ion 217, 290
  green 22, 170
  infrared 238
  nanoscale 4
  oxazine dye 217, 290
  pulsed 49
  tunable 240
laser ablation method 47, 58
latent fingerprint 4, 21, 22, 42, 44, 46
LCD screen 127–129, 141, 143
linear attenuation coefficient 107, 115, 297, 298, 305
lithium niobate crystal 196, 200, 201, 207, 216, 217, 289
logarithmic interpolation method 107
longitudinal plasmon band (LPB) 24, 25
LPB *see* longitudinal plasmon band
luminance 94, 98
luminescence 10, 45
LUMO 92, 94

magnetic field 153, 158, 254, 276, 277
Manley–Rowe relation 223, 226
mass attenuation coefficient 99
matching condition 223
  birefringence phase 169, 222, 223
  critical phase 288
  quasi-phase 223
MATLAB software 99, 102, 104, 109, 113
Maxwell's equation 12, 14, 46, 151–153, 158–160, 171, 253, 270, 271
mercury 4, 19–20
mercury detection 19, 20, 40
metal 5, 6, 8, 14, 17, 29, 30, 37, 39, 41, 46, 50, 51, 54, 55, 70, 71, 259, 260
  alkali 257
  dielectric constant of 6, 7, 9
  free electron oscillation of 11, 13
  heavy 19, 47, 54, 64, 78
  high work function of 260
  noble 4, 5, 54
  rare 57
momentum conservation 118, 171, 192, 220, 259, 265
Monte Carlo approach 104
Monte Carlo simulation 113, 121–123, 125, 131
muscle 120, 121, 133
  skeletal 123, 130, 131, 141

nanocomposites 47, 64, 76
nanodots 7, 8
nanomaterials 4, 48, 51
nanoparticles 4–5, 9, 49–50, 53–54, 59, 65–66, 77, 80, 82
  metal 4, 9, 10, 16, 17, 22, 28–30, 39
  ZnO 47, 48, 50–56, 58–62, 64–66, 68–71, 261
nanorods 5, 7, 24, 31, 36–37, 39–41
nanostructures 3, 5, 8, 9, 29, 46, 48, 256
nanowires 7, 8, 46, 255
negative uniaxial crystal 157
nonlinear coefficient 172, 196, 205, 207, 209, 212, 219
nonlinear effect 151, 215, 222, 281
  second-order 170, 215, 281
nonlinear polarization 170, 172, 174, 177, 181, 183, 184, 192, 194, 195, 197, 199, 200, 202, 205–208, 221, 286, 289, 290
nonlinear susceptibility 170, 216, 286

OA *see* optic axis
OLED *see* organic light-emitting diode
optic axis (OA) 158, 159, 167, 181, 183, 184, 188, 190, 194, 197, 199–201, 203, 216, 217, 272, 273, 288, 289
optimization model 99, 102, 106, 112
organic light-emitting diode (OLED) 16, 87, 88, 90, 92, 94, 96–98, 263
oscillation 170, 242, 243, 272, 295
  collective 5
  conduction electron 26
  continuous wave 248, 296
  dipole 41
oscillator 238, 240, 245, 246, 248–250, 295, 296
  double resonant 246, 249, 295, 296
  parametric 238, 239, 242, 248, 296
  single resonant 242, 243, 296
oscilloscope 142

PDMS *see* polydimethylsiloxane
photocatalyst 52–54, 77, 83
polydimethylsiloxane (PDMS) 89, 96
positive uniaxial crystal 157, 158, 277
power efficiency 87, 94, 204, 212, 214, 229, 238–240
Poynting vector 15, 154
printing method 87, 95
probability density function 123, 132, 148, 269, 270
probe 20, 119–121, 130, 140
  absorption 58–60
  fluorescent 9
  refraction 147, 266
process 57, 63, 90, 102, 103, 110, 111, 169, 188, 202, 206, 207, 220, 299
  amplification 225
  degenerate parametric amplifier 229
  fluorescence enhancement 29
  fluorescence quenching 29
  for-loop 102, 103, 297, 298, 302, 303
  high-yield 90
  if-clause 102
  nanomaterial production 48
  nonradiative energy transfer 39
  phototoxicity 64
  pinocytosis 62
  purification 57
  recombination 84, 94
  release film stamp 90
  simulation 131
  second-order 171
pump 181, 207, 219, 223, 229, 231–233, 236, 237, 240, 241, 244, 248, 249, 295, 296
  continuous 243
  fixed 219
pump photon 223, 226, 231
pump power 226, 232, 241, 244, 245, 248, 295, 296

QPM *see* quasi-phase matching
quantum yield 10, 16, 17, 34, 36, 76
quasi-phase matching (QPM) 169, 192, 202, 203, 206, 208, 210–214, 216, 241, 283–285

radiation 113, 240, 272, 305
radiation intensity 99, 102, 106, 107, 109, 305
radioactive waste 99, 102, 106, 113
Raman scattering 4
reactive oxygen species 51, 52, 62, 63, 71, 261
reflectivity 240–242
refractive index 17, 23, 34, 35, 156, 158, 246, 248, 249, 275, 285, 290, 310, 311
response surface method 17, 34, 36, 37
rhodamine 10, 16, 20, 31–34, 36–40, 43, 54, 73, 76, 78–83, 85
Ruby laser 216, 288

scattering 67, 101, 102, 121, 122, 125, 265, 273
Schottky barrier effects 47
Schottky barrier height 54, 56
Schottky junction 67
second harmonic generation (SHG) 169–186, 188, 190–192, 194, 196, 198, 200, 202, 204, 206, 208, 212–214, 216, 217, 288, 290–292
second harmonic power efficiency 202, 203, 213
second harmonic wave 191, 198
semiconductor 50, 52, 54, 70–71, 258–260